Advances in CANCER RESEARCH

Volume 107

Advances in CANCER RESEARCH

Volume 107

Edited by

George F. Vande Woude
Van Andel Research Institute
Grand Rapids
Michigan, USA

George Klein
Microbiology and Tumor Biology Center
Karolinska Institute
Stockholm, Sweden

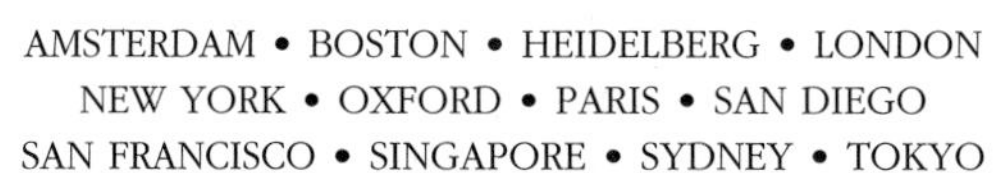
AMSTERDAM • BOSTON • HEIDELBERG • LONDON
NEW YORK • OXFORD • PARIS • SAN DIEGO
SAN FRANCISCO • SINGAPORE • SYDNEY • TOKYO
Academic Press is an imprint of Elsevier

Academic Press is an imprint of Elsevier
525 B Street, Suite 1900, San Diego, CA 92101-4495, USA
30 Corporate Drive, Suite 400, Burlington, MA 01803, USA
32 Jamestown Road, London, NW1 7BY, UK
Linacre House, Jordan Hill, Oxford OX2 8DP, UK
Radarweg 29, PO Box 211, 1000 AE Amsterdam, The Netherlands

First edition 2010

ISBN: 978-0-12-374770-9
ISSN: 0065-230X

For information on all Academic Press publications
visit our website at www.elsevierdirect.com

Printed and bound in USA

10 11 12 10 9 8 7 6 5 4 3 2 1

Contents

Contributors

Numbers in parentheses indicate the pages on which the authors' contributions begin.

Ami Albihn, Department of Microbiology, Tumor and Cell Biology (MTC), Karolinska Institutet, Stockholm, Sweden (163)

Aniruddha Choudhury, Department of Oncology and Pathology, Karolinska University Hospital, Cancer Center Karolinska R8:01, Stockholm, Sweden (57)

Harry A. Drabkin, Department of Medicine and Division of Hematology-Oncology, Medical University of South Carolina, Charleston, South Carolina, USA (39)

Robert M. Gemmill, Department of Medicine and Division of Hematology-Oncology, Medical University of South Carolina, Charleston, South Carolina, USA (39)

Nechama Haran Ghera, Department of Immunology, The Weizmann Institute of Science, Rehovot, Israel (1)

Marie Arsenian Henriksson, Department of Microbiology, Tumor and Cell Biology (MTC), Karolinska Institutet, Stockholm, Sweden (163)

Dai Iwakiri, Department of Tumor Virology, Institute for Genetic Medicine, Hokkaido University, Sapporo, Japan (119)

C. Christian Johansson, Department of Oncology and Pathology, Karolinska University Hospital, Cancer Center Karolinska R8:01, Stockholm, Sweden (57)

John Inge Johnsen, Childhood Cancer Research Unit, Department of Woman & Children's Health, Karolinska Institutet, Stockholm, Sweden (163)

Rolf Kiessling, Department of Oncology and Pathology, Karolinska University Hospital, Cancer Center Karolinska R8:01, Stockholm, Sweden (57)

Kristin R. Lamont, Departments of Urology, Biochemistry and Molecular Biology, Mayo Clinic College of Medicine, Rochester, Minnesota, USA (137)

Alvaro Lladser, Department of Oncology and Pathology, Karolinska University Hospital, Cancer Center Karolinska R8:01, Stockholm, Sweden (57)

Dimitrios Mougiakakos, Department of Oncology and Pathology, Karolinska University Hospital, Cancer Center Karolinska R8:01, Stockholm, Sweden (57)

Amir Sharabi, Department of Immunology, The Weizmann Institute of Science, Rehovot, Israel (1)

Kenzo Takada, Department of Tumor Virology, Institute for Genetic Medicine, Hokkaido University, Sapporo, Japan (119)

Donald J. Tindall, Departments of Urology, Biochemistry and Molecular Biology, Mayo Clinic College of Medicine, Rochester, Minnesota, USA (137)

Breaking Tolerance in a Mouse Model of Multiple Myeloma by Chemoimmunotherapy

Amir Sharabi and Nechama Haran Ghera

Department of Immunology, The Weizmann Institute of Science Rehovot, Israel

---

A unique mouse model of multiple myeloma (MM), namely 5T2MM-bearing mouse, was useful for elucidating the pathophysiological mechanisms underlying the disease. Increased accumulation of suppressive $CD4^{+}CD25^{High}Foxp3^{+}$ regulatory T cells (Tregs) was observed in the thymus and lymphoid peripheral organs during disease progression. Adoptive transfer of Tregs, but not other thymocytes, from 5T2MM-bearing mice led to increased progression of disease manifestations in young syngeneic mice. Depletion of Tregs, a proposed strategy in cancer immunotherapy, was tested using cyclophosphamide (CYC), an alkylating agent with selective cytotoxicity. Both low- and high-dose CYC, administered to sick mice with hind limb paralysis, caused the paralysis to disappear, the plasma tumor cells in the bone marrow (BM) cavity to be replaced by normal cell populations, and the survival of the mice to be significantly prolonged. Low-dose CYC, which selectively depletes Tregs, decreased MM incidence, in contrast to high-dose CYC, which was generally cytotoxic, and did not reduce MM incidence. In contrast, low-dose CYC induced Tregs to become susceptible to apoptosis by down-regulating Bcl-xL and CTLA-4 in these cells, and by decreasing the production of IL-2 by

Advances in CANCER RESEARCH

0065-230X/10 $35.00
DOI: 10.1016/S0065-230X(10)07001-6

effector CD4 cells. This treatment consequently triggered the recovery of IFN-γ-producing natural killer T cells and the maturation of dendritic cells. Transient gradual depletion of Tregs in low-dose CYC-treated 5T2MM mice was maintained beyond 45 days. Thus, less frequent injections of low-dose CYC enabled us to recruit compatible immune-derived cells that would reduce tumor load and delay or prevent tumor recurrence, hence breaking immune tolerance toward MM tumor cells.

I. INTRODUCTION

Multiple myeloma (MM) is a progressive B-lineage neoplasia characterized by proliferation of clonal malignant plasma cells in the bone marrow (BM). The tumor cells secrete an immunoglobulin, usually monoclonal IgE or IgA in the serum and/or light chains in the urine. The progression of the disease may include anemia, lytic bone lesions, renal dysfunction, hypercalcemia, hypogammaglobulinemia, and peripheral neuropathy. Immune dysfunction is an important feature of the disease and leads to infections that are a major cause of morbidity and mortality. Moreover, it may promote tumor growth and resistance to chemotherapy. MM is characterized by numerous defects in the immune system including impaired lymphocyte functions, steroid-related immunosuppression, and neutropenia secondary to chemotherapy (Bergsagel and Kuehl, 2005). A reduced level of polyclonal immunoglobulins is a consistent feature of active MM, reflecting the suppression of $CD19^+$ B lymphocytes that correlate inversely with the disease stage (Rawstron *et al.*, 1998). The relationship between myeloma plasma cells and the BM microenvironment is critical for maintaining the disease. Tumor cells and stromal cells interact via adhesion molecules and cytokine networks to simultaneously promote tumor cell survival, drug resistance, angiogenesis, and disordered bone metabolism. In addition, a number of immunologically active compounds are increased including transforming growth factor (TGF)-β, interleukin (IL)-10, IL-6, vascular endothelial growth factor (VEGF), Fas ligand, Mucin 1 (MUC-1), Cyclooxygenase (COX)-2, and related prostanoids and metalloproteinases (Pratt *et al.*, 2007).

Various drugs having immunomodulatory effects have been used in MM treatment. Thalidomide, shown to have potent anti-inflammatory, antiangiogenic, and immunomodulatory properties, was reported to have anti-MM activity as well (Bartlett *et al.*, 2004; Rajkumar *et al.*, 2002; Singhal *et al.*, 1999). Lenalidomine is another immunomodulatory drug used recently (Richardson *et al.*, 2006) in a NKT cell target combinatorial immunotherapy approach (Chang *et al.*, 2006).

Animal models mimicking human MM are useful for better understanding the pathophysiological mechanisms involved in the progression of the disease and for developing new therapeutic strategies. A series of murine models were described by Radl *et al.* (1988), in which MM arose spontaneously in aging mice of the C57BL/KaLwRij strain with a frequency of 0.5%. A series of tumors have been propagated *in vivo* by intravenous transfer of the diseased BM into young syngeneic mice. This series of MM tumors represents the human form of the disease since their clinical characteristics involve selective localization in the BM, serum M component, angiogenesis, and adhesion and chemokine profiles that are similar to human myeloma (Asosingh *et al.*, 2000; Vanderkerken *et al.*, 1997). The BM microenvironment consists of extracellular matrix protein and BM stromal cells, osteoblasts, and osteoclasts that play a crucial role in the pathogenesis of MM cell growth and survival (Hideshima *et al.*, 2007).

T cell tolerance to tumor-associated antigens plays a significant role in immune evasion by tumors (Drake *et al.*, 2006; Zou, 2006). Naturally occurring and adaptive regulatory T cells (Tregs) are anergic cells with suppressive capabilities that constitute 5–10% of CD4 cells. These cells are induced early during tumor development and were shown to contribute to tumor tolerance (Peng *et al.*, 2002; Zhou and Levitsky, 2007). The mechanisms underlying these effects include inhibiting the activity of a variety of immune cells that are tumor specific such as effector CD4 cells, CD8 cells, dendritic cells (DCs), natural killer (NK) cells, natural killer T (NKT) cells, and B cells (Chen *et al.*, 2005; Ghiringhelli *et al.*, 2006; Lim *et al.*, 2005; Nishikawa *et al.*, 2005; Piccirillo and Shevach, 2001; Thornton and Shevach, 1998; Turk *et al.*, 2004). Phenotypically, these suppressor cells are characterized by their expression of certain surface and intracellular molecules, which include the following: the IL-2 receptor alpha chain (e.g., CD25), cytotoxic T lymphocyte-associated protein 4 (CTLA-4), and glucocorticoid-induced TNFR-related protein (GITR). Recently, the lack of CD127 expression was shown to predict functional Tregs in normal humans (Liu *et al.*, 2006), but it is relatively unstudied in tumor Tregs. The transcription factor Forkhead-box-p3 (Foxp3) is a more specific marker of Tregs (Hori *et al.*, 2003). Recently, it was demonstrated that Bcl-xL plays a role in the induction and suppressive function of Tregs, in addition to its antiapoptotic effect (Sharabi *et al.*, 2009).

The presence of Tregs in tumors is associated with a poor prognosis (Curiel *et al.*, 2004). Patients with many different types of cancers had increased numbers of Tregs in their blood, tumor mass, and draining lymph nodes. Increased numbers of Tregs in lung and ovarian cancers were first reported by Woo *et al.* (2001). Later it was demonstrated that high frequencies of Tregs are allocated not only at the proximity of tumors but also in peripheral blood, thus suggesting that an increased number of Tregs is a generalized phenomenon (Liyanage *et al.*, 2002). It is thought that active proliferation of Tregs rather than redistribution from other compartments is responsible for the tumor-associated increase in the numbers of Tregs (Wolf *et al.*, 2006).

Increased numbers of Tregs were found in patients with MM as well (Beyer and Schultze, 2006; Beyer *et al.*, 2006; Feyler *et al.*, 2009). Interestingly, *in vitro* expansion of Tregs could be induced in the presence of MM-specific antigens (Han *et al.*, 2008). The increased number of Tregs was associated with reduced immune effector functions (Han *et al.*, 2008), and was suggestive of the progression of malignant transformation (Beyer *et al.*, 2006).

Therapeutic approaches for breaking tolerance to tumor cells have been tried; the depletion of Tregs is the most studied strategy (Ercolini *et al.*, 2005; Ghiringhelli *et al.*, 2004; Shimizu *et al.*, 1999). Specific depletion of Tregs by anti-CD25 antibodies improved endogenous immune-mediated tumor rejection (Shimizu *et al.*, 1999) by enabling the development of tumor-specific CD8 cells and NK cells that reacted against tumors (Shimizu *et al.*, 1999). Nevertheless, despite the tumor antigen-specific immunity (Tanaka *et al.*, 2002), the tumors were not completely rejected (Jones *et al.*, 2002). Cyclophosphamide (CYC) was found to have specific effects on T cells, with tumor-inhibiting properties (Proietti *et al.*, 1998). This alkylating agent was shown to have beneficial effects in the treatment of MM, and to be associated with increased survival rates (Rivers *et al.*, 1963). It was reported that the beneficial effects of CYC were due to the removal of suppressor T cells rather than to the reduction in tumor burden (McCune *et al.*, 1998).

The use of various doses of CYC for depleting Tregs in different types of solid tumors has been reported. In this regard, low doses of CYC had a specific effect in depleting Tregs (Awwad and North, 1989; Ghiringhelli *et al.*, 2004). High-dose CYC also depleted Tregs but was less effective than the low-dose CYC in rejecting the tumor (Castano *et al.*, 2008). Thus, apparently the beneficial effects of low-dose CYC on tumor rejection may predominantly be immune mediated and less cytotoxic mediated. Indeed, the resulting depletion of Tregs by low-dose CYC augmented the immune response to cancer immunotherapy (Machiels *et al.*, 2001), unlike the high-dose CYC, which caused general immune cell depletion, and as a consequence, the concomitant depletion of CD4 cells and CD8 effector T cells that are required for developing an effective antitumor immunity (Castano *et al.*, 2008). Further, low-dose CYC inhibited angiogenesis and vasculogenesis (Kerbel and Kamen, 2004), and impeded tumor cell repopulation kinetics (Wu and Tannock, 2003). In agreement, mathematical analysis of the evolutionary dynamics of tumor populations predicted that the control of tumors by chemotherapy could be achieved using progressively lower doses and increasingly long intervals between doses (Gatenby *et al.*, 2009). Hence, it is suggested that a desirable effect of a chemotherapeutic compound would result in a tumor volume that is either stable or slowly increases for a prolonged period of time.

II. INCREASED TREGS IN MOUSE MODELS OF MM

In recent years, the role of Tregs in tumor development has been extensively studied. $CD4^{+}CD25^{+}Foxp3^{+}$ Tregs suppress T cell proliferation, downregulate proinflammatory cytokines, and are involved in tumor tolerance, which is one of the main obstacles to overcome for improving antitumor immunity. Elevated Treg levels in rodents and humans with solid tumors and hematological malignancies, including human MM have been observed (Beyer and Schultze, 2006; Curiel *et al.*, 2004; Liyanage *et al.*, 2002; Marshall *et al.*, 2004; Ormandy *et al.*, 2005), and their functional role in reducing antitumor responses has been demonstrated in rodents (Onizuka *et al.*, 1999, Shimizu *et al.*, 1999; Sutmuller *et al.*, 2001; Turk *et al.*, 2004). In humans, the contribution of Tregs to tumor tolerance was strongly suggested by the significant correlation between Treg levels and the poor survival of ovarian cancer patients as well as tumor relapse in other malignancies such as breast cancer and non-small lung cancers (Bates *et al.*, 2006; Curiel *et al.*, 2004; Petersen *et al.*, 2006). Onizuka *et al.* (1999) were the first to suggest that $CD4^{+}CD8^{+}$ T cells played an important role in inhibiting tumor immunity, causing regression induced by $CD25^{+}$ cell depletion; similar conclusions were presented by Shimizu *et al.* (1999). These studies on depletion of $CD4^{+}CD25^{+}$ T cells and adoptive transfer of $CD4^{+}CD25^{+}$ T cells strongly suggest that the effectiveness of an antitumor therapy is greatly enhanced by removal of $CD4^{+}CD25^{+}$ T cell suppression activity. Sutmuller *et al.* (2001) were able to demonstrate that antibody-mediated depletion of $CD25^{+}$ T cells followed by vaccination with the GM-CSF-transfused melanocyte cell line resulted in enhanced tumor rejection. Experiments with adoptively transfused Tregs provided a direct link between Treg cells and reduced tumor immunity (Antony *et al.*, 2005). Thus, it is essential to reveal the mechanism leading to Treg expansion for developing strategies to eliminate them and to improve the results of cancer immunotherapy (Zou, 2005).

Several mechanisms describing Treg induction or recruitment to the tumor site have been described in the literature. It has been suggested that Tregs are induced at the tumor site and further affect the tumor microenvironment and draining lymph nodes (Kim *et al.*, 2006; Zou, 2005). Indeed, it was recently shown that Tregs were induced at the tumor site as a result of IL-10 and TGF-β secretion by tumor cells (Jarnicki *et al.*, 2006; Larmonier *et al.*, 2007; Liyanage *et al.*, 2006). Additionally, Tregs were shown to specifically recruit to the tumor by chemotaxis that was mediated by the release of CCL22 and CCL17 by the tumor cells (Mizukami *et al.*, 2008). The thymus is recognized as the main site of Treg development (Itoh *et al.*, 1999; Kim *et al.*, 2007; Sakaguchi, 2005; Shevach, 2000). Treg development in the thymus has been discussed as a possible mechanism contributing to Treg accumulation in malignancy (Beyer and Schultze, 2006) and thymus output was indirectly tested in human MM patients (Beyer *et al.*, 2006). However, Treg development in the thymus during malignancy has not been directly explored. Unique mouse models (5TM series) that mimic human MM disease served as a tool to examine levels in the periphery as well as developmental processes that may occur in the thymus to increase Treg ratios in MM development. The 5T33 (IgG2bκ)MM and 5T2MM (IgG2aκ) are the best characterized tumors (Radl *et al.*, 1988; Vanderkerken *et al.*, 1997). The 5T2MM model closely resembles the most common form of human MM in its selective localization to the BM, the presence of serum M component, the development of osteolytic bone disease, and the moderate progressive course of the disease. 5T2MM cells grow exclusively *in vivo* and can only be maintained *in vitro* for a very short period when coculture with BM stromal cells. In contrast, the 5T33 MM model represents an aggressive, rapidly progressive variant and cells can easily be maintained *in vitro* (Manning *et al.*, 1992).

Accumulation of suppressive functional $CD4^{+}CD25^{High}Foxp3^{+}$ Tregs was observed in peripheral organs during disease progression in both 5T2MM and 5T33MM mouse models. Treg levels were tested in spleen, lymph nodes, BM, and peripheral blood at different time points (28, 42, 66, 90, and 104 days) following 5T2MM cell injection. At the first two time points, 5T2MM-bearing mice were asymptomatic; the clinical phase involving hind limb paralysis appeared around 60 days and became more severe with increased latency. Treg frequency significantly increased at an early stage (28 d) in spleen and lymph nodes and remained constant during disease progression. Treg ratios increased similarly in lymph nodes surrounding the main sites of tumor infiltration (inguinal, caudal, and lumbar nodes) and lymph nodes distal to the main tumor site (superficial nodes, auxiliary nodes, and branchial nodes). In contrast, Treg frequency in BM remained normal in early stages but increased markedly only in the more progressive phases of the disease, about 90 days onwards after tumor cell challenge. In peripheral blood, a mild but significant increase was observed before paralysis onset (at 42 days) and remained constant during disease progression (Laronne-Bar-On *et al.*, 2008). These observations concerning elevated Treg levels during MM progression coincide with similar findings in MM patients (Beyer *et al.*, 2006; Feyler *et al.*, 2009).

A. Changes in Thymus Structure and Composition

Since Tregs normally develop in the thymus (Itoh *et al.*, 1999; Kim *et al.*, 2007; Sakaguchi, 2005; Shevach, 2000), it was essential to examine whether thymic processes were involved in increased Treg frequency in the periphery of MM-bearing mice. Thymus atrophy, manifested by a significant reduction in thymus weight and cellularity (~5.5), was observed in both 5T33M and 5T2MM mouse models during the disease's progression (Laronne-Bar-On *et al.*, 2008). A distortion of the normal distinction between cortical and medullary areas was observed. No thymus atrophy was observed in MM-bearing mice during the asymptomatic phase (40 days after tumor cell injection). The atrophy was correlated with the clinical phase of hind limb paralysis caused by spinal cord compression (from 60 days onwards post-tumor cell challenge) and further increased with disease progression and/or severity. Only thymus atrophy that occurred in paralyzed mice was associated with increased Treg-to-effector T cell proportions in MM-diseased mice. Although thymus cellularity was reduced, Treg numbers were not severely decreased in MM-bearing mice whereas numbers of effector T cells were dramatically reduced. The $CD4^+CD8^+$ double positive (DP) population, normally the largest thymocyte subset, significantly decreased, whereas the $CD4^-CD8^-$ double negative (DN) population increased. The proportion of the most mature population of $CD4^+CD8^+$ single positive (SP) cells significantly increased in the thymus, suggesting that a change in kinetics rather than a developmental block at the DN stage was responsible for changes in the subpopulation proportions.

Thymus atrophy was reported in cancer patients and tumor-bearing animals (Thomas *et al.*, 1985; Zhang, 1989). In a mouse mammary tumor model, thymic atrophy progressed with tumor growth; in mice with a large tumor mass, the thymus became involuted to less than a 10th of its normal size and its architecture was totally disrupted. Phenotypic analysis of the thymus from tumor bearers revealed a dramatic decrease in the percentage of DP immature thymocytes compared with those in normal controls. Severely altered levels of subpopulations of the $CD4^-CD8^-$ precursors suggested an early block in the maturation of DN cells. An impaired thymic stromal microenvironment in tumor-bearing mice, increasingly disorganized and altered, coincided with tumor growth (Adkins *et al.*, 2000; Lopez *et al.*, 2002). Similar observations were described in the thymus in mice bearing Lewis lung carcinoma and in ascitic growth of a spontaneous transplantable T cell lymphoma (Kaiserlian *et al.*, 1984; Shanker *et al.*, 2000).

Prolonged infusion of recombinant VEGF, a factor secreted by various tumors including MM, caused profound thymic atrophy (Ohm *et al.*, 2003). A dramatic reduction in $CD4^+CD8^+$ thymocytes and a decreased number of the earliest occurring progenitors in the thymus was observed. Thus, pathophysiologically relevant concentrations of VEGF may block the differentiation and/or migration of these progenitors, resulting in thymic atrophy. Cessation of VEGF infusion resulted in the restoration of the normal composition and cellularity of the thymus. Thus, continuous administration of recombinant VEGF mimics the profound thymic atrophy observed in tumor-bearing mice, and inhibits the production of T cells. VEGF acts on thymic progenitors rather than directly on the thymus itself. VEGF infusion results in defective seeding of the thymus by BM-derived progenitors. These earliest thymocytes fail to replace maturing T cells emigrating to the periphery and consequently, all thymocytes are depleted.

B. Increased Frequency of Treg Development in the Thymus of MM-Bearing Mice

The thymus is normally the main site of Treg development. Following the observation of significant changes in thymus characteristics in MM-bearing mice, it was important to determine whether Treg development was altered in the thymus of MM-bearing mice. The frequency of mature $CD4^+$ SP thymocytes expressing CD25 significantly increased in 5T2MM-bearing mice approximately twofold, and most of the $CD4^+$ SP cells that expressed CD25, coexpressed Foxp3. Foxp3 is a transcription factor that identifies functional Tregs (Hori *et al.*, 2003). There was no significant difference in the percentage of $Foxp3^+$ among $CD4^+$ SP $CD25^+$ cells in controls or in 5T2MM-bearing mice. Interestingly, CD25 expression was increased already at the DP stage. Although most $CD25^+$ DP cells did not express Foxp3, the frequency of $CD25^+$ $Foxp3^+$ cells significantly increased during this stage, and this increase was accompanied by a decrease in the ratios of $CD25^+$ $Foxp3^-$ DP cells (Laronne-Bar-On *et al.*, 2008). These results are in accordance with previous data suggesting the commitment of the Treg linage as early as the DP stage (Bayer *et al.*, 2007; Cabarrocas *et al.*, 2006; Pennington *et al.*, 2006).

Increased $CD25^+Foxp3^+$ expression in the DP stage implies that increased Treg ratios among mature thymocytes result from changes in the developmental processes in the thymus of MM-bearing mice. To exclude the possibility that increased Treg ratios reflect Treg recirculation from the periphery to the thymus (Bosco *et al.*, 2006; Zhan *et al.*, 2007), we compared the naïve phenotype of Tregs in the thymus and in the periphery of 5T2MM-bearing mice. The mouse naïve T cells can be distinguished by the marked expression of CD62L. Tregs in the periphery might have been activated, thereby losing their naïve phenotype. Actually, the percentage of peripheral $CD62L^{high}$ Tregs significantly decreased in the 5T2MM-diseased mice compared with controls, indicating loss of the naïve phenotype. Tregs in the thymus of the same mice retained a naïve phenotype and a statistically insignificant increase was observed (compared with the controls). Effector T cells in the periphery and thymus of 5T2MM-bearing mice exhibited similar trends. The Treg memory phenotype was also tested in the thymus and periphery using the memory marker CD44. The percentage of Tregs expressing CD44 did not change in the periphery (spleen and lymph nodes) or in the thymus of 5T2MM-bearing mice. Analysis of CD44 and CD62L coexpression revealed a similar decrease in CD62L^{high} expressing cells as was found in the total Treg population, suggesting a shift from a CD44^{high} CD62L^{high} (central memory) to CD44^{high} CD62L^{low} (effector memory) phenotype among peripheral Tregs (Laronne-Bar-On *et al.*, 2008). These results indicate that Tregs have a distinct naïve phenotype in the thymus of 5T2MM mice and suggest that Treg recirculation from the periphery is not the cause of increased ratios of Tregs in the thymus.

Since atrophy in MM mice was associated with reduced cellularity, it was interesting to follow Treg levels and their physiological activity in the involuted thymuses. Treg absolute numbers in the thymus of diseased mice were not altered when compared with controls. In contrast, the number of effector T CD25^{-} cells was reduced ~2.5-fold. Treg numbers decreased only in the severely atrophied thymuses up to ~3-fold, compared with a more dramatic ~9.5-fold reduction of effector T cell numbers (Laronne-Bar-On *et al.*, 2008).

The effect of thymus atrophy on peripheral effector T cell numbers and function in MM are largely unknown (Raitakari *et al.*, 2003). Low effector T cell frequency occurring in the peripheral blood of 5T2MM mice may be associated with thymus atrophy. Effector T cell depletion but not activation or proliferation in the periphery is significantly correlated with thymus atrophy. However, since thymus atrophy is associated with diseased severity, it cannot be concluded that effector T cell depletion results from thymus atrophy or from other processes associated with the disease. Accumulating data suggest that increased Treg-to-effector T cell ratios in the thymus of MM-diseased mice did not result from altered thymocyte survival or increased Treg proliferation. Importantly, the balance between Tregs and effectors has been stressed as critical for deciding between immune response and suppression (Belkaid and Rouse, 2005; Pennington *et al.*, 2006). The data showing increased Treg to effector T cell proportions among immature thymocytes suggest that an effector immune balance exists in the periphery of MM mice. The reviewed data suggest a thymic contribution to increased Treg ratios among CD4^{+} cells, as was found in the mouse MM models and in human MM patients (Beyer *et al.*, 2006), in addition to peripheral mechanisms reported to contribute to Treg accumulation at the tumor site.

C. Adoptive Transfer of Thymocytes from 5T2MM-Diseased Mice Affects the Severity of MM Manifestations in 5T2MM-Injected Mice

Patients with MM commonly develop bone disease, including bone pain, osteolytic lesions, pathologic fractures, and hypercalcemia. Bone destruction in MM results from asynchronous bone turnover. Normal osteoclasts are induced by osteoclast-activating factors produced by myeloma cells or the cells in the microenvironment; however, the process is not accompanied by increased bone formation by osteoblasts (Callander and Roodman, 2001; Terpos *et al.*, 2007; Yeh and Berenson, 2006). The 5T2MM mouse model also involves bone lesions as a primary sign of the disease (Dingli and Russell, 2007; Vanderkerken *et al.*, 1997). The 5T2MM cells localize primarily to the BM, replacing the normal BM cells and causing bone lesions. The mice develop hind limb paralysis as a result of spinal cord compression. An adoptive transfer assay (Deng *et al.*, 2006) was carried out to determine whether thymocytes from 5T2MM-diseased mice could support *in vivo* tumor progression. Mice challenged with 5T2MM cells, still in the asymptomatic phase (42 days after 5T2MM cell challenge), received thymocytes from paralyzed 5T2MM-bearing mice, or from healthy mice. The severity of disease manifestations was apparent. Eighty percent of mice injected with thymocytes from diseased mice developed severe bone destruction and massive tumor growth around the spine, and had infiltration into the surrounding muscles in contrast to 20% in mice injected with control thymocytes, which developed less severe bone destruction. Adoptive transfer of Treg thymocytes from thymus of 5T2MM-diseased mice and thymocytes excluding Tregs (Treg depleted) presented an early onset of disease only following the transfer of Treg thymocytes. Thus, Tregs alone, but not other thymocyte populations, could account for the tumor progressive effect of 5T2MM-derived thymocytes (Laronne-Bar-On *et al.*, 2008).

III. TREG DEPLETION BY CYC IMPROVES ANTITUMOR IMMUNITY

The role of Tregs in tumor development has been extensively studied in recent years. Tregs suppress T cell proliferation, downregulate proinflammatory cytokines, and are involved in tumor tolerance to self-antigens. In addition, Tregs are thought to dampen T cell immunity to tumor-associated antigens and to be the main obstacle to successful immunotherapy. Much data suggest that early-stage cancers are eliminated by immune surveillance, whereas established tumors are more likely to induce immune tolerance (Pardoll, 2003). A multitude of tumor-derived factors contribute to tumor microenvironmental immune tolerance and to immunosuppression; this helps elucidate the lack of effective immune surveillance in later stages of tumor development. Functional Tregs are increased in peripheral blood and in the tumor microenvironment of patients suffering from different types of cancer. A correlation between increased rates of Tregs and disease progression was observed in cancer patients and in rodent models of solid tumors and hematological malignancies. In humans, the contribution of Tregs to tumor tolerance was strongly suggested by the significant correlation between Treg levels and the poor survival of ovarian cancer patients, progression of pancreatic ductal adenocarcinoma, and tumor relapse in patients with breast cancer and non-small cell lung cancer (Bates *et al.*, 2006; Curiel *et al.*, 2004; Hiraoka *et al.*, 2006; Petersen *et al.*, 2006). Thus, Treg-mediated immunosuppression could be a crucial evasion mechanism that prevents the elimination of cancerous cells by the immune system. Experiments with adoptively transferred Tregs provided a direct link between Treg cells and reduced tumor immunity. Hence, new strategies in cancer immunotherapy, aimed at reducing Tregs, have been proposed (Ruter *et al.*, 2009). Five general strategies to reduce Treg functions have been used: (1) depletion of Tregs; (2) blockade of Treg functions; (3) blockade of Treg trafficking; (4) blockade of Treg differentiation; and (5) combining depletion of Tregs with tumor vaccines.

In our previous studies concerned with the effect of erythropoietin on MM development, using the 5T33MM mouse model, we found that erythropoietin acted as an immunomodulating agent, promoting specific T cell-dependent immune response (Mittelman *et al.*, 2001). The 5T33MM mouse model represents an aggressive rapidly progressive variant that survives for about 4 weeks. Since we were interested in following the pathophysiological mechanism involved in MM development and prevention, for our further studies we chose the 5T2MM mouse model, which has a moderate, progressive course of disease, lasting about 3 months. We observed a correlation between increased ratios of $CD4^{+}CD25^{High}Foxp3^{+}$ Tregs and disease progression (Laronne-Bar-On *et al.*, 2008). The obvious next phase was to study the effect of $CD4^{+}CD25^{High}$ $Foxp3^{+}$ Treg depletion on the progression of the disease. CYC was used to deplete Tregs.

CYC is an alkylating agent widely used in chemotherapeutic regimes because of its broad antitumor spectrum and its selective cytotoxicity (Brode and Cooke, 2008). CYC is known to reverse immunological tolerance and to facilitate adoptive immunotherapy through inhibition of suppressor T cell activity. High doses of CYC are required for effective tumor chemotherapy, which might lead to immunosuppression. Strikingly, low-dose CYC can selectively decrease Tregs; therefore, it can be useful for immunomodulation. CYC was also shown to increase the production of inflammatory cytokines (IL-1, TNF-α/β, IFN-γ), and tumor-induced immuno-suppressive factors (TGF-β, IL-10, VEGF).

Low-dose CYC was shown to decrease Treg numbers and to inhibit their suppressive function (Ikezawa *et al.*, 2005; Lutsiak *et al.*, 2005), as well as to enhance apoptosis and decrease Treg homeostatic proliferation (Lutsiak *et al.*, 2005). A single administration of low-dose CYC was shown to deplete Tregs in colon carcinoma-bearing rats, thereby delaying tumor growth. In rats bearing established tumors, treatment with a single dose of CYC, followed by an immunotherapy strategy, restored antitumor activity of effector T cells (Ghiringhelli *et al.*, 2004). Inhibitory effects of low-dose CYC on tumor were determined in mice that spontaneously develop prostate carcinoma also through the depletion of Tregs (Wada *et al.*, 2009). Treatment of a mammary tumor model in the neu-N line with immunomodulating doses of CYC in sequence with neu-targeted vaccine revealed high avidity-specific $CD8^+$ T cell activity associated with more effective eradication of neu-expressing tumors *in vivo* (Ercolini *et al.*, 2005). The mechanism by which CYC chemotherapy enhances the vaccine-induced specific T cells is through depletion of Tregs. Adoptive transfer of $CD4^+CD25^+$ Tregs was shown to inhibit the antitumor immune response induced by CYC administered with vaccine. This is the first report demonstrating the unmasking of high-avidity $CD8^+$ T cell responses against a naturally expressed tissue-specific tumor antigen in a murine model of tolerance.

Another model showed a direct functional link between the transfer of $CD4^+CD25^+$ T cells and reduced therapeutic efficiency of adoptively transferred tumor-antigen-specific effector T cells in a mouse melanoma model. Thus, the optimal vaccine effect against melanoma antigen could be achieved only when $CD4^+CD25^+$ Tregs were depleted by CYC treatment (Antony *et al.*, 2005). Single administration of low-dose CYC was shown to potentiate the antitumor effect of DC vaccine in mice bearing B16 melanoma or C26 colon carcinoma. Increased proportions of IFN-γ by removing suppressor T cells induced a bystander effect (Gorelik *et al.*, 1994; Liu *et al.*, 2007; Machiels *et al.*, 2001). Schiavoni *et al.* (2000) showed that CYC acts by removing suppressor T cells followed by production of type I IFN, thus increasing $CD44^{hi}$ $CD4^+$ and $CD44^{hi}$ $CD8^+$ T cells (memory phenotype). CYC was also shown to have an antiangiogenic component. Scheduled CYC administration for shorter intervals without interruption (defined metronomic regime; Kerbel and Kamen, 2004) resulted in apoptosis of vascular endothelial cells within the tumor bed. The therapeutic advantage of slowing or suppressing the growth of tumors was demonstrated in mice bearing Lewis lung carcinoma cells or L1210 leukemia cells (Browder *et al.*, 2000).

The metronomic low-dose CYC regime used in advanced cancer patients was shown to induce a profound and selective reduction of circulating Tregs and the reduction of tumor-induced tolerance. CYC treatment led to the restoration of peripheral T cell proliferation and innate killing activities, favoring a better control of tumor progression. This metronomic CYC regime dramatically enhanced T and NK cell effector function through its suppressive effect on Treg number and function (Ghiringhelli *et al.*, 2007).

A. Effects of a Single Low- and High-Dose CYC on 5T2MM Progression

Norths' pioneering studies in the 1980s suggested that suppressive T cell function could be selectively inhibited in tumor hosts receiving low-dose CYC treatment (North, 1982). Extensive studies on Treg biology presented evidence that different mechanisms govern the antitumor effect of low- and high-dose CYC (Brode and Cooke, 2008; Lutsiak *et al.*, 2005; Motoyoshi *et al.*, 2006). A single injection of different doses of CYC (50, 100, and 200 mg/kg body weight) administered to 5T2MM-bearing mice in the early clinical phase (70 days after cell challenge) prolonged their survival very significantly in comparison with the control group (5T2MM with diluent treatment). The tumor load at the timing of CYC treatment, reflected in the serum protein level (using a standard electrophoretic technique), was 0.95–1.52 g/dl and administering CYC reduced it to the control level (0.13–0.2 g/dl) within 2 days after injecting CYC. The hind limb paralysis involving a nerve compression syndrome such as spinal cord compression, observed in the 70-day clinical phase of 5T2MM-bearing mice, disappeared 14 days after administering the three different CYC doses. The tumor cells in the hind limbs were replaced by normal BM cell populations for several months. The prolonged survival of cell populations following a single CYC injection, irrespective of its dose level, might be related to the CYC-induced disappearance of plasma tumor cells from the BM. Homing of MM cells in the BM is important for their interaction with stromal cells, which induce a microenvironment for their survival as well as growth signals (Hideshima *et al.*, 2007). The main difference between administering low- and high-doses of CYC to the 5T2MM mice lies in the ultimate development of disease (Fig. 1). A high incidence of diseased mice (80%) was observed in those 5T2MM mice treated with a high CYC dose (200 mg/kg). Since the cytotoxic high-dose CYC is less selective to all lymphocytes, including populations with antitumor properties, the residual 5T2MM cells apparently recovered during their prolonged latency, ultimately yielding a high MM incidence. Both groups treated with a low CYC dose (50 or 100 mg/kg) developed a similar lower MM incidence (53% and 59%). Low-dose CYC treatment is associated with selective transient depletion of Tregs in the diseased mice, leading to restoration of peripheral T cell proliferation and immune functions.

The clinical effect of a single injection of low-dose CYC was shown to depend on tumor load. Administering CYC at different intervals of the 5T2MM tumor cell injection affected the final MM incidence, though prolonged survival was observed irrespective of the tumor load level. The single CYC treatment was given to mice harboring 5T2MM cells for 47, 70, or 94 days. The levels of M paraproteins associated with MM development

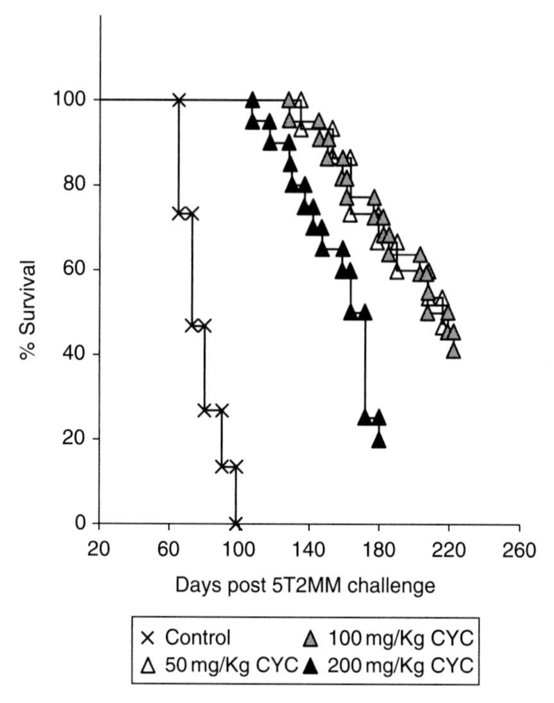

Fig. 1 Myeloma progression following treatment with a single low- or high-dose CYC. Seventy days after 5T2MM cell challenge (i.v. injection of 3×10^5cells/mouse) the mice were divided into four treatment groups as follows: i.p. injection of two different low doses, 50 mg/kg ($n=15$), or 100mg/kg ($n=22$), a high dose CYC, 200 mg/kg ($n=20$), or the diluent ($n=15$). A follow-up of the mice's survival was carried out for 240 days. A 100% (15/15) of the control group developed MM at 82±11 days latency. Mice injected with 50 mg/kg CYC exhibited a reduced incidence of 53% (8/15) and a prolonged survival of 176±22 days, and 59% (13/22) treated with 100 mg/kg CYC at 181±26 days mean survival; mice injected with the high CYC dose developed a high MM incidence of 80% (16/20) with a prolonged survival of 146±28 days.

were 0.57–0.98 g/dl in mice in the asymptomatic phase (47 d), 1.1–1.52 g/dl in the early clinical phase (70 d), (all mice with early hind limb paralysis), and 2.55–2.99 g/dl in very sick mice (94 days). In the control group (injected with diluent) all mice (100%) developed MM at a mean latency of 73±9 days. In mice carrying the lowest tumor load (at 47 d), reduced tumor incidence was observed (66%); a further tumor increase (83%) was noted in the early clinical phase (70 d) and 100% in the very sick mice (94 days post MM cell challenge). The efficiency of a single low-dose of CYC in reducing MM progression depended on the tumor load, as reflected in the serum paraprotein level at the time of drug administration. There was a 40–60% MM incidence with a lower tumor burden, and 80–100% incidence with a further increased tumor burden. Nevertheless, substantially prolonged survival was observed among 5T2MM mice treated with CYC in comparison with those injected with the diluent, irrespective of the tumor burden.

B. Cellular Component of the Immune System in MM

The number and function of T cell subsets were reported to be abnormal in patients with MM. The CD4:CD8 ratio is inverted, and the Th1:Th2 ratio among $CD4^+$ cells is abnormal (Mills and Cawley, 1983; Ogawara *et al.*, 2005). T cells from MM patients were shown to function aberrantly (Brown *et al.*, 1998; Frassanito *et al.*, 2001). In addition, the levels of expression of CD28 and CTLA-4 costimulatory molecules required for T cell activation and inhibition, respectively, were downregulated in T cells derived from MM patients (Mozaffari *et al.*, 2004). B cell activity was suppressed in patients with an active stage of MM because the cells secreted reduced levels (hypogammaglobulinemia) of polyclonal immunoglobulin, which was inversely correlated with the disease stage (Rawstron *et al.*, 1998). The elevated levels of TGF-β (Urashima *et al.*, 1996), in addition to the impaired accessory signals from Th cells, contributed to dysfunctional B cells. Defective NK cells have also been noted in patients with MM (Jarahian *et al.*, 2007). This is of major importance since NK cells have antimyeloma activity (Carbone *et al.*, 2005; Frohn *et al.*, 2002). Circulating DCs from MM patients were shown to be dysfunctional because the cells failed to upregulate costimulatory molecules required for activation (Brimnes *et al.*, 2006; Brown *et al.*, 2001). It was suggested that reduced function of DCs indicates the progression of the disease (Brown *et al.*, 2001). Further, DCs from MM patients had reduced phagocytic capacity (Ratta *et al.*, 2002). In addition, monocyte-derived DCs exhibited downregulated expression of activation markers and impaired presentation capacity to T cells (Wang *et al.*, 2006). Impaired activity of DCs may be linked to the upregulation of Tregs (Onishi *et al.*, 2008). Cytokines such as IL-6, TGF-β, IL-10, and VEGF, which were actively produced by myeloma cells (Brown *et al.*, 2001), and were found to be in the tumor microenvironment as well as in the serum (Wang *et al.*, 2006), played a role in preventing the development of functional DCs.

C. CYC Effects on Molecules Essential for the Survival and Function of Tregs

There are several molecules that phenotypically characterize Tregs and enable their suppressive function. Foxp3 is a master gene that identifies functional Tregs (Hori *et al.*, 2003). It was reported that injecting a low-dose

CYC results in downregulated expression of Foxp3 in Tregs, which could cause a loss of suppressive activity (Lutsiak *et al.*, 2005). It is possible that CYC downregulate the expression of Foxp3 in Tregs because CYC was shown to result in the upregulation of OX40 (CD134) primarily in Tregs (Hirschhorn-Cymerman *et al.*, 2009) and OX40 engagement on Tregs can reduce Foxp3 levels (Kitamura *et al.*, 2009; Vu *et al.*, 2007). TGF-β is elevated in patients with MM (Cook *et al.*, 1999), and this immunosuppressive cytokine plays a significant role in many aspects of Treg activity. For example, it can maintain the expression of Foxp3 in Tregs (Marie *et al.*, 2005), it induces responder T cells to be sensitive to suppression (Fahlén *et al.*, 2005), and when it is membrane-bound, it may mediate suppression (Nakamura *et al.*, 2001). Although treatment with CYC may result in the upregulation of TGF-β and enhance the induction of functional Tregs, plasma cells from MM patients were resistant to the inhibitory effects of TGF-β on B cell proliferation and immunoglobulin production (Urashima *et al.*, 1996).

Bcl-xL is an antiapoptotic molecule known to play a role in the development, differentiation, and clonal selection of B cells (Amanna *et al.*, 2003; Takahashi *et al.*, 1999). Upregulation of Bcl-xL expression was demonstrated in patients with MM (Gauthier *et al.*, 1996; Tu *et al.*, 1998). Further, the expression of Bcl-xL was associated with the progression of MM and impaired the response to treatment in those patients with elevated levels of this antiapoptotic molecule (Tu *et al.*, 1998). Recently, we showed that Bcl-xL plays a role in the induction of Tregs (Sharabi *et al.*, 2010a). Bcl-xL was involved in the induction of Foxp3 in Tregs and in enabling their suppressive function. We also found that the reduced numbers of Tregs in 5T2MM-bearing mice following treatment with a low-dose CYC could be accomplished by downregulating the expression of Bcl-xL in Tregs and increasing their apoptosis.

CTLA-4 is an inhibitory T cell molecule essential for T cell homeostasis and tolerance induction (Chambers, 2001; Salomon and Bluestone, 2001). It is constitutively expressed in Tregs. We found that administration of a low-dose CYC to 5T2MM-bearing mice resulted in a significant reduction of CTLA-4 expression in Tregs. Recently, it was demonstrated that deficient expression of CTLA-4 may hinder the *in vivo* development and suppressive function of Tregs (Wing *et al.*, 2008). In addition, the downregulation of CTLA-4 may decrease the expression of Bcl-xL (Sharabi *et al.*, 2010a), thus interfering further the development of functional Tregs.

Tregs highly consume IL-2 for their homeostasis and since they cannot produce this cytokine, they depend on effector T cell production (Fontenot *et al.*, 2005). We showed that treatment of 5T2MM-bearing mice with a low-dose CYC resulted in a significant decreased production of IL-2 in CD4 effector cells (Sharabi *et al.*, 2010b). Therefore, it is possible that deficient expression of IL-2 might interrupt Treg maintenance.

D. Adoptive Transfer of Tregs Shortly After Administering CYC to 5T2MM-Bearing Mice

The involvement of Tregs in the pathogenesis of MM has been frequently manifested by the increased number of Tregs associated with the progression of MM, and also by improved disease manifestations after depletion of Tregs. Since specific downregulation of Tregs can be accomplished by injection of low-dose CYC, we conducted a series of experiments in 5T2MM-bearing mice aimed at highlighting other beneficial aspects of CYC, in addition to depletion of Tregs, which might explain its ameliorative effects on MM. In these experiments, 5T2MM-bearing mice with full-blown MM were treated with a single injection of low-dose CYC, and 24 h later, when the cytotoxic effects of CYC were substantially diminished (Sladek *et al.*, 1984), these mice were injected by means of adoptive transfer of two types of cells, for example, the treated mice received either Tregs or effector T cells. Thus, we found that amelioration of MM manifestations, observed in diseased mice in response to low-dose CYC, was abrogated when the mice were injected with Tregs. In contrast, CYC-treated mice that were adoptively transferred with effector T cells preserved the ameliorative effects of CYC on MM.

E. The Effect of CYC on NKT Cells and DCs

Patients with MM have reduced numbers of NKT cells, and IFN-γ production by freshly isolated NKT cells was deficient in patients with progressive myeloma (Dhodapkar *et al.*, 2003). In agreement with previous reports showing that a reciprocal relationship exists between NKT and Tregs (Smyth and Godfrey, 2000) and that Tregs could suppress the function of NKT cells (Azuma *et al.*, 2003; Nishikawa *et al.*, 2003), we noted that depletion of Tregs in 5T2MM-bearing mice that were treated with low-dose CYC was accompanied by significantly upregulated numbers of IFN-γ-producing NKT cells. The main role of NKT cells is to protect against tumors and pathogens (Kronenberg, 2005; Smyth *et al.*, 2002). It has been well documented that NKT cells produce large amounts of IFN-γ upon activation (Arase *et al.*, 1992), and that the antitumor properties of NKT cells are linked to this capability (Liu *et al.*, 2005; Smyth *et al.*, 2002).

Using 5T2MM-bearing mice, we demonstrated the reversibility of aberrant differentiation and function of DCs, observed in patients with MM (Brown *et al.*, 2001; Ratta *et al.*, 2002). DCs from mice with MM and treated with low-dose CYC did not expand but instead differentiated further and acquired a mature phenotype, for example, the DCs upregulated the expression of MHC class II and costimulatory molecules (Cederbom *et al.*, 2000; Höltl *et al.*, 2005; Larmonier *et al.*, 2007; Misra *et al.*, 2004). The latter effect is of great importance since tumor cells may evade immune responses by losing the expression of HLA molecules (Seliger *et al.*, 2000). It is possible that the elevated production of IFN-γ in the treated mice contributed to the differentiation process of DCs (Beatty and Paterson, 2001). Remarkably, treatment of 5T2MM-bearing mice with high-dose CYC, as oppose to treatment with low-dose CYC, neither affected substantially the number of NKT cells nor the production of IFN-γ by these cells, and did not result in maturation of DCs (Sharabi *et al.*, 2010b).

F. A Window of Opportunity

Because patients with MM are considered to have competent immune systems, it is reasonable to speculate that each patient's system would be capable of dealing with the disease by generating an antitumor immune response, provided that the inhibitory and regulatory pathways of the immune system are removed or at least put on hold. Tregs are major suppressors of the immune response; therefore, these cells may serve as a convenient target through which the development of MM can be manipulated. In this review, we focused on CYC and showed that using low doses of this drug may, on the one hand, result in depletion of Tregs, and on the other hand, still maintain functional immune-derived cells that would contribute to the amelioration of MM. Hence, the number and function of NKT cells could be recovered, the production of IFN-γ was enhanced, and DCs could continue their differentiation and become mature and ready for activation. Once the concept of low-dose CYC was proven feasible for potentially enabling an effective immune response against myeloma cells, it was essential to find the most effective protocol of treatment that would optimally achieve satisfactory and durable antimyeloma effects.

IV. OPTIMAL TIME SCHEDULES OF CYC TREATMENT AFFECTING MM PROGRESSION

Studies in animal models of cancer showed that tumor rejection can be facilitated by inhibiting the function of Tregs, which play a key role in tumor-induced tolerance. Administration of either low- or high-dose CYC to 5T2MM-bearing mice in their early clinical phase of the disease prolonged dramatically their survival. The main difference between the single injection of low- or high-dose CYC was the ultimate high MM incidence following high-dose CYC treatment in comparison with low-dose CYC. Since the cytotoxic effect of high-dose CYC was substantially less selective and without resulting in the recovery of immune-derived cells with anticancer properties, it may have enabled the growth of residual tumor cells, yielding ultimately a high MM incidence. Treatment with low-dose CYC was associated with selective transient depletion of Treg in the diseased mice, leading to restoration of peripheral T cell proliferation and immune functions. It seemed of interest to test whether reduced MM development during prolonged latency could be accomplished by repeated injections of low-dose CYC at intervals that would coincide with the timing before Treg restoration occurred.

A. The Clinical Effect of a Single Injection Versus Repeated Injections of Low-Dose CYC at Different Time Intervals

The kinetics of suppressor T cell depletion following low-dose CYC administration was described in several studies (summarized in Table I). The results differ according to whether normal mice, tumor-bearing rodents, or advanced cancer patients were tested. In normal mice Treg reduction began after 1–2 days, with the lowest decrease at 4–6 days, and this was restored to a normal level at 10–14 days (in one study the levels were monitored for 4 weeks and were still low). In colon cancer-bearing rats the decrease in Tregs began at day 1, the lowest level was at 7 days and it was restored to normal at 28 days. In 5T2MM-bearing mice a gradual decrease (tested at 14, 25, and 42 days post CYC administration) was observed at all testing points, including 42 days. In advanced cancer patients, 1 month after administering CYC, a dramatic selective Treg depletion was observed and 2 months after starting treatment, pretreatment Treg levels were observed.

Our studies involving the kinetics of $CD4^{+}CD25^{High}Foxp3^{+}$ Tregs following administration of low-dose CYC to 5T2MM-bearing mice showed that Treg depletion was maintained beyond 45 days. Populations involved in antitumor immune responses could effectively be recruited during this period, before renewal of Tregs. Thus, it was of interest to test the possible influence of the "timing window" period on MM progression involving repeated CYC treatments at 45-day intervals. To this end, mice bearing 5T2MM cells for 70 days were treated with three repeated CYC injections at 21- or 45-day intervals. Results are shown in Fig. 2. All treated mice had hind limb paralysis and their paraprotein level was 1.1–1.75 g/dl; the control group developed 100% MM (75$\pm$ 8 days mean survival) versus 71% MM (188$\pm$14 days mean survival) at 21-day interval CYC treatments and

Table I Kinetics of Tregs Following the Single Administration of CYC

Animal model/Human patients	Dose of i.p. CYC (mg/Kg)	First day of Treg reduction	Peak day of Treg depletion	Day of Treg normalization	References
Normal C57BL/6	100	1	4	10	Lutsiak *et al.* (2005)
Naïve neu-N	100	2	n.d.	14	Ercolini *et al.* (2005)
Colon carcinoma-bearing rats	25	1	7	28	Ghiringhelli *et al.* (2004)
Normal C_3H/HeN	20	1	4[a]	28[b]	Motoyoshi *et al.* (2006)
Normal C_3H/HeN	200	1	4	28[b]	Motoyoshi *et al.* (2006)
5T2MM-bearing mice	100	n.d.	14[a]	42[a]	Sharabi *et al.* (2010)
5T2MM-bearing mice	200	n.d.	14[a]	42[a]	Sharabi *et al.* (2010)
Patients with advanced cancer	100[c]	n.d.	30[d]	60[f]	Ghiringhelli *et al.* (2007)
Patients with advanced cancer	200[c]	n.d.	30[e]	n.d	Ghiringhelli *et al.* (2007)

[a]Depletion of 50% of baseline levels.
[b]Levels of Tregs remained reduced.
[c]Administered orally, daily, every 2 weeks, for a month.
[d]Selective reduction in Tregs number and function.
[e]Nonselective cell reduction.
[f]Days after treatment cessation.

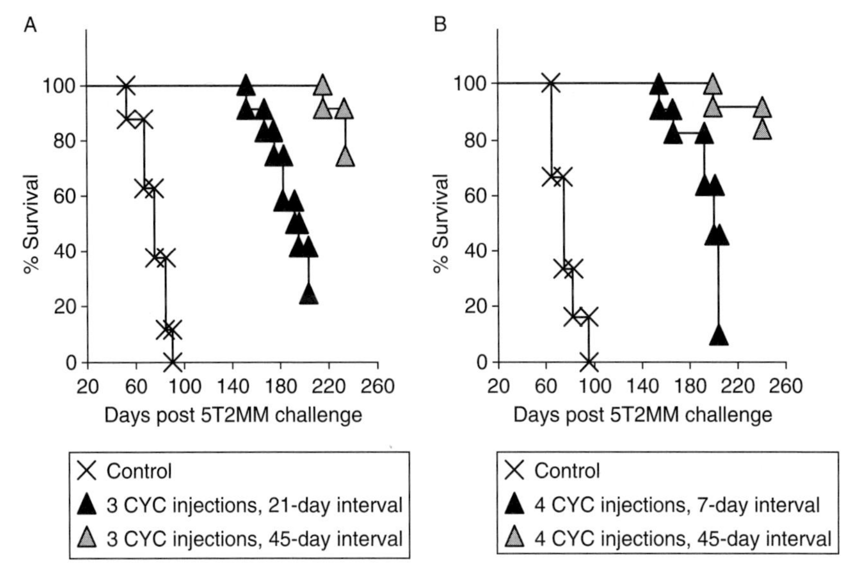

Fig. 2 Effect of repeated treatments of low-dose CYC (100 mg/kg). (A) Sixty days after 5T2MM cell injection, the mice were treated with 3 CYC injections administered at intervals of 21 days ($n=12$) or 45 days ($n=12$). The control group ($n=16$) was injected with the diluent at intervals of 21 days. All mice (16/16) in the control group developed MM at 75±8 days mean latency. MM incidence in mice receiving repeated injections of CYC at intervals of 21 days was 75% (9/12) at 188±14 days latency versus 25% (3/12) at 228±9 days mean survival at 45-day repeated treatments. (B) 5T2MM-bearing mice were treated 60 days after tumor cell challenge with four CYC injections either at 7-day intervals ($n=10$) or at 45-day intervals ($n=12$): 100% (12/12) in the control group developed MM at 76±9 days mean survival; 90% (9/10) that received four weekly injections developed MM at a mean latency of 191±11 days versus 16% (2/12) at 45-day intervals (sick at 201 and 240 days).

25% MM (228±8 days mean latency) when the interval between the repeated treatments was prolonged to 45 days (Fig. 2A). CYC administration markedly prolonged the survival of the CYC-treated mice, but the repeated treatments often (21 days vs. 45 days) did not improve the effectiveness of the drug, since MM incidence was much higher in spite of more frequent CYC treatment.

A similar experiment involving four repeated CYC treatments at intervals of 7 or 45 days also did not improve CYC effectiveness. The four weekly treatments at 7-day intervals resulted in 90% MM development at a 191± 11 day mean survival rate versus 16% MM at a 45-day interval. The prolonged survival in both groups was very significant (Fig. 2B). Thus, more frequent administration of CYC at intervals of 7 or 21 days did not improve the therapeutic effect versus a longer 45-day interval that was more beneficial. The prolonged maintenance of transient depletion of Tregs following repeated injection of low doses of CYC might enhance the reduction of MM incidence by tipping the balance toward effector T cells for a durable period coinciding with previous observations that depletion of Tregs promoted anti-T-cell responses (O'Garra and Vieira, 2004; Piccirillo and Thornton, 2004). The latter effect is of major importance since the number of Tregs is increased in patients and in mice with MM progression (Curiel *et al.*, 2004; Hiraoka *et al.*, 2006; Laronne-Bar-On *et al.*, 2008; Liyanage *et al.*, 2002; Marshall *et al.*, 2004; Ormandy *et al.*, 2005).

B. Prolonged Maintenance of Treg Depletion

The effect of chemotherapy by administering CYC depends on the timing and dose of CYC, while considering the transient depletion of Tregs. Our observations concerning the efficacy of a long time interval between repeated low-dose CYC treatments served as the basis for testing the prolonged maintenance of Treg depletion (thereby increasing immune antitumor responses) for developing MM. The protocol for this experiment is presented in Fig. 3A.

The initial antitumor treatment involved the administration of a cytotoxic high-dose CYC (200 mg/kg body weight) to mice that received 5T2MM cells 70 days earlier. The tumor load was eradicated (indicated by the normalization of serum preparation level), and hind limb paralysis disappeared within 14 days following CYC treatment. MM incidence in the treated mouse group B was 71% (15/21) within a mean latency of 157±17 days. In the control group A, of the mice bearing 5T2MM cells injected with diluent, 100% (23/23) developed the disease within 82±17 days of mean latency. Bone lesions mostly in femur and/or tibia developed in 43% (10/23) of the control group in about 80–114 days following tumor cell injection. No bone lesions were observed in mice treated with CYC. In mice treated with a high dose of CYC, this chemotherapeutic administration kills both tumor cells but also induces systemic immune suppression, thereby damping the therapeutic efficacy of immunotherapy. To further control the proliferation of 5T2MM residual cells (escaping the high-dose CYC cytotoxic effect), repeated low doses of CYC (100 mg/kg) were administered at 45-day intervals. We tested the effect of three different time schedules, 80-, 60-, and 45-day intervals following the administration of the initial high-dose CYC. In group C, 80 days following high-dose CYC treatment, two additional low doses of CYC were administered at 45-day intervals, yielding 30% (3/10) MM development at a mean latency of 200±27 days. In group D, 60 days after the initial treatment, two additional repeated low-dose CYC injections at 45-day intervals resulted in 20% (2/10) MM development at 191±4 days mean latency. Mice in group E were treated with three repeated low doses of

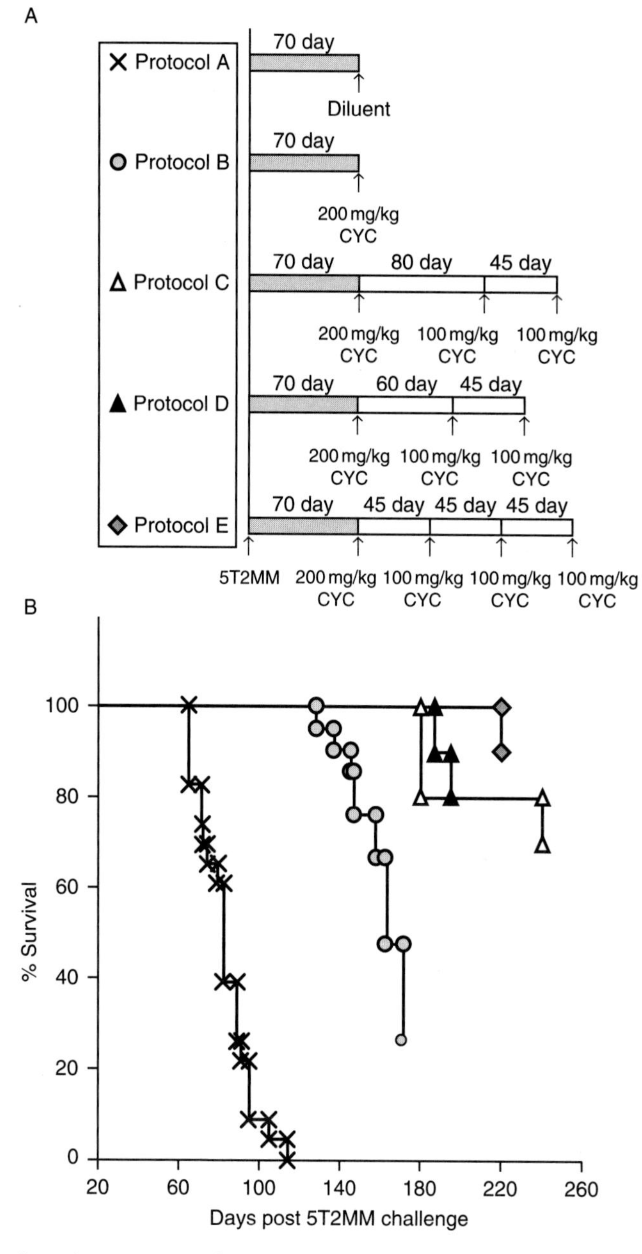

Fig. 3 Prolonged maintenance of transient Treg depletion for myeloma prevention by administering a single high-dose CYC followed by repeated low-doses of CYC. Seventy days after 5T2MM cell injection, 74 mice were divided into five groups. The control group A (n=23) was injected with diluent—all mice (23/23) developed MM at 82±17 days latency. Group B (n=21) administered with 200 mg/kg CYC yielded 71% MM (15/21) at 157±17 days mean survival.

CYC at 45-day intervals, starting 45 days after the high-dose CYC initial treatment, resulting in 10% (1/10) MM development at 220 days (Fig. 3). Thus, durable transient depletion of Treg cells in 5T2MM-bearing mice with low-dose CYC enhances the function of Treg depletion by tipping the balance toward effector T cells, thereby reducing tumor load to minimal residual disease during prolonged latency.

The prevention of bone lesions developing in CYC-treated mice was remarkable. Summing up results from several experiments involving 150 mice treated with high or low doses of CYC indicated that only 6% (9/150) bone lesions were observed. In 5 mice from these groups, CYC was administered quite late, around 80–92 days after the 5T2MM cell challenge. In control mice challenged with 5T2MM cells, hind limb paralysis was observed at about 60–70 days but bone lesions, due to uncontrolled osteoblast bone resorption, appeared later, from about 80 days onwards in 90% of mice surviving for 90–120 days. The development of lytic bone disease is due to an imbalance, with increased osteoclasts and decreased osteoblasts. MM cells trigger osteoclast activity by secreting an osteoblast stimulating factor and an angiogenesis factor, which result in the development of osteolytic lesions involving bone resorption and the formation of new blood cells (Yaccoby *et al.*, 2002).

Bone disease in MM patients is a major cause of morbidity. Bisphosphanates are potent inhibitors of osteolytic bone resorption. They were found to reduce the incidence of skeletal-related events, thus preventing the development of MM bone disease *in vivo*. Angiogenesis is also an active and important process in MM disease progression since the BM is richly vascularized. An important open question is whether treatment with bisphosphanates would influence the tumor burden and MM progression. Dallas *et al.* (1999) used the 5T3MM mouse model to examine the effect of a potent bisphosphanate ibandronate on myeloma-associated bone destruction. Treatment with ibandronate significantly reduced the development of osteolytic lesions in myeloma-bearing mice, but it was not effective in preventing mice from developing hind limb paralysis and did not prolong the survival of myeloma-bearing mice. Treatment of 5T2MM-bearing mice with another potent heterocyclic bisphosphanate, zoledronic acid, prevented the development of lytic bone lesions. A moderate decrease in tumor burden (a 31–35% decrease in serum paraprotein), angiogenesis, and prolonged survival (about 15 days) was also observed (Croucher *et al.*, 2003).

At different intervals after the initial high-dose CYC treatment (80, 60, or 45 days), low-dose CYC (100 mg/kg) was administered repeatedly at 45-day intervals (protocol schedules are presented in Fig. 3A). In group C (n=10), 30% (3/10) developed MM at 200±27 days. In group D (n=10), 20% (2/10) developed MM at 191±4 days and in group E (n=10), 10% (1/10) developed MM at 220 days.

C. Residual Tumor Cells

Initial antitumor treatment may reduce the tumor mass to minimal residual disease, thereby altering the balance of the disease. We evaluated the therapeutic efficacy of immunotherapy involving prolonged Treg depletion, by recruiting antitumor immune response expressed in tumor load size. 5T2MM-bearing mice were reduced to minimal residual disease by injecting a cytotoxic high dose of CYC followed by prolonged administration of low doses of CYC at long intervals. The CYC-induced immunomodulation resulted in remarkably low MM incidence and prolonged survival. An important question was whether prolonged CYC treatment eradicated all tumor cells. We approached this enigma by transferring BM from CYC-treated mice that did not develop overt disease for a prolonged period to young normal syngeneic recipients and followed MM development in these BM recipients for 220 days. BM (2×10^7 cells) was transferred i.v. from one donor to one recipient. The results are summarized in Table II.

Experiment I involved three groups. Group A, a control group, provided evidence that the transferred BM collected from sick mice not receiving any additional CYC treatment reflects the tumor load in these 5T2MM-bearing mice. BM was collected individually from five mice 80 days after 5T2MM cell injection and was transferred to normal recipients. All five BM recipients developed overt disease at a mean latency of 61 ± 5days. Group B—low-dose CYC (100 mg/kg) was administered to 5T2MM-bearing mice 60 days after tumor cell injection (all mice had hind limb paralysis) and after 170 days the BM of mice grossly normal were transferred to syngeneic young recipients. None of these recipients (0/10) developed MM within a 220-day follow-up period. Group C: 5T2MM-bearing mice (for 60 days) were treated with a high dose of CYC (200 mg/kg) and 170 days later, BM from grossly normal mice was transferred to young normal recipients. All BM recipients (10/10) developed MM at a mean latency of 111 ± 14 days. The high-dose CYC reduced the tumor load only transiently (similar results are shown in Fig. 1), but during their prolonged survival the cells regained their tumor growth potential. In the control group (injected with 3×10^5 BM cells from sick mice) bearing only MM cells without any further CYC treatment, their BM activity was replaced by tumor cells and therefore transferring their BM included a high tumor load and all BM recipients developed the disease within a short period of 61 ± 5 days. This situation represents the acute phase of the disease. CYC administration irrespective of CYC dose levels triggered the disappearance of plasma tumor cells from the BM area (replaced by the normal BM population) and markedly prolonged their survival (150–220 days vs. 61–95 days survival of the controls), thereby reverting the disease development to a chronic phase.

Table II Residual MM Cells in BM of Grossly "Normal-Appearing" 5T2MM-Bearing Donor Mice Following Treatment with CYC[a]

Number of experiment	Timing of CYC injection of donor mice (days post 5T2MM injection)	Dose of i.p. CYC (mg/Kg)	Timing of BM transfer from donor mice (days post CYC injection)	MM incidence in BM of recipient mice (n/n, %)	Mean (±SD) latency of recipient mice (days)
Experiment I	60	0	80	5/5, 100%	61±5
	60	100	170	0/10, 0%	220
	60	200	170	10/10, 100%	111±14
Experiment II	66	100	196	4/10, 40%	170±20
	70	100	240	1/10, 10%	142

[a]BM (2×10^7/mouse) from 5T2MM injected mice treated with CYC that did not develop overt disease for a prolonged period was transferred to young normal syngeneic recipients (from one donor to one recipient) and followed for MM development in the BM recipient for 220 days.

The high CYC dose destroys all T lymphocyte populations, in contrast with the low CYC dose that transiently depletes Treg cells and thereby facilitates antitumor immune responses as long as Treg cells are blocked. BM collected from mice 170 days after being treated with low doses of CYC might therefore have a decreased tumor load.

Experiment II involved two groups of 5T2MM-bearing mice treated either at 66 or 70 days after a tumor cell challenge (at the clinical phase) with a single low dose (100 mg/kg) of CYC, and BM was collected at 196 or 240 days afterwards, when the mice looked grossly normal. The development of MM manifestations occurred in 4/10 recipients in Group A at 170$\pm$20 days mean latency and in 1/10 recipients in Group B at 142 days. Thus, the residual tumor load after a long latency following the repeated low-dose CYC treatment seems to be very much reduced, thereby delaying or preventing tumor recurrence. The BM donors might still carry dormant solitary tumor cells that are quiescent and/or in growth arrest (G0/G1 phase) or as small avascular foci. Among the prolonged surviving mice (200–250 days following the initial 5T2MM cell challenge), 30 mice developed undifferentiated lymphoid tumors. These tumors would also grow after subcutaneous grafts (in contrast to 5T2MM tumor cells that grow only following i.v. cell transfer). The spleen was always the main site of lymphoma development, usually involving an enlarged spleen (two- to eightfold weight): the involvement of lymph nodes was observed in 50% of these sick mice and sometimes small foci were observed in the liver; however, their BM was always normal. In several mice (9/30) besides the lymphoma, small foci of plasma tumor cells were observed in the spleen and lymph nodes.

V. CONCLUDING REMARKS

A major impediment to cancer immunotherapy is tumor-induced suppression and tumor evasion of antitumor immune response, which ultimately render the host tolerant to tumor-associated antigens. In recent years, the role of Tregs in tumor development has been extensively studied. A direct link between Tregs and reduced immunity has been demonstrated, strongly suggesting that the effectiveness of antitumor therapy could be greatly enhanced by removal of Treg suppressive activity. A mouse model mimicking human MM was useful to perceive those mechanisms involved in the progression and prevention of the disease. The clinical phase of the disease in 5T2MM-bearing mice involves hind limb paralysis coinciding with increased tumor load and initiation of bone lesions. Suppressive functional Tregs accumulate in the spleen, LNs, BM, peripheral blood, and thymus of sick mice, and contribute to the development of MM. Eradication of Tregs in this context is therefore desired. The use of CYC may be beneficial for treating MM, since it may selectively deplete Tregs depending on timing and dose. High-dose CYC is cytotoxic and causes general lymphodepletion, whereas low-dose CYC selectively depletes Tregs, induces immunostimulation and antiangiogenesis, and enhances effector cell functions. A single low- or high-dose CYC administered to 5T2MM-bearing mice in their early clinical phase prolonged their survival very significantly in comparison with the control group. More specifically, the tumor load was eradicated, hind limb paralysis disappeared, and the tumor cells homing in the BM cavity were replaced by normal BM cell populations. Thus, this treatment changed the acute phase (100% control mice challenged with 5T2MM cells, with a surviving rate of 80–120 days) into a chronic phase (surviving rate of 160–240 days). Administering a single low-dose CYC reduced the disease incidence (38–60%) in contrast with the high-dose CYC, which resulted in higher incidence rates (70–85%). The efficiency of a single low-dose CYC in reducing MM progression was found to depend on tumor load. Kinetic studies on transient Treg depletion showed that low-dose CYC injected in 5T2MM-bearing mice maintained Treg depletion beyond 45 days. Cell populations with antitumor activity could be recovered while Treg renewal was still blocked. More frequent injections of low-dose CYC at 7- or 21-day intervals did not improve the therapeutic effect since these treated mice developed a high incidence of MM. In contrast, mice treated at 45-day intervals developed a significantly lower MM incidence, thus tipping the balance toward effector T cells for a more prolonged period of time. To further control the proliferation of residual tumor cells that escaped the cytotoxic high-dose CYC, we injected additional low-dose CYC at 45-day intervals. These repeated CYC treatments prolonged the transient Treg depletion, thereby facilitating antitumor immune responses to decrease tumor load to minimal residual disease. The low incidence of bone lesions following CYC injection might be due to the disappearance of plasma tumor cells from the BM. More specifically, they detach from the BM microenvironment, which leads to bone resorption and bone lesions.

In summary, the data presented here and supported by evidence from previous studies indicate that beneficial treatment of mice affected with MM may be accomplished by repeated injections of low-dose CYC at long time intervals corresponding to the transient Treg depletion. Consequently, compatible immune cells such as effector T cells, NKT cells, and DCs may possibly be recovered and play a role in breaking immune tolerance against the tumor cells. We believe this approach should be translated to a clinical setting in future therapy for MM in humans.

ACKNOWLEDGMENTS

We are grateful to Dr. Jay A. Levy of the University of California, San Francisco, for his continuous support, valuable discussions and helpful suggestions. Special thanks to Mrs. Tania Meri for patient assistance in the preparation of the manuscript.

REFERENCES

Adkins, B., Charyulu, V., Sun, Q. L., Lobo, D., and Lopez, D. M. (2000). Early block in maturation is associated with thymic involution in mammary tumor-bearing mice. *J. Immunol.* **164,** 5635–5640.

Amanna, I. J., Dingwall, J. P., and Hayes, C. E. (2003). Enforced bcl-xL gene expression restored splenic B lymphocyte development in BAFF-R mutant mice. *J. Immunol.* **170,** 4593–4600.

Antony, P. A., Piccirillo, C. A., Akpinarli, A., Finkelstein, S. E., Speiss, P. J., Surman, D. R., Palmer, D. C., Chan, C. C., Klebanoff, C. A., Overwijk, W. W., Rosenberg, S. A., and Restifo, N. P. (2005). CD8+ T cell immunity against a tumor/self-antigen is augmented by CD4+ T helper cells and hindered by naturally occurring T regulatory cells. *J. Immunol.* **174,** 2591–2601.

Arase, H., Arase, N., Ogasawara, K., Good, R. A., and Onoé, K. (1992). An NK1.1+ CD4+8− single-positive thymocyte subpopulation that expresses a highly skewed T-cell antigen receptor V beta family. *Proc. Natl. Acad. Sci. USA* **89,** 6506–6510.

Asosingh, K., Radl, J., Van Riet, I., Van Camp, B., and Vanderkerken, K. (2000). The 5TMM series: A useful *in vivo* mouse model of human multiple myeloma. *Hematol. J.* **1,** 351–356.

Awwad, M., and North, R. J. (1989). Cyclophosphamide-induced immunologically mediated regression of a cyclophosphamide-resistant murine tumor: A consequence of eliminating precursor L3T4+ suppressor T-cells. *Cancer Res.* **49,** 1649–1654.

Azuma, T., Takahashi, T., Kunisato, A., Kitamura, T., and Hirai, H. (2003). Human CD4+ CD25+ regulatory T cells suppress NKT cell functions. *Cancer Res.* **63,** 4516–4520.

Bartlett, J. B., Dredge, K., and Dalgleish, A. G. (2004). The evolution of thalidomide and its IMiD derivatives as anticancer agents. *Nat. Rev. Cancer* **4,** 314–322.

Bates, G. J., Fox, S. B., Han, C., Leek, R. D., Garcia, J. F., Harris, A. L., and Banham, A. H. (2006). Quantification of regulatory T cells enables the identification of high-risk breast cancer patients and those at risk of late relapse. *J. Clin. Oncol.* **24,** 5373–5380.

Bayer, A. L., Yu, A., and Malek, T. R. (2007). Function of the IL-2R for thymic and peripheral CD4+CD25+ Foxp3+ T regulatory cells. *J. Immunol.* **178,** 4062–4071.

Beatty, G. L., and Paterson, Y. (2001). Regulation of tumor growth by IFN-gamma in cancer immunotherapy. *Immunol. Res.* **24,** 201–210.

Belkaid, Y., and Rouse, B. T. (2005). Natural regulatory T cells in infectious disease. *Nat. Immunol.* **6,** 353–360.

Bergsagel, P. L., and Kuehl, W. M. (2005). Molecular pathogenesis and a consequent classification of multiple myeloma. *J. Clin. Oncol.* **23,** 6333–6338.

Beyer, M., and Schultze, J. L. (2006). Regulatory T cells in cancer. *Blood* **108,** 804–811.

Beyer, M., Kochanek, M., Giese, T., Endl, E., Weihrauch, M. R., Knolle, P. A., Classen, S., and Schultze, J. L. (2006). *In vivo* peripheral expansion of naive CD4+CD25high FoxP3+ regulatory T cells in patients with multiple myeloma. *Blood* **107,** 3940–3949.

Bosco, N., Agenes, F., Rolink, A. G., and Ceredig, R. (2006). Peripheral T cell lymphopenia and concomitant enrichment in naturally arising regulatory T cells: The case of the pre-Talpha gene-deleted mouse. *J. Immunol.* **177,** 5014–5023.

Brimnes, M. K., Svane, I. M., and Johnsen, H. E. (2006). Impaired functionality and phenotypic profile of dendritic cells from patients with multiple myeloma. *Clin. Exp. Immunol.* **144,** 76–84.
Brode, S., and Cooke, A. (2008). Immune-potentiating effects of the chemotherapeutic drug cyclophosphamide. *Crit. Rev. Immunol.* **28,** 109–126.
Browder, T., Butterfield, C. E., Kraling, B. M., Shi, B., Marshall, B., O'Reilly, M. S., and Folkman, J. (2000). Antiangiogenic scheduling of chemotherapy improves efficacy against experimental drug-resistant cancer. *Cancer Res.* **60,** 1878–1886.
Brown, R. D., Pope, B., Yuen, E., Gibson, J., and Joshua, D. E. (1998). The expression of T cell related costimulatory molecules in multiple myeloma. *Leuk. Lymphoma* **31,** 379–384.
Brown, R. D., Pope, B., Murray, A., Esdale, W., Sze, D. M., Gibson, J., Ho, P. J., Hart, D., and Joshua, D. (2001). Dendritic cells from patients with myeloma are numerically normal but functionally defective as they fail to up-regulate CD80 (B7-1) expression after huCD40LT stimulation because of inhibition by transforming growth factor-beta1 and interleukin-10. *Blood* **98,** 2992–2998.
Cabarrocas, J., Cassan, C., Magnusson, F., Piaggio, E., Mars, L., Derbinski, J., Kyewski, B., Gross, D. A., Salomon, B. L., Khazaie, K., Saoudi, A., and Liblau, R. S. (2006). Foxp3+ CD25+ regulatory T cells specific for a neo-self-antigen develop at the double-positive thymic stage. *Proc. Natl. Acad. Sci. USA* **103,** 8453–8458.
Callander, N. S., and Roodman, G. D. (2001). Myeloma bone disease. *Semin. Hematol.* **38,** 276–285.
Carbone, E., Neri, P., Mesuraca, M., Fulciniti, M. T., Otsuki, T., Pende, D., Groh, V., Spies, T., Pollio, G., Cosman, D., Catalano, L., Tassone, P., *et al.* (2005). HLA class I, NKG2D, and natural cytotoxicity receptors regulate multiple myeloma cell recognition by natural killer cells. *Blood* **105,** 251–258.
Castano, A. P., Mroz, P., Wu, M. X., and Hamblin, M. R. (2008). Photodynamic therapy plus low-dose cyclophosphamide generates antitumor immunity in a mouse model. *Proc. Natl. Acad. Sci. USA* **105,** 5495–5500.
Cederbom, L., Hall, H., and Ivars, F. (2000). CD4+CD25+ regulatory T cells down-regulate co-stimulatory molecules on antigen-presenting cells. *Eur. J. Immunol.* **30,** 1538–1543.
Chambers, C. A. (2001). The expanding world of co-stimulation: the two-signal model revisited. *Trends Immunol.* **22,** 217–223.
Chang, D. H., Liu, N., Klimek, V., Hassoun, H., Mazumder, A., Nimer, S. D., Jagannath, S., and Dhodapkar, M. V. (2006). Enhancement of ligand-dependent activation of human natural killer T cells by lenalidomide: Therapeutic implications. *Blood* **108,** 618–621.
Chen, M. L., Pittet, M. J., Gorelik, L., Flavell, R. A., Weissleder, R., von Boehmer, H., and Khazaie, K. (2005). Regulatory T cells suppress tumor-specific CD8 T cell cytotoxicity through TGF-beta signals *in vivo*. *Proc. Natl. Acad. Sci. USA* **102,** 419–424.
Cook, G., Campbell, J. D., Carr, C. E., Boyd, K. S., and Franklin, I. M. (1999). Transforming growth factor beta from multiple myeloma cells inhibits proliferation and IL-2 responsiveness in T lymphocytes. *J. Leukoc. Biol.* **66,** 981–988.
Croucher, P. I., De Hendrik, R., Perry, M. J., Hijzen, A., Shipman, C. M., Lippitt, J., Green, J., Van Marck, E., Van Camp, B., and Vanderkerken, K. (2003). Zoledronic acid treatment of 5T2MM-bearing mice inhibits the development of myeloma bone disease: Evidence for decreased osteolysis, tumor burden and angiogenesis, and increased survival. *J. Bone Miner. Res.* **18,** 482–492.
Curiel, T. J., Coukos, G., Zou, L., Alvarez, X., Cheng, P., Mottram, P., Evdemon-Hogan, M., Conejo-Garcia, J. R., Zhang, L., Burow, M., Zhu, Y., Wei, S., *et al.* (2004). Specific recruitment of regulatory T cells in ovarian carcinoma fosters immune privilege and predicts reduced survival. *Nat. Med.* **10,** 942–949.
Dallas, S. L., Garrett, I. R., Oyajobi, B. O., Dallas, M. R., Boyce, B. F., Bauss, F., Radl, J., and Mundy, G. R. (1999). Ibandronate reduces osteolytic lesions but not tumor burden in a murine model of myeloma bone disease. *Blood* **93,** 1697–1706.

Deng, S., Moore, D. J., Huang, X., Mohiuddin, M., Lee, M. K.t., Velidedeoglu, E., Lian, M. M., Chiaccio, M., Sonawane, S., Orlin, A., Wang, J., Chen, H., *et al.* (2006). Antibody-induced transplantation tolerance that is dependent on thymus-derived regulatory T cells. *J. Immunol.* **176,** 2799–2807.

Dhodapkar, M. V., Geller, M. D., Chang, D. H., Shimizu, K., Fujii, S., Dhodapkar, K. M., and Krasovsky, J. (2003). A reversible defect in natural killer T cell function characterizes the progression of premalignant to malignant multiple myeloma. *J. Exp. Med.* **197,** 1667–1676.

Dingli, D., and Russell, S. J. (2007). Mouse models and the RANKL/OPG axis in myeloma bone disease. *Leukemia* **21,** 2090–2093.

Drake, C. G., Jaffee, E., and Pardoll, D. M. (2006). Mechanisms of immune evasion by tumors. *Adv. Immunol.* **90,** 51–81.

Ercolini, A. M., Ladle, B. H., Manning, E. A., Pfannenstiel, L. W., Armstrong, T. D., Machiels, J. P., Bieler, J. G., Emens, L. A., Reilly, R. T., and Jaffee, E. M. (2005). Recruitment of latent pools of high-avidity CD8(+) T cells to the antitumor immune response. *J. Exp. Med.* **201,** 1591–1602.

Fahlén, L., Read, S., Gorelik, L., Hurst, S. D., Coffman, R. L., Flavell, R. A., and Powrie, F. (2005). T cells that cannot respond to TGF-beta escape control by CD4(+)CD25(+) regulatory T cells. *J. Exp. Med.* **201,** 737–746.

Feyler, S., von Lilienfeld-Toal, M., Jarmin, S., Marles, L., Rawstron, A., Ashcroft, A. J., Owen, R. G., Selby, P. J., and Cook, G. (2009). CD4(+)CD25(+)FoxP3(+) regulatory T cells are increased whilst CD3(+)CD4(-)CD8(-)alphabetaTCR(+) Double Negative T cells are decreased in the peripheral blood of patients with multiple myeloma which correlates with disease burden. *Br. J. Haematol.* **144,** 686–695.

Fontenot, J. D., Rasmussen, J. P., Gavin, M. A., and Rudensky, A. Y. (2005). A function for interleukin 2 in Foxp3-expressing regulatory T cells. *Nat. Immunol.* **6,** 1142–1151.

Frassanito, M. A., Cusmai, A., and Dammacco, F. (2001). Deregulated cytokine network and defective Th1 immune response in multiple myeloma. *Clin. Exp. Immunol.* **125,** 190–197.

Frohn, C., Höppner, M., Schlenke, P., Kirchner, H., Koritk, P., and Luhm, J. (2002). Anti-myeloma activity of natural killer lymphocytes. *Br. J. Haematol.* **119,** 660–664.

Gatenby, R. A., Silva, A. S., Gillies, R. J., and Frieden, B. R. (2009). Adaptive therapy. *Cancer Res.* **69,** 4894–4903.

Gauthier, E. R., Piché, L., Lemieux, G., and Lemieux, R. (1996). Role of bcl-X(L) in the control of apoptosis in murine myeloma cells. *Cancer Res.* **56,** 1451–1456.

Ghiringhelli, F., Larmonier, N., Schmitt, E., Parcellier, A., Cathelin, D., Garrido, C., Chauffert, B., Solary, E., Bonnotte, B., and Martin, F. (2004). CD4+CD25+ regulatory T cells suppress tumor immunity but are sensitive to cyclophosphamide which allows immunotherapy of established tumors to be curative. *Eur. J. Immunol.* **34,** 336–344.

Ghiringhelli, F., Ménard, C., Martin, F., and Zitvogel, L. (2006). The role of regulatory T cells in the control of natural killer cells: Relevance during tumor progression. *Immunol. Rev.* **214,** 229–238.

Ghiringhelli, F., Menard, C., Puig, P. E., Ladoire, S., Roux, S., Martin, F., Solary, E., Le Cesne, A., Zitvogel, L., and Chauffert, B. (2007). Metronomic cyclophosphamide regimen selectively depletes CD4+CD25+ regulatory T cells and restores T and NK effector functions in end stage cancer patients. *Cancer Immunol. Immunother.* **56,** 641–648.

Gorelik, L., Prokhorova, A., and Mokyr, M. B. (1994). Low-dose melphalan-induced shift in the production of a Th2-type cytokine to a Th1-type cytokine in mice bearing a large MOPC-315 tumor. *Cancer Immunol. Immunother.* **39,** 117–126.

Han, S., Wang, B., Cotter, M. J., Yang, L. J., Zucali, J., Moreb, J. S., and Chang, L. J. (2008). Overcoming immune tolerance against multiple myeloma with lentiviral calnexin-engineered dendritic Cells. *Mol. Ther.* **16,** 269–279.

Hideshima, T., Mitsiades, C., Tonon, G., Richardson, P. G., and Anderson, K. C. (2007). Understanding multiple myeloma pathogenesis in the bone marrow to identify new therapeutic targets. *Nat. Rev. Cancer* **7,** 585–598.

Hiraoka, N., Onozato, K., Kosuge, T., and Hirohashi, S. (2006). Prevalence of FOXP3+ regulatory T cells increases during the progression of pancreatic ductal adenocarcinoma and its premalignant lesions. *Clin. Cancer Res.* **12,** 5423–5434.

Hirschhorn-Cymerman, D., Rizzuto, G. A., Merghoub, T., Cohen, A. D., Avogadri, F., Lesokhin, A. M., Weinberg, A. D., Wolchok, J. D., and Houghton, A. N. (2009). OX40 engagement and chemotherapy combination provides potent antitumor immunity with concomitant regulatory T cell apoptosis. *J. Exp. Med.* **206,** 1103–1116.

Höltl, L., Ramoner, R., Zelle-Rieser, C., Gander, H., Putz, T., Papesh, C., Nussbaumer, W., Falkensammer, C., Bartsch, G., and Thurnher, M. (2005). Allogeneic dendritic cell vaccination against metastatic renal cell carcinoma with or without cyclophosphamide. *Cancer Immunol. Immunother.* **54,** 663–670.

Hori, S., Nomura, T., and Sakaguchi, S. (2003). Control of regulatory T cell development by the transcription factor Foxp3. *Science* **299,** 1057–1061.

Ikezawa, Y., Nakazawa, M., Tamura, C., Takahashi, K., Minami, M., and Ikezawa, Z. (2005). Cyclophosphamide decreases the number, percentage and the function of CD25+ CD4+ regulatory T cells, which suppress induction of contact hypersensitivity. *J. Dermatol. Sci.* **39,** 105–112.

Itoh, M., Takahashi, T., Sakaguchi, N., Kuniyasu, Y., Shimizu, J., Otsuka, F., and Sakaguchi, S. (1999). Thymus and autoimmunity: Production of CD25+CD4+ naturally anergic and suppressive T cells as a key function of the thymus in maintaining immunologic self-tolerance. *J. Immunol.* **162,** 5317–5326.

Jarahian, M., Watzl, C., Issa, Y., Altevogt, P., and Momburg, F. (2007). Blockade of natural killer cell-mediated lysis by NCAM140 expressed on tumor cells. *Int. J. Cancer.* **120,** 2625–2634.

Jarnicki, A. G., Lysaght, J., Todryk, S., and Mills, K. H. (2006). Suppression of antitumor immunity by IL-10 and TGF-beta-producing T cells infiltrating the growing tumor: Influence of tumor environment on the induction of CD4+ and CD8+ regulatory T cells. *J. Immunol.* **177,** 896–904.

Jones, E., Dahm-Vicker, M., Simon, A. K., Green, A., Powrie, F., Cerundolo, V., and Gallimore, A. (2002). Depletion of CD25+ regulatory cells results in suppression of melanoma growth and induction of autoreactivity in mice. *Cancer Immun.* **2,** 1–12.

Kaiserlian, D., Savino, W., Hassid, J., and Dardenne, M. (1984). Studies of the thymus in mice bearing the Lewis lung carcinoma. III. Possible mechanisms of tumor-induced thymic atrophy. *Clin. Immunol. Immunopathol.* **32,** 316–325.

Kerbel, R. S., and Kamen, B. A. (2004). The anti-angiogenic basis of metronomic chemotherapy. *Nat. Rev. Cancer* **4,** 423–436.

Kim, R., Emi, M., and Tanabe, K. (2006). Cancer immunosuppression and autoimmune disease: Beyond immunosuppressive networks for tumour immunity. *Immunology* **119,** 254–264.

Kim, J. M., Rasmussen, J. P., and Rudensky, A. Y. (2007). Regulatory T cells prevent catastrophic autoimmunity throughout the lifespan of mice. *Nat. Immunol.* **8,** 191–197.

Kitamura, N., Murata, S., Ueki, T., Mekata, E., Reilly, R. T., Jaffee, E. M., and Tani, T. (2009). OX40 costimulation can abrogate Foxp3+ regulatory T cell-mediated suppression of antitumor immunity. *Int. J. Cancer* **125,** 630–638.

Kronenberg, M. (2005). Toward an understanding of NKT cell biology: Progress and paradoxes. *Annu. Rev. Immunol.* **23,** 877–900.

Larmonier, N., Marron, M., Zeng, Y., Cantrell, J., Romanoski, A., Sepassi, M., Thompson, S., Chen, X., Andreansky, S., and Katsanis, E. (2007). Tumor-derived CD4(+)CD25(+) regulatory T cell suppression of dendritic cell function involves TGF-beta and IL-10. *Cancer Immunol. Immunother.* **56,** 48–59.

Laronne-Bar-On, A., Zipori, D., and Haran-Ghera, N. (2008). Increased regulatory versus effector T cell development is associated with thymus atrophy in mouse models of multiple myeloma. *J. Immunol.* **181,** 3714–3724.

Lim, H. W., Hillsamer, P., Banham, A. H., and Kim, C. H. (2005). Cutting edge: Direct suppression of B cells by CD4+ CD25+ regulatory T cells. *J. Immunol.* **175,** 4180–4183.

Liu, K., Idoyaga, J., Charalambous, A., Fujii, S., Bonito, A., Mordoh, J., Wainstok, R., Bai, X. F., Liu, Y., and Steinman, R. M. (2005). Innate NKT lymphocytes confer superior adaptive immunity via tumor-capturing dendritic cells. *J. Exp. Med.* **202,** 1507–1516.

Liu, W., Putnam, A. L., Xu-Yu, Z., Szot, G. L., Lee, M. R., Zhu, S., Gottlieb, P. A., Kapranov, P., Gingeras, T. R., Fazekas de St Groth, B., Clayberger, C., Soper, D. M., *et al.* (2006). CD127 expression inversely correlates with FoxP3 and suppressive function of human CD4+ T reg cells. *J. Exp. Med.* **203,** 1701–1711.

Liu, J. Y., Wu, Y., Zhang, X. S., Yang, J. L., Li, H. L., Mao, Y. Q., Wang, Y., Cheng, X., Li, Y. Q., Xia, J. C., Masucci, M., and Zeng, Y. X. (2007). Single administration of low dose cyclophosphamide augments the antitumor effect of dendritic cell vaccine. *Cancer Immunol. Immunother.* **56,** 1597–1604.

Liyanage, U. K., Goedegebuure, P. S., Moore, T. T., Viehl, C. T., Moo-Young, T. A., Larson, J. W., Frey, D. M., Ehlers, J. P., Eberlein, T. J., and Linehan, D. C. (2006). Increased prevalence of regulatory T cells (Treg) is induced by pancreas adenocarcinoma. *J. Immunother.* **29,** 416–424.

Liyanage, U. K., Moore, T. T., Joo, H. G., Tanaka, Y., Herrmann, V., Doherty, G., Drebin, J. A., Strasberg, S. M., Eberlein, T. J., Goedegebuure, P. S., and Linehan, D. C. (2002). Prevalence of regulatory T cells is increased in peripheral blood and tumor microenvironment of patients with pancreas or breast adenocarcinoma. *J. Immunol.* **169,** 2756–2761.

Lopez, D. M., Charyulu, V., and Adkins, B. (2002). Influence of breast cancer on thymic function in mice. *J. Mammary Gland Biol. Neoplasia* **7,** 191–199.

Lutsiak, M. E., Semnani, R. T., De Pascalis, R., Kashmiri, S. V., Schlom, J., and Sabzevari, H. (2005). Inhibition of CD4(+)25+ T regulatory cell function implicated in enhanced immune response by low-dose cyclophosphamide. *Blood* **105,** 2862–2868.

Machiels, J. P., Reilly, R. T., Emens, L. A., Ercolini, A. M., Lei, R. Y., Weintraub, D., Okoye, F. I., and Jaffee, E. M. (2001). Cyclophosphamide, doxorubicin, and paclitaxel enhance the antitumor immune response of granulocyte/macrophage-colony stimulating factor-secreting whole-cell vaccines in HER-2/neu tolerized mice. *Cancer Res.* **61,** 3689–3697.

Manning, L. S., Berger, J. D., O'Donoghue, H. L., Sheridan, G. N., Claringbold, P. G., and Turner, J. H. (1992). A model of multiple myeloma: Culture of 5T33 murine myeloma cells and evaluation of tumorigenicity in the C57BL/KaLwRij mouse. *Br. J. Cancer* **66,** 1088–1093.

Marie, J. C., Letterio, J. J., Gavin, M., and Rudensky, A. Y. (2005). TGF-beta1 maintains suppressor function and Foxp3 expression in CD4+CD25+ regulatory T cells. *J. Exp. Med.* **201,** 1061–1067.

Marshall, N. A., Christie, L. E., Munro, L. R., Culligan, D. J., Johnston, P. W., Barker, R. N., and Vickers, M. A. (2004). Immunosuppressive regulatory T cells are abundant in the reactive lymphocytes of Hodgkin lymphoma. *Blood* **103,** 1755–1762.

McCune, T. R., Thacker, L. R., II, Peters, T. G., Mulloy, L., Rohr, M. S., Adams, P. A., Yium, J., Light, J. A., Pruett, T., Gaber, A. O., Selman, S. H., Jonsson, J., *et al.* (1998). Effects of tacrolimus on hyperlipidemia after successful renal transplantation: A Southeastern Organ Procurement Foundation multicenter clinical study. *Transplantation* **65,** 87–92.

Mills, K. H., and Cawley, J. C. (1983). Abnormal monoclonal antibody-defined helper/suppressor T-cell subpopulations in multiple myeloma: Relationship to treatment and clinical stage. *Br. J. Haematol.* **53,** 271–275.

Misra, N., Bayry, J., Lacroix-Desmazes, S., Kazatchkine, M. D., and Kaveri, S. V. (2004). Cutting edge: Human CD4+CD25+ T cells restrain the maturation and antigen-presenting function of dendritic cells. *J. Immunol.* **172,** 4676–4680.

Mittelman, M., Neumann, D., Peled, A., Kanter, P., and Haran-Ghera, N. (2001). Erythropoietin induces tumor regression and antitumor immune responses in murine myeloma models. *Proc. Natl. Acad. Sci. USA* **98,** 5181–5186.

Mizukami, Y., Kono, K., Kawaguchi, Y., Akaike, H., Kamimura, K., Sugai, H., and Fujii, H. (2008). CCL17 and CCL22 chemokines within tumor microenvironment are related to accumulation of Foxp3+ regulatory T cells in gastric cancer. *Int. J. Cancer* **122,** 2286–2293.

Motoyoshi, Y., Kaminoda, K., Saitoh, O., Hamasaki, K., Nakao, K., Ishii, N., Nagayama, Y., and Eguchi, K. (2006). Different mechanisms for anti-tumor effects of low- and high-dose cyclophosphamide. *Oncol. Rep.* **16,** 141–146.

Mozaffari, F., Hansson, L., Kiaii, S., Ju, X., Rossmann, E. D., Rabbani, H., Mellstedt, H., and Osterborg, A. (2004). Signalling molecules and cytokine production in T cells of multiple myeloma-increased abnormalities with advancing stage. *Br. J. Haematol.* **124,** 315–324.

Nakamura, K., Kitani, A., and Strober, W. (2001). Cell contact-dependent immunosuppression by CD4(+)CD25(+) regulatory T cells is mediated by cell surface-bound transforming growth factor beta. *J. Exp. Med.* **194,** 629–644.

Nishikawa, H., Kato, T., Tanida, K., Hiasa, A., Tawara, I., Ikeda, H., Ikarashi, Y., Wakasugi, H., Kronenberg, M., Nakayama, T., Taniguchi, M., Kuribayashi, K., *et al.* (2003). CD4+ CD25+ T cells responding to serologically defined autoantigens suppress antitumor immune responses. *Proc. Natl. Acad. Sci. USA* **100,** 10902–10906.

Nishikawa, H., Jäger, E., Ritter, G., Old, L. J., and Gnjatic, S. (2005). CD4+ CD25+ regulatory T cells control the induction of antigen-specific CD4+ helper T cell responses in cancer patients. *Blood* **106,** 1008–1011.

North, R. J. (1982). Cyclophosphamide-facilitated adoptive immunotherapy of an established tumor depends on elimination of tumor-induced suppressor T cells. *J. Exp. Med.* **155,** 1063–1074.

O'Garra, A., and Vieira, P. (2004). Regulatory T cells and mechanisms of immune system control. *Nat. Med.* **10,** 801–805.

Ogawara, H., Handa, H., Yamazaki, T., Toda, T., Yoshida, K., Nishimoto, N., Al-ma'Quol, W. H., Kaneko, Y., Matsushima, T., Tsukamoto, N., Nojima, Y., Matsumoto, M., *et al.* (2005). High Th1/Th2 ratio in patients with multiple myeloma. *Leuk. Res.* **29,** 135–140.

Ohm, J. E., Gabrilovich, D. I., Sempowski, G. D., Kisseleva, E., Parman, K. S., Nadaf, S., and Carbone, D. P. (2003). VEGF inhibits T-cell development and may contribute to tumor-induced immune suppression. *Blood* **101,** 4878–4886.

Onishi, Y., Fehervari, Z., Yamaguchi, T., and Sakaguchi, S. (2008). Foxp3+ natural regulatory T cells preferentially form aggregates on dendritic cells *in vitro* and actively inhibit their maturation. *Proc. Natl. Acad. Sci. USA* **105,** 10113–10118.

Onizuka, S., Tawara, I., Shimizu, J., Sakaguchi, S., Fujita, T., and Nakayama, E. (1999). Tumor rejection by *in vivo* administration of anti-CD25 (interleukin-2 receptor alpha) monoclonal antibody. *Cancer Res.* **59,** 3128–3133.

Ormandy, L. A., Hillemann, T., Wedemeyer, H., Manns, M. P., Greten, T. F., and Korangy, F. (2005). Increased populations of regulatory T cells in peripheral blood of patients with hepatocellular carcinoma. *Cancer Res.* **65,** 2457–2464.

Pardoll, D. (2003). Does the immune system see tumors as foreign or self? *Annu. Rev. Immunol.* **21,** 807–839.

Peng, L., Kjaergäard, J., Plautz, G. E., Awad, M., Drazba, J. A., Shu, S., and Cohen, P. A. (2002). Tumor-induced L-selectin high suppressor T cells mediate potent effector T cell blockade and cause failure of otherwise curative adoptive immunotherapy. *J. Immunol.* **169,** 4811–4821.

Pennington, D. J., Silva-Santos, B., Silberzahn, T., Escorcio-Correia, M., Woodward, M. J., Roberts, S. J., Smith, A. L., Dyson, P. J., and Hayday, A. C. (2006). Early events in the thymus affect the balance of effector and regulatory T cells. *Nature* **444,** 1073–1077.

Petersen, R. P., Campa, M. J., Sperlazza, J., Conlon, D., Joshi, M. B., Harpole, D. H., Jr., and Patz, E. F., Jr. (2006). Tumor infiltrating Foxp3+ regulatory T-cells are associated with recurrence in pathologic stage I NSCLC patients. *Cancer* **107,** 2866–2872.

Piccirillo, C. A., and Shevach, E. M. (2001). Cutting edge: Control of CD8+ T cell activation by CD4+CD25+ immunoregulatory cells. *J. Immunol.* **167,** 1137–1140.

Piccirillo, C. A., and Thornton, A. M. (2004). Cornerstone of peripheral tolerance: Naturally occurring CD4+CD25+ regulatory T cells. *Trends Immunol.* **25,** 374–380.

Pratt, G., Goodyear, O., and Moss, P. (2007). Immunodeficiency and immunotherapy in multiple myeloma. *Br. J. Haematol.* **138,** 563–579.

Proietti, E., Greco, G., Garrone, B., Baccarini, S., Mauri, C., Venditti, M., Carlei, D., and Belardelli, F. (1998). Importance of cyclophosphamide-induced bystander effect on T cells for a successful tumor eradication in response to adoptive immunotherapy in mice. *J. Clin. Invest.* **101,** 429–441.

Radl, J., Croese, J. W., Zurcher, C., Van den Enden-Vieveen, M. H., and de Leeuw, A. M. (1988). Animal model of human disease. Multiple myeloma. *Am. J. Pathol.* **132,** 593–597.

Raitakari, M., Brown, R. D., Gibson, J., and Joshua, D. E. (2003). T cells in myeloma. *Hematol. Oncol.* **21,** 33–42.

Rajkumar, S. V., Hayman, S., Gertz, M. A., Dispenzieri, A., Lacy, M. Q., Greipp, P. R., Geyer, S., Iturria, N., Fonseca, R., Lust, J. A., Kyle, R. A., and Witzig, T. E. (2002). Combination therapy with thalidomide plus dexamethasone for newly diagnosed myeloma. *J. Clin. Oncol.* **20,** 4319–4323.

Ratta, M., Fagnoni, F., Curti, A., Vescovini, R., Sansoni, P., Oliviero, B., Fogli, M., Ferri, E., Della Cuna, G. R., Tura, S., Baccarani, M., and Lemoli, R. M. (2002). Dendritic cells are functionally defective in multiple myeloma: The role of interleukin-6. *Blood* **100,** 230–237.

Rawstron, A. C., Davies, F. E., Owen, R. G., English, A., Pratt, G., Child, J. A., Jack, A. S., and Morgan, G. J. (1998). B-lymphocyte suppression in multiple myeloma is a reversible phenomenon specific to normal B-cell progenitors and plasma cell precursors. *Br. J. Haematol.* **100,** 176–183.

Richardson, P. G., Blood, E., Mitsiades, C. S., Jagannath, S., Zeldenrust, S. R., Alsina, M., Schlossman, R. L., Rajkumar, S. V., Desikan, K. R., Hideshima, T., Munshi, N. C., Kelly-Colson, K., *et al.* (2006). A randomized phase 2 study of lenalidomide therapy for patients with relapsed or relapsed and refractory multiple myeloma. *Blood* **108,** 3458–3464.

Rivers, S. L., Whittington, R. M., and Patno, M. E. (1963). Comparison of effect of cyclophosphamide and a placebo in treatment of multiple myeloma. *Cancer Chemother. Rep.* **29,** 115–119.

Ruter, J., Barnett, B. G., Kryczek, I., Brumlik, M. J., Daniel, B. J., Coukos, G., Zou, W., and Curiel, T. J. (2009). Altering regulatory T cell function in cancer immunotherapy: A novel means to boost the efficacy of cancer vaccines. *Front. Biosci.* **14,** 1761–1770.

Sakaguchi, S. (2005). Naturally arising Foxp3-expressing CD25+CD4+ regulatory T cells in immunological tolerance to self and non-self. *Nat. Immunol.* **6,** 345–352.

Salomon, B., and Bluestone, J. A. (2001). Complexities of CD28/B7: CTLA-4 costimulatory pathways in autoimmunity and transplantation. *Annu. Rev. Immunol.* **19,** 225–252.

Schiavoni, G., Mattei, F., Di Pucchio, T., Santini, S. M., Bracci, L., Belardelli, F., and Proietti, E. (2000). Cyclophosphamide induces type I interferon and augments the number of CD44(hi) T lymphocytes in mice: Implications for strategies of chemoimmunotherapy of cancer. *Blood* **95,** 2024–2030.

Seliger, B., Maeurer, M. J., and Ferrone, S. (2000). Antigen-processing machinery breakdown and tumor growth. *Immunol. Today* **21,** 455–464.

Shanker, A., Singh, S. M., and Sodhi, A. (2000). Ascitic growth of a spontaneous transplantable T cell lymphoma induces thymic involution. 1. Alterations in the CD4/CD8 distribution in thymocytes. *Tumour Biol.* **21,** 288–298.

Sharabi, A., Lapter, S., and Mozes, E. (2010a). Bcl-xL is required for the development of functional regulatory CD4 cells in lupus-afflicted mice following treatment with a tolerogenic peptide. *J. Autoimmun.* **34,** 87–95.

Sharabi, A., Laronne-Bar-On, A., Meshorer, A., and Haran-Ghera, N. (2010b). Low-dose cyclophosphamide at wide intervals reduces incidence and prolongs survival in a mouse model of multiple myeloma. *Cancer Immunol. Immunother.* (In Press).

Shevach, E. M. (2000). Regulatory T cells in autoimmmunity. *Annu. Rev. Immunol.* **18,** 423–449.

Shimizu, J., Yamazaki, S., and Sakaguchi, S. (1999). Induction of tumor immunity by removing CD25+CD4+ T cells: A common basis between tumor immunity and autoimmunity. *J. Immunol.* **163,** 5211–5218.

Singhal, S., Mehta, J., Desikan, R., Ayers, D., Roberson, P., Eddlemon, P., Munshi, N., Anaissie, E., Wilson, C., Dhodapkar, M., Zeddis, J., and Barlogie, B. (1999). Antitumor activity of thalidomide in refractory multiple myeloma. *N. Engl. J. Med.* **341,** 1565–1571.

Sladek, N. E., Powers, J. F., and Grage, G. M. (1984). Half-life of oxazaphosphorines in biological fluids. *Drug Metab. Dispos.* **12,** 553–559.

Smyth, M. J., and Godfrey, D. I. (2000). NKT cells and tumor immunity—A double-edged sword. *Nat. Immunol.* **1,** 459–460.

Smyth, M. J., Crowe, N. Y., Hayakawa, Y., Takeda, K., Yagita, H., and Godfrey, D. I. (2002). NKT cells—Conductors of tumor immunity? *Curr. Opin. Immunol.* **14,** 165–171.

Sutmuller, R. P., van Duivenvoorde, L. M., van Elsas, A., Schumacher, T. N., Wildenberg, M. E., Allison, J. P., Toes, R. E., Offringa, R., and Melief, C. J. (2001). Synergism of cytotoxic T lymphocyte-associated antigen 4 blockade and depletion of CD25(+) regulatory T cells in antitumor therapy reveals alternative pathways for suppression of autoreactive cytotoxic T lymphocyte responses. *J. Exp. Med.* **194,** 823–832.

Takahashi, Y., Cerasoli, D. M., Dal Porto, J. M., Shimoda, M., Freund, R., Fang, W., Telander, D. G., Malvey, E. N., Mueller, D. L., Behrens, T. W., and Kelsoe, G. (1999). Relaxed negative selection in germinal centers and impaired affinity maturation in bcl-xL transgenic mice. *J. Exp. Med.* **190,** 399–410.

Tanaka, H., Tanaka, J., Kjaergaard, J., and Shu, S. (2002). Depletion of CD4+ CD25+ regulatory cells augments the generation of specific immune T cells in tumor-draining lymph nodes. *J. Immunother.* **25,** 207–217.

Terpos, E., Sezer, O., Croucher, P., and Dimopoulos, M. A. (2007). Myeloma bone disease and proteasome inhibition therapies. *Blood* **110,** 1098–1104.

Thomas, E., Smith, D. C., Lee, M. Y., and Rosse, C. (1985). Induction of granulocytic hyperplasia, thymic atrophy, and hypercalcemia by a selected subpopulation of a murine mammary adenocarcinoma. *Cancer Res.* **45,** 5840–5844.

Thornton, A. M., and Shevach, E. M. (1998). CD4+CD25+ immunoregulatory T cells suppress polyclonal T cell activation *in vitro* by inhibiting interleukin 2 production. *J. Exp. Med.* **188,** 287–296.

Tu, Y., Renner, S., Xu, F., Fleishman, A., Taylor, J., Weisz, J., Vescio, R., Rettig, M., Berenson, J., Krajewski, S., Reed, J. C., and Lichtenstein, A. (1998). BCL-X expression in multiple myeloma: Possible indicator of chemoresistance. *Cancer Res.* **58,** 256–262.

Turk, M. J., Guevara-Patino, J. A., Rizzuto, G. A., Engelhorn, M. E., Sakaguchi, S., and Houghton, A. N. (2004). Concomitant tumor immunity to a poorly immunogenic melanoma is prevented by regulatory T cells. *J. Exp. Med.* **200,** 771–782.

Urashima, M., Ogata, A., Chauhan, D., Hatziyanni, M., Vidriales, M. B., Dedera, D. A., Schlossman, R. L., and Anderson, K. C. (1996). Transforming growth factor-beta1: Differential effects on multiple myeloma versus normal B cells. *Blood* **87,** 1928–1938.

Vanderkerken, K., De Raeve, H., Goes, E., Van Meirvenne, S., Radl, J., Van Riet, I., Thielemans, K., and Van Camp, B. (1997). Organ involvement and phenotypic adhesion profile of 5T2 and 5T33 myeloma cells in the C57BL/KaLwRij mouse. *Br. J. Cancer* **76,** 451–460.

Vu, M. D., Xiao, X., Gao, W., Degauque, N., Chen, M., Kroemer, A., Killeen, N., Ishii, N., and Chang Li, X. (2007). OX40 costimulation turns off Foxp3+ Tregs. *Blood* **110,** 2501–2510.

Wada, S., Yoshimura, K., Hipkiss, E. L., Harris, T. J., Yen, H. R., Goldberg, M. V., Grosso, J. F., Getnet, D., Demarzo, A. M., Netto, G. J., Anders, R., Pardoll, D. M., *et al.* (2009). Cyclophosphamide augments antitumor immunity: Studies in an autochthonous prostate cancer model. *Cancer Res.* **69,** 4309–4318.

Wang, S., Hong, S., Yang, J., Qian, J., Zhang, X., Shpall, E., Kwak, L. W., and Yi, Q. (2006). Optimizing immunotherapy in multiple myeloma: Restoring the function of patients' monocyte-derived dendritic cells by inhibiting p38 or activating MEK/ERK MAPK and neutralizing interleukin-6 in progenitor cells. *Blood* **108,** 4071–4077.

Wing, K., Onishi, Y., Prieto-Martin, P., Yamaguchi, T., Miyara, M., Fehervari, Z., Nomura, T., and Sakaguchi, S. (2008). CTLA-4 control over Foxp3+ regulatory T cell function. *Science* **322,** 271–275.

Wolf, D., Rumpold, H., Koppelstätter, C., Gastl, G. A., Steurer, M., Mayer, G., Gunsilius, E., Tilg, H., and Wolf, A. M. (2006). Telomere length of *in vivo* expanded CD4(+)CD25 (+) regulatory T-cells is preserved in cancer patients. *Cancer Immunol. Immunother.* **55,** 1198–1208.

Woo, E. Y., Chu, C. S., Goletz, T. J., Schlienger, K., Yeh, H., Coukos, G., Rubin, S. C., Kaiser, L. R., and June, C. H. (2001). Regulatory CD4(+)CD25(+) T cells in tumors from patients with early-stage non-small cell lung cancer and late-stage ovarian cancer. *Cancer Res.* **61,** 4766–4772.

Wu, L., and Tannock, I. F. (2003). Repopulation in murine breast tumors during and after sequential treatments with cyclophosphamide and 5-fluorouracil. *Cancer Res.* **63,** 2134–2138.

Yaccoby, S., Pearse, R. N., Johnson, C. L., Barlogie, B., Choi, Y., and Epstein, J. (2002). Myeloma interacts with the bone marrow microenvironment to induce osteoclastogenesis and is dependent on osteoclast activity. *Br. J. Haematol.* **116,** 278–290.

Yeh, H. S., and Berenson, J. R. (2006). Myeloma bone disease and treatment options. *Eur. J. Cancer* **42,** 1554–1563.

Zhan, Y., Bourges, D., Dromey, J. A., Harrison, L. C., and Lew, A. M. (2007). The origin of thymic CD4+CD25+ regulatory T cells and their co-stimulatory requirements are determined after elimination of recirculating peripheral CD4+ cells. *Int. Immunol.* **19,** 455–463.

Zhang, M. (1989). The relationships between thymus, other immune organs and various diseases in children (analysis of 621 cases). *Zhonghua Bing Li Xue Za Zhi* **18,** 92–95.

Zhou, G., and Levitsky, H. I. (2007). Natural regulatory T cells and de novo-induced regulatory T cells contribute independently to tumor-specific tolerance. *J. Immunol.* **178,** 2155–2162.

Zou, W. (2005). Immunosuppressive networks in the tumour environment and their therapeutic relevance. *Nat. Rev. Cancer* **5,** 263–274.

Zou, W. (2006). Regulatory T cells, tumour immunity and immunotherapy. *Nat. Rev. Immunol.* **6,** 295–307.

Obesity, Cholesterol, and Clear-Cell Renal Cell Carcinoma (RCC)

Harry A. Drabkin and Robert M. Gemmill

Department of Medicine and Division of Hematology-Oncology
Medical University of South Carolina, Charleston
South Carolina, USA

Multiple epidemiologic studies have linked the development of renal cancer to obesity. In this chapter, we begin with a review of selected population studies, followed by recent mechanistic discoveries that further link lipid deregulation to the RCC development. The upregulation of leptin and downregulation of adiponectin pathways in obesity fit well with our molecular understanding of RCC pathogenesis. In addition, two forms of hereditary RCC involve proteins, Folliculin and TRC8, that are positioned to coordinately regulate lipid and protein biosynthesis. Both of these biosynthetic pathways have important downstream consequences on HIF-1/2α levels and angiogenesis, key aspects in the disease pathogenesis. The role of lipid biology and its interface with protein translation regulation represents a new dimension in RCC research with potential therapeutic implications. © 2010 Elsevier Inc.

I. OBESITY, CHOLESTEROL, AND RCC

A subset of human cancer has been linked to obesity in multiple population/case control studies. Most often, this subset has included cancers of the colon, breast, esophagus (adenocarcinoma), uterus, ovary, kidney, and pancreas (Flegal *et al.*, 2007). This discussion is focused on the relationship between kidney cancer and lipids. We begin with a review of selected population studies, followed by recent mechanistic discoveries that further link lipid deregulation to RCC development.

Advances in CANCER RESEARCH

0065-230X/10 $35.00
DOI: 10.1016/S0065-230X(10)07002-8

Renal cell carcinoma comprises 5% of epithelial cancers in the United States, with more than 38,000 new cases each year (Jemal *et al.*, 2006). Most tumors are clear cell RCCs, and of these a majority contains mutations or epigenetic silencing of the von Hippel-Lindau gene with upregulation of hypoxia-inducible factor (HIF) α subunits and a constitutively activated hypoxic response (Kaelin, 2005). RCC is more common in males than females by about 2:1 (Lipworth *et al.*, 2006). One of the larger studies linking RCC to obesity was conducted by Samanic *et al.* (2006) involving 362,552 Swedish men followed on average for 19 years between 1971 and 1992. This study was notable because it used actual measurements of weight and body size (body mass index, i.e., BMI = kg/m^2), as opposed to questionnaire-derived data, and was linked to the population-based Swedish cancer registry. However, the average age of individuals at the time of entrée into the cohort was only 34 years, whereas to reach the average age of RCC diagnosis would require almost 30 years of follow-up (Setiawan *et al.*, 2007).

Nevertheless, after adjusting for age and smoking, being either overweight (BMI 25.0–29.9) or obese (BMI > 30.0) was associated with a statistically significant increase in kidney cancer compared with nonoverweight controls (BMI 18.5–24.9). For overweight and obese individuals, the relative risk (RR) was 1.28 and 1.82, respectively ($p < 0.001$). Among never-smokers, this relationship was greater (RR = 1.94 and 3.49, respectively, $p < 0.001$). Moreover, for a subset of 107,815 individuals who underwent a 6-year follow-up exam, the risk of RCC progressively increased with incremental elevations in the BMI. In this study, other obesity-linked cancers included esophageal (adenocarcinoma), melanoma, cancers of the lower GI tract, liver, and prostate. Obesity is particularly associated with inflammation in the liver. In this study, the RR of developing hepatocellular carcinoma was 3.13 overall, and 4.83 in the subset of never-smokers. In contrast, there was no association between overweight/obesity and cancers of the renal pelvis (transitional cell carcinoma), lung, and malignant hematologic diseases.

In the United States, the NIH-AARP Diet and Health Study (Adams *et al.*, 2008) utilized a self-administered questionnaire sent in 1995–1996 to 3.5 million AARP (American Association of Retired Persons) members, whose ages ranged from 50 to 71 years. Adequate responses were obtained from 566,402 and a second questionnaire with 320,618 responses was used to define the cohort (male:female = 1.38). Compared to individuals with a normal BMI, progressive degrees of overweight/obesity were associated with a statistically significant increase in RCC after adjusting for age and other factors. In the most obese group (BMI > 35), the RR was 2.47 and 2.59 for men and women, respectively.

Similar findings were observed in the "multiethnic cohort," comprised (in descending frequency) of Japanese Americans, Whites, Latinos, African

Americans, native Hawaiians, and others living in California and Hawaii (Setiawan *et al.*, 2007). This study began in 1993 and involved 161,126 individuals with an average age of ~59 years who were followed by questionnaire for an average of 8.3 years. In men and women, after adjusting for smoking and other factors, obesity was associated with an increased RR of RCC (1.76 and 2.27, respectively). Moreover, risk progressively increased for individuals in the higher weight quartiles. In this study, the risk for women was greater than men. Overweight, but not obese, men had a marginal, nonstatistically significant, increased risk (RR = 1.14), whereas both overweight and obese women had a significantly increased risk.

In the Million Women Study (Reeves *et al.*, 2007), which recruited 1.2 million English and Scottish women followed on average 5.4 years for cancer incidence and 7.4 years for cancer mortality, a significant correlation was found between kidney cancer and increased BMI. At recruitment during 1996–2001, their ages ranged from 50 to 64. RCC was the third highest cancer associated with increased BMI following endometrial and esophageal adenocarcinoma. In the EPIC trial (European Prospective Investigation into Cancer and Nutrition) involving 348,550 individuals, women with a weight or BMI in the top quintile had a twofold increased risk of RCC after adjusting for smoking and other risk factors (Pischon *et al.*, 2006). In men, there was no significant association. Similarly, in a case-control study from Iowa, increased weight was associated with increased RCC risk in women (Chiu *et al.*, 2006). In this same group of participants, diets richest in animal and saturated fats, oleic acid, and cholesterol were associated with statistically significant increases in RCC (1.9–2.6-fold, depending on the factor) (Brock *et al.*, 2009). Thus, being overweight or obese increases the probability of developing RCC. Depending on the study, this risk applies to both men and women.

II. MECHANISTIC FACTORS LINKING OBESITY AND LIPID DEREGULATION TO RCC

Previous excellent reviews (Chow and Devesa, 2008; Decastro and McKiernan, 2008; Klinghoffer *et al.*, 2009; Pascual and Borque, 2008) have discussed the role of hormones, including insulin signaling, insulin-like growth factor and other factors, which will not be repeated here. Rather, we focus on the potential roles of leptin and adiponectin, as well as two hereditary cancer genes, which encode proteins that provide a regulatory link between the lipid and biosynthetic pathways. We have included discussions on the clear-cell phenotype, transcriptional regulation of lipid biosynthesis, and also VHL function, since this is the most frequently mutated gene in both hereditary and spontaneous RCC. Together, these data make a compelling case that alterations in lipid homeostasis play a role in RCC development.

A. The Role of Leptin and Adiponectin

Leptin is an adipose tissue-derived hormone that regulates food intake and energy expenditure. Normally, leptin promotes satiety following feeding. Leptin levels positively correlate with increasing BMI. This increased expression results from leptin resistance that develops in obese individuals leading to high levels concurrent with the lack of response. Leptin has been shown to be a growth factor in cancer cell lines and elevated circulating leptin levels have been identified in patients with various types of cancer. In RCC patients, elevated leptin levels in blood were associated with both higher leptin receptor levels in the tumor cells and increased venous invasion (Horiguchi *et al.*, 2006a). Furthermore, the leptin receptor was expressed in each of six human RCC cell lines (Horiguchi *et al.*, 2006b). *In vitro*, leptin promotes collagen gel invasion by mouse renal cancer cells as well as non-transformed MDCK kidney cells (Attoub *et al.*, 2000). These results suggest that leptin signaling plays a role in kidney cancer development (see Fig. 1).

RCC is characteristically associated with a rich vascularity, and anti-VEGF therapy in cancer has had its greatest effect in this disease. Interestingly, the leptin receptor is also expressed in endothelial cells and leptin is a potent angiogenic factor *in vitro* and *in vivo* (Bouloumie *et al.*, 1998; Sierra-Honigmann *et al.*, 1998). In addition, tumor-driven lymphangiogenesis causes upregulation of the leptin receptor (Clasper *et al.*, 2008). Endothelial cells exposed to physiologic concentrations of leptin activate AKT, ERK1/2 and increase their migration. Interestingly, thiazolidinediones,

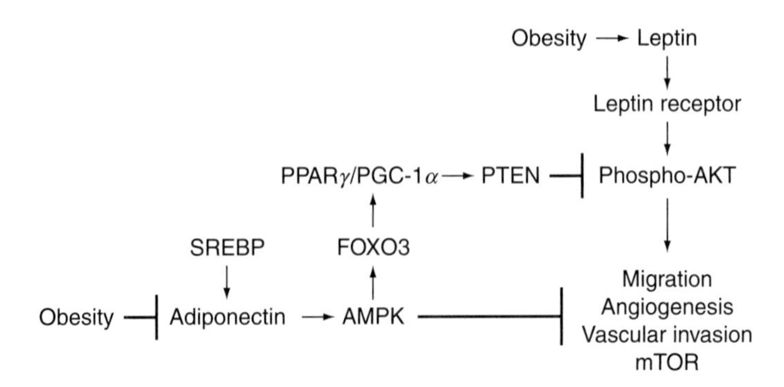

Fig. 1 Leptin–adiponectin interactions that may play a role in RCC.

which act as PPARγ and AMPK agonists, block leptin-induced AKT activation and migration by upregulating PTEN (Goetze *et al.*, 2002). This appears to be a direct effect of PPARγ binding to the PTEN promoter (reviewed in Teresi and Waite, 2008). Thus, PPARγ agonists may represent a nonspecific approach to downregulate leptin-induced angiogenesis.

Adiponectin is normally produced exclusively by adipose tissue and is an abundant circulating plasma protein (reviewed in Fang and Sweeney, 2006). In contrast to leptin, adiponectin levels are reduced in obese individuals and cancer patients. Several studies have demonstrated that adiponectin levels are reduced in patients with RCC, and also that adiponectin levels inversely correlate with tumor size (Horiguchi *et al.*, 2008; Pinthus *et al.*, 2008; Spyridopoulos *et al.*, 2009). Furthermore, a reduction in levels of the adiponectin receptor, AdipoR2, was associated with increased metastases (Pinthus *et al.*, 2008). These results suggest that adiponectin has tumor suppressor activity in RCC. In this regard, adiponectin exerts much of its reported effects by activating AMP Kinase (AMPK) (Guerre-Millo, 2008; Lim *et al.*, 2009), a known tumor suppressor (Shackelford and Shaw, 2009). In angiogenesis, the role of adiponectin is less clear with reports of both stimulatory and inhibitor activities (see Barresi *et al.*, 2009; Brakenhielm *et al.*, 2004; Ouchi *et al.*, 2004 and references therein). However, AMPK phosphorylates and activates FOXO3, which in turn upregulates the PPARγ coactivator, PGC-1α (Greer *et al.*, 2007). Thus, leptin and adiponectin may have opposite effects on angiogenesis, at least in some contexts. A summary of these interactions is shown in Fig. 1.

The regulation of adiponectin mRNA level is multifactorial, involving C/EBP, PPARγ, TNFα, and the SREBPs (sterol regulatory element binding proteins-1/2) (Kita *et al.*, 2005), the activity of which is decreased in sterol-overloaded cells. Two evolutionarily conserved SREBP response elements were identified in the mouse adiponectin promoter (Seo *et al.*, 2004). While mutation of either single site had no effect, mutation of both sites abolished basal expression. Induction of adiponectin by SREBP-1c was potentiated by the E-box protein, E47, and these factors were shown to interact directly at the adiponectin promoter (Doran *et al.*, 2008).

B. Regulation of Lipid Biosynthesis—Role of SREBPs

The transcriptional regulation of cholesterol and fatty acid biosynthesis is under the positive control of SREBP-1 and 2, which are synthesized as inactive 125-kDa precursor proteins tethered to the endoplasmic reticulum (ER) (Sato *et al.*, 1994; Yokoyama *et al.*, 1993). Activation (during sterol deficiency) requires intramembrane proteolytic processing in the Golgi, which releases their N-terminal bHLH domains to relocate in the nucleus and activate target genes (Fig. 2). This process is controlled by sterol levels, growth factors, and mTOR signaling (Horton *et al.*, 2002; Porstmann *et al.*, 2008), although SREBP-independent effects of mTOR on lipid biosynthetic processes (e.g., LDL levels) have also been described (Sharpe and Brown, 2008; Yoon *et al.*, 2007). When membrane cholesterol is high, the precursors are retained in the ER in a complex with SCAP (SREBP-cleavage activating protein) (Nohturfft *et al.*, 1996). In turn, SCAP contains a sterol-sensing domain that mediates its interaction with the ER-anchor proteins, INSIG-1/2 (*INS*ulin-*I*nduced Gene) (Yang *et al.*, 2002). When sterol levels are low, the SCAP–INSIG interaction is lost and SCAP escorts the SREBPs in COPII vesicles to the Golgi, where they are cleaved by the Site 1 and 2 proteases (Rawson *et al.*, 1997; Sakai *et al.*, 1998). Only a handful of proteins contain sterol-sensing domains. Among others, these include SCAP, HMG-CoA reductase, and TRC8. Mutations in the sterol-sensing domain of SCAP cause constitutive SREBP processing (Hua *et al.*, 1996). The sterol-sensing domain of HMG-CoA reductase regulates its association with INSIG in response to levels of lanosterol, a cholesterol precursor. In turn, ISIG binds the E3-ubiquitin ligase, gp78, which polyubiquitinates HMG-CoA reductase resulting in its degradation (Lee *et al.*, 2006). Thus, sterols regulate the retention of SREBPs in the ER and the degradation of HMG-CoA reductase.

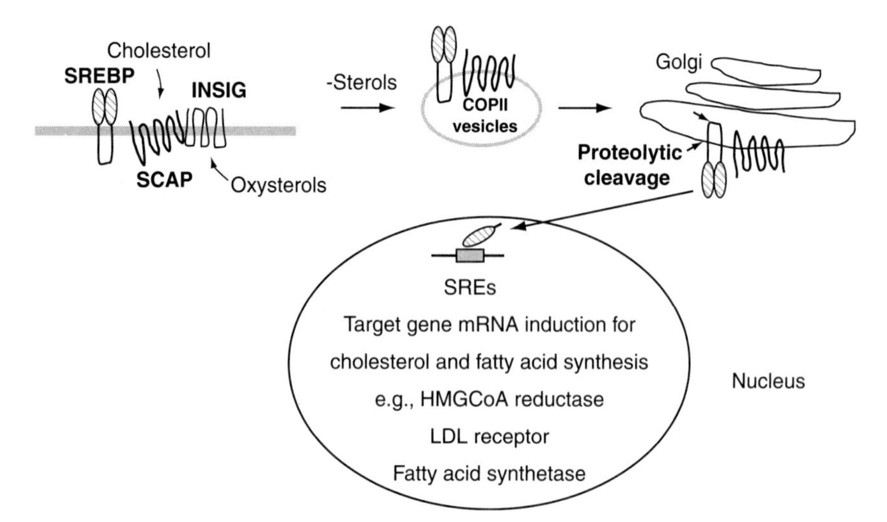

Fig. 2 Activation of SREBP precursors in response to sterol deficiency.

C. The Clear-Cell Phenotype in RCC

In RCC, the clear-cell phenotype is associated with intense oil-red O staining, a fat-soluble dye that marks neutral triglycerides and other neutral lipids, including esterified sterols. Gebhard *et al.* (1987) initially reported that RCCs contain elevated levels of cholesterol and cholesterol esters (8- and 35-fold, respectively). The cholesterol esters were predominantly comprised of oleate, which suggested they were locally produced. Likewise, by electron microscopy, the intracellular accumulations were free of membranes. However, the number of samples examined was limited and the question of whether increased LDL-mediated uptake might be responsible in part for the cholesterol accumulation is important (see below). Using magnetic resonance spectroscopy (MRS), Tugnoli *et al.* (2004) reported that normal kidney contained high levels of triglycerides and free cholesterol, whereas RCCs were characterized by cholesterol esterified with oleate, confirming the earlier work of Gebbhard *et al.* Of note, the clear-cell RCC lipid profile was distinct from those of papillary carcinomas and renal oncocytomas (Righi *et al.*, 2007; Tugnoli *et al.*, 2004).

The formation of cholesterol esters by acyl-coenzyme A:cholesterol acyl transferase (ACAT) protects cells from the toxic effects of high levels of free cholesterol. Gebhard *et al.* found approximately twofold higher levels of ACAT in RCCs compared with uninvolved kidney (Gebhard *et al.*, 1987). In contrast, levels of HMG-CoA reductase, the rate-limiting step in cholesterol biosynthesis, were reduced and cholesterol ester hydrolase activity appeared normal, which suggested that increased ACAT activity was responsible for the cholesterol ester accumulation. There are two ACAT enzymes in mammalian cells. Matsumoto *et al.* (2008) reported that ACAT activity was upregulated by 5.7-fold in RCCs versus normal kidney and ACAT-1 protein was 2.9-fold higher, whereas ACAT-2 was undetectable. Of note, cells containing excess cholesterol respond by increasing the production of cholesterol esters and turning down biosynthesis. This was observed, for example, in cells made deficient for cholesterol ester transfer protein (Izem and Morton, 2007).

ACAT-1 mRNA levels are controlled by multiple factors, although the SREBPs appear not to be involved. In macrophages, interferon-γ, urotensin II, and dexamethasone upregulate ACAT-1 mRNA, while adiponectin inhibits its expression (Chang *et al.*, 2009). Leptin has been reported to increase ACAT-1 activity and inhibit cholesterol efflux (Hongo *et al.*, 2009). Thus ACAT-1 activity could be increased with obesity, which would be consistent with the hypothesis that leptin facilitates RCC development, while adiponectin has an inhibitory role. In breast cancer cells, estrogen receptor-negative tumors were associated with elevated levels of ACAT-1 mRNA, cholesterol esters, and LDL-cholesterol uptake, while cholesterol biosynthesis was reduced.

These findings would be reminiscent of the abnormalities in RCC if LDL-mediated uptake was, indeed, upregulated. ACAT-1 (and ACAT-2) are integral ER-membrane proteins and the activity of ACAT-1 is stimulated by cholesterol in a sigmoidal manner. ACAT-2 is regulated by Cdx2 (in the gut) and HNF-1α. Of note, loss of VHL is associated with downregulation of HNF-1α, and this effect is HIF-independent (Hughes *et al.*, 2007). Potentially, this may explain the absence of ACAT-2 in RCC (Matsumoto *et al.*, 2008). In addition, HNF-1α physically and functionally interacts with HNF-4α (Rowley *et al.*, 2006), which in turn can bind and be inhibited by SREBP-1 (Kanayama *et al.*, 2007).

Regardless of whether increased ACAT is responsible for the clear-cell phenotype, its inhibition has been shown to negatively affect tumor cell growth and invasion. In NIH-3T3 and U87 glioma cells driven by activated cholecystokinin receptor 2 (CCK2R), a G-protein-coupled receptor that stimulates the formation of cholesterol esters, ACAT inhibition with Sah58-035 inhibited proliferation by 34% and invasion by 73% (Paillasse *et al.*, 2009). In contrast, the addition of cholesterol oleate stimulated proliferation and invasion. Cholesterol ester stimulation by CCK2R was dependent on the atypical protein kinase C, zeta (PKCζ) as well as ERK-1/2. Of note, VHL, the major "gate-keeper" in RCC is known to target atypical PKCs for ubiquitination and proteasome degradation (Okuda *et al.*, 2001), thus providing a potential link to cholesterol esterol formation. CCK2R activation also leads to proteosome-mediated degradation of PPARγ (Chang *et al.*, 2006), which would be expected to enhance the effects of leptin upregulation (and adiponectin inhibition) on AKT. This occurs through transactivation of EGFR and ERK-1/2 activation. G-protein-coupled receptors known to transactivate EGFR in RCC cells include protease-activated receptor-1 (PAR1/coagulation factor 2 receptor), which is activated by thrombin (Bergmann *et al.*, 2006), and the bradykinin B2 receptor (Mukhin *et al.*, 2006). However, whether PPARγ levels are affected by these or related receptors and contribute to cholesterol ester formation in RCC cells is unknown. A summary of these interactions is shown in Fig. 3.

Lastly, normal kidney proximal tubular cells upregulate cholesterol biosynthesis following stress, which produces a "cytoresistant state" to further injury. If perturbed by statins (or cholesterol depletion) the cells undergo apoptosis (Zager and Kalhorn, 2000). Similarly, lovastatin induces apoptosis in renal mesangial cells (Ghosh *et al.*, 1997), which are induced during development by the ureteral bud to form glomeruli and proximal/distal tubules. Thus, it is tempting to speculate that upregulated cholesterol biosynthesis plays a role in the pathogenesis of RCC, possibly by allowing stressed cells to survive. Included in the endogenous cholesterol/lipid biosynthetic pathway are the isoprenoids, farnesyl and geranylgeranyl

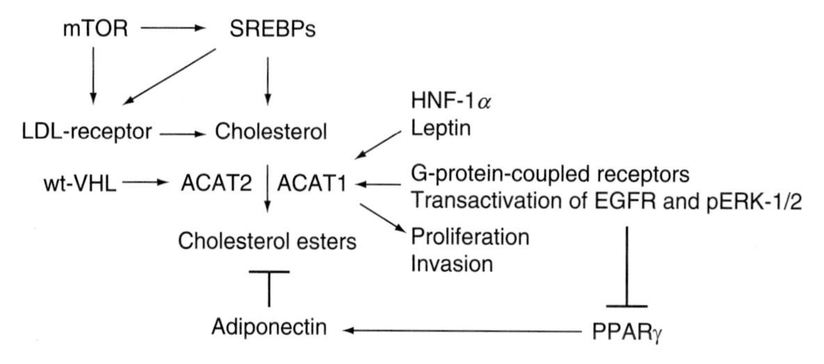

Fig. 3 Cholesterol ester regulation.

pyrophosphate. These moieties are critical regulators of various small GTPases that are critical in tumor development, invasion, and metastasis such as Ras, Rho, and RHEB, among others.

D. VHL—The "Gate-Keeper" Gene in RCC

The term "gate-keeper" was utilized by Vogelstein and Kinzler in 1997 to refer to "genes that directly regulate the growth of tumors by inhibiting growth or promoting death." In the course of tumor development, a loss of gate-keeper function is required and the alterations are rate-limiting. VHL is believed to function as a gate-keeper in RCC, since its function is lost in the most common form of hereditary RCC (VHL syndrome) and the majority of spontaneous RCCs contain loss of function of VHL mutations or epigenetic silencing.

As part of an E3-ubiquitin ligase complex, pVHL targets substrates for polyubiquitylation and destruction by the proteasome (Kaelin, 2004). Undoubtedly the best characterized substrates are the alpha subunits of HIF, HIF-1/2α, which have been strongly implicated in RCC development (Li *et al.*, 2007). Although RCC cells with wild-type VHL are able to proliferate *in vitro*, their *in vivo* tumorigenic potential is substantially impaired in a VHL and HIF-dependent manner (Zimmer *et al.*, 2004). Additional levels of control that affect HIF protein levels include HIF prolyl hydroxylation, which increases its binding to VHL, and HIF-1/2α protein translation initiation, which is regulated by the activity of mTOR. Mutations in the Krebs cycle enzymes, fumarate hydratase and succinate dehydrogenase, negatively affect prolyl hydroxylase function and result in elevated levels of HIF-1/2α proteins (Isaacs *et al.*, 2005; Pollard *et al.*, 2005). Mutations affecting the Tuberous Sclerosis (TSC) complex cause upregulation of HIFα subunits since the TSC complex is an upstream inhibitor of mTOR (Brugarolas *et al.*, 2003). Mutations in fumarate hydratase, succinate dehydrogenase, and TSC1/2 result in hereditary tumors that include RCC, although their mutation in spontaneous tumors is rare.

Experimentally, HIF deregulation is associated with *in vivo* tumorigenesis, aneupleudy and escape from growth arrest, yet some biologic consequences of VHL mutations are HIF-independent (Hughes *et al.*, 2007). These include impaired fibronectin deposition, upregulated $\alpha 5/\beta 1$ integrin expression, and loss of the differentiation markers, leucine aminopeptidase, and HNF-1α (as mentioned above). Other functions of VHL include the regulation of NF-κB (An and Rettig, 2005; Yang *et al.*, 2007) and stabilization of p53 (Roe *et al.*, 2006).

III. HEREDITARY RCC GENES AFFECTING LIPID AND PROTEIN BIOSYNTHETIC PATHWAYS

A. Birt-Hogg-Dubé/Folliculin

The Birt–Hogg–Dubé (BHD) syndrome is an autosomal dominant disorder (with variable penetrance) characterized by benign fibrofolliculomas and trichodiscomas of the hair follicles as well as renal tumors, colonic polyps, thyroid medullary carcinoma, and multifocal pulmonary cysts, which may rupture resulting in spontaneous pneumothorax. The renal tumors are primarily chromophobic, but clear-cell and chromophobe/oncocytic tumors have also been reported. Genetic linkage studies led to the isolation of the BHD gene, folliculin (Nickerson *et al.*, 2002), and subsequently folliculin was linked to AMPK through folliculin interacting protein (FNIP1) (Baba *et al.*, 2006). Recently, BHD mutations were shown to result in activation of mTOR (Hasumi *et al.*, 2009). At least in chromophobe tumors, HIF-2α protein levels were consistently upregulated (Kim *et al.*, 2006). In the mouse, BHD mutation is associated with polycystic kidneys—a phenotype that is partially reversed by treatment with the mTOR inhibitor, rapamycin (Baba *et al.*, 2008). Thus, BHD/folliculin loss of function mutations appears to result in renal tumors, at least in part, by activating mTOR and upregulating HIFα protein translation.

Of note, AMPK was originally identified by its phosphorylation and inhibition of HMG-CoA reductase, the rate-limiting step in cholesterol biosynthesis (Beg *et al.*, 1979). Thus, folliculin loss would be expected to upregulate both cholesterol biosynthesis and protein translation initiation through reduced activation of AMPK. AMPK also inhibits SREBP-1 at both the mRNA and protein levels (Tomita *et al.*, 2005; Zhou *et al.*, 2001); BHD mutations would be expected to have the opposite effects.

B. TRC8/RNF139

The hereditary RCC gene, TRC8, was isolated from a family with hereditary renal/thyroid cancer and a constitutional 3;8 chromosome translocation (Gemmill *et al.*, 1998). Subsequent independent cases of TRC8 rearrangements have been identified involving RCC as well as an ovarian dysgerminoma (Gimelli *et al.*, 2009; Poland *et al.*, 2007). TRC8 encodes a multimembrane spanning ER protein with E3-ubiquitin ligase activity (Gemmill *et al.*, 1998; Lorick *et al.*, 1999). From genetic interaction studies in *Drosophila*, knockdown of either d*TRC8* or d*VHL* caused an identical appearing mid-line defect, while the combined overexpression of both genes in the wing resulted in a unique phenotype, consistent with effects on interacting pathways (Gemmill *et al.*, 2002). In mammalian cells, overexpression of TRC8 inhibits growth and destabilizes the membrane-bound precursor forms of SREBP-1 and SREBP-2 without inducing processing to the nuclear forms (Brauweiler *et al.*, 2007; Lee *et al.*, 2010). Multiple SREBP target genes have reduced expression when TRC8 is expressed, consistent with the loss of SREBP precursors (Lee *et al.*, 2010). TRC8-mediated growth inhibition and SREBP loss are dependent upon a functional RING domain, implicating ubiquitination and the 26S proteasome in both effects. The growth inhibition can be partially overcome by expressing a constitutively active (nuclear) form of SREBP-1, which bypasses regulation by TRC8 (Brauweiler *et al.*, 2007). Mechanistically, TRC8 protein levels are sterol-responsive, increasing upon sterol starvation and decaying when cells are sterol replete. It also binds and stimulates ubiquitylation of the ER anchor protein, INSIG, an effect that appears dependent upon a sterol sensing domain in the amino terminus of TRC8. TRC8 knockdown has opposite effects, leading to increased levels of both precursor and nuclear forms of SREBPs along with increased SREBP target gene expression, at least in sterol-deprived cells (Lee *et al.*, 2010).

Previously, TRC8 was shown to physically interact with the MPN (Mpr1p, Pad1p N-terminal) protein–protein interaction domain of JAB1/CSN5, a subunit of the COP9 signalosome (Gemmill *et al.*, 2005). A subsequent screen demonstrated that TRC8 physically interacted with two additional MPN domain proteins, eIF3f and eIF3h, which are subunits of the eIF3 translation initiation complex (Lee *et al.*, 2010). This was confirmed by genetic interaction studies in flies, demonstrating restoration of growth in dTrc8-inhibited wing tissues following hemizygous loss of eIF3f or eIF3h. In mammalian cells, eIF3 was coimmunoprecipitated with ectopic TRC8.

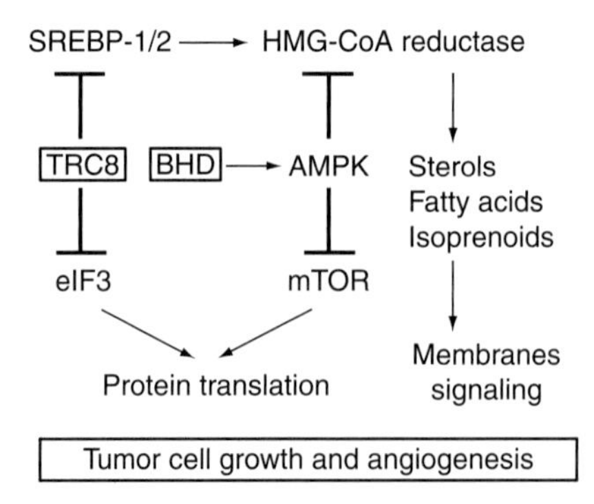

Fig. 4 The hereditary RCC genes, TRC8 and BHD/Folliclin, are positioned to coordinately regulate lipid and protein biosynthesis, tumor growth, and angiogenesis.

Furthermore, TRC8 overexpression inhibited polysome loading, indicative of impaired protein translation, and in *Drosophila* dTrc8 overexpression resulted in a *Minute* phenotype, characteristically associated with ribosomal protein mutations. Together, these results indicate that TRC8 provides a regulatory link between lipid and protein biosynthetic pathways. Thus, two hereditary RCC genes encode proteins that are poised to coordinately regulate lipid and protein biosynthesis (Fig. 4).

There are additional data that suggest links between lipid biosynthesis and RCC biology. Among the NCI panel of 60 cell lines, RCC as a tumor type is the most sensitive to growth inhibition by statins, which target HMG-CoA reductase. Furthermore, statins or 25-hydroxycholesterol, which blocks SREBP nuclear processing, inhibits angiogenesis induced by either VEGF (Schiefelbein *et al.*, 2008) or IL-8 (Yao *et al.*, 2006). This is associated with downregulation of Rho-GTP and involves isoprenoids, since the addition of farnesyl and geranylgeranyl pyrophosphate overcomes the effect of 25-hydroxycholesterol.

IV. CONCLUDING REMARKS

Multiple epidemiologic studies have linked the development of renal cancer to obesity. The upregulation of leptin and downregulation of adiponectin pathways in obesity fit with our molecular understanding of RCC pathogenesis. In addition, two forms of hereditary RCC involve proteins, Folliculin and TRC8, positioned to coordinately regulate lipid and protein biosynthesis, both of which have important downstream consequences on HIF-1/2α levels and angiogenesis. The role of lipid biology and its interface with protein translation regulation represents a new dimension in RCC research with potential therapeutic implications.

REFERENCES

Adams, K. F., Leitzmann, M. F., Albanes, D., Kipnis, V., Moore, S. C., Schatzkin, A., and Chow, W. H. (2008). Body size and renal cell cancer incidence in a large US cohort study. *Am. J. Epidemiol.* **168,** 268–277.

An, J., and Rettig, M. B. (2005). Mechanism of von Hippel-Lindau protein-mediated suppression of nuclear factor kappa B activity. *Mol. Cell. Biol.* **25,** 7546–7556.

Attoub, S., Noe, V., Pirola, L., Bruyneel, E., Chastre, E., Mareel, M., Wymann, M. P., and Gespach, C. (2000). Leptin promotes invasiveness of kidney and colonic epithelial cells via phosphoinositide 3-kinase-, rho-, and rac-dependent signaling pathways. *FASEB J.* **14,** 2329–2338.

Baba, M., Hong, S. B., Sharma, N., Warren, M. B., Nickerson, M. L., Iwamatsu, A., Esposito, D., Gillette, W. K., Hopkins, R. F., 3rd, Hartley, J. L., *et al.* (2006). Folliculin encoded by the BHD gene interacts with a binding protein, FNIP1, and AMPK, and is involved in AMPK and mTOR signaling. *Proc. Natl. Acad. Sci. USA* **103,** 15552–15557.

Baba, M., Furihata, M., Hong, S. B., Tessarollo, L., Haines, D. C., Southon, E., Patel, V., Igarashi, P., Alvord, W. G., Leighty, R., *et al.* (2008). Kidney-targeted Birt-Hogg-Dube gene inactivation in a mouse model: Erk1/2 and Akt-mTOR activation, cell hyperproliferation, and polycystic kidneys. *J. Natl. Cancer Inst.* **100,** 140–154.

Barresi, V., Tuccari, G., and Barresi, G. (2009). Adiponectin immunohistochemical expression in colorectal cancer and its correlation with histological grade and tumour microvessel density. *Pathology* **41,** 533–538.

Beg, Z. H., Stonik, J. A., and Brewer, H. B., Jr. (1979). Characterization and regulation of reductase kinase, a protein kinase that modulates the enzymic activity of 3-hydroxy-3-methylglutaryl-coenzyme A reductase. *Proc. Natl. Acad. Sci. USA* **76,** 4375–4379.

Bergmann, S., Junker, K., Henklein, P., Hollenberg, M. D., Settmacher, U., and Kaufmann, R. (2006). PAR-type thrombin receptors in renal carcinoma cells: PAR1-mediated EGFR activation promotes cell migration. *Oncol. Rep.* **15,** 889–893.

Bouloumie, A., Drexler, H. C., Lafontan, M., and Busse, R. (1998). Leptin, the product of Ob gene, promotes angiogenesis. *Circ. Res.* **83,** 1059–1066.

Brakenhielm, E., Veitonmaki, N., Cao, R., Kihara, S., Matsuzawa, Y., Zhivotovsky, B., Funahashi, T., and Cao, Y. (2004). Adiponectin-induced antiangiogenesis and antitumor activity involve caspase-mediated endothelial cell apoptosis. *Proc. Natl. Acad. Sci. USA* **101,** 2476–2481.

Brauweiler, A., Lorick, K. L., Lee, J. P., Tsai, Y. C., Chan, D., Weissman, A. M., Drabkin, H. A., and Gemmill, R. M. (2007). RING-dependent tumor suppression and G2/M arrest induced by the TRC8 hereditary kidney cancer gene. *Oncogene* **26,** 2263–2271.

Brock, K. E., Gridley, G., Chiu, B. C., Ershow, A. G., Lynch, C. F., and Cantor, K. P. (2009). Dietary fat and risk of renal cell carcinoma in the USA: A case-control study. *Br. J. Nutr.* **101,** 1228–1238.

Brugarolas, J. B., Vazquez, F., Reddy, A., Sellers, W. R., and Kaelin, W. G., Jr. (2003). TSC2 regulates VEGF through mTOR-dependent and -independent pathways. *Cancer Cell* **4,** 147–158.

Chang, A. J., Song, D. H., and Wolfe, M. M. (2006). Attenuation of peroxisome proliferator-activated receptor gamma (PPARgamma) mediates gastrin-stimulated colorectal cancer cell proliferation. *J. Biol. Chem.* **281,** 14700–14710.

Chang, T. Y., Li, B. L., Chang, C. C., and Urano, Y. (2009). Acyl-coenzyme A:cholesterol acyltransferases. *Am. J. Physiol.* **297,** E1–E9.

Chiu, B. C., Gapstur, S. M., Chow, W. H., Kirby, K. A., Lynch, C. F., and Cantor, K. P. (2006). Body mass index, physical activity, and risk of renal cell carcinoma. *Int. J. Obes.* **30,** 940–947.

Chow, W. H., and Devesa, S. S. (2008). Contemporary epidemiology of renal cell cancer. *Cancer J. (Sudbury, MA)* **14,** 288–301.

Clasper, S., Royston, D., Baban, D., Cao, Y., Ewers, S., Butz, S., Vestweber, D., and Jackson, D. G. (2008). A novel gene expression profile in lymphatics associated with tumor growth and nodal metastasis. *Cancer Res.* **68,** 7293–7303.

Decastro, G. J., and McKiernan, J. M. (2008). Epidemiology, clinical staging, and presentation of renal cell carcinoma. *Urol. Clin. North Am.* **35,** 581–592, vi.

Doran, A. C., Meller, N., Cutchins, A., Deliri, H., Slayton, R. P., Oldham, S. N., Kim, J. B., Keller, S. R., and McNamara, C. A. (2008). The helix-loop-helix factors Id3 and E47 are novel regulators of adiponectin. *Circ. Res.* **103,** 624–634.

Fang, X., and Sweeney, G. (2006). Mechanisms regulating energy metabolism by adiponectin in obesity and diabetes. *Biochem. Soc. Trans.* **34,** 798–801.

Flegal, K. M., Graubard, B. I., Williamson, D. F., and Gail, M. H. (2007). Cause-specific excess deaths associated with underweight, overweight, and obesity. *JAMA* **298,** 2028–2037.

Gebhard, R. L., Clayman, R. V., Prigge, W. F., Figenshau, R., Staley, N. A., Reesey, C., and Bear, A. (1987). Abnormal cholesterol metabolism in renal clear cell carcinoma. *J. Lipid Res.* **28,** 1177–1184.

Gemmill, R. M., West, J. D., Boldog, F., Tanaka, N., Robinson, L. J., Smith, D. I., Li, F., and Drabkin, H. A. (1998). The hereditary renal cell carcinoma 3;8 translocation fuses FHIT to a patched-related gene, TRC8. *Proc. Natl. Acad. Sci. USA* **95,** 9572–9577.

Gemmill, R. M., Bemis, L. T., Lee, J. P., Sozen, M. A., Baron, A., Zeng, C., Erickson, P. F., Hooper, J. E., and Drabkin, H. A. (2002). The TRC8 hereditary kidney cancer gene suppresses growth and functions with VHL in a common pathway. *Oncogene* **21,** 3507–3516.

Gemmill, R. M., Lee, J. P., Chamovitz, D. A., Segal, D., Hooper, J. E., and Drabkin, H. A. (2005). Growth suppression induced by the TRC8 hereditary kidney cancer gene is dependent upon JAB1/CSN5. *Oncogene* **24,** 3503–3511.

Ghosh, P. M., Mott, G. E., Ghosh-Choudhury, N., Radnik, R. A., Stapleton, M. L., Ghidoni, J. J., and Kreisberg, J. I. (1997). Lovastatin induces apoptosis by inhibiting mitotic and post-mitotic events in cultured mesangial cells. *Biochim. Biophys. Acta* **1359,** 13–24.

Gimelli, S., Beri, S., Drabkin, H. A., Gambini, C., Gregorio, A., Fiorio, P., Zuffardi, O., Gemmill, R. M., Giorda, R., and Gimelli, G. (2009). The tumor suppressor gene TRC8/RNF139 is disrupted by a constitutional balanced translocation t(8;22)(q24.13;q11.21) in a young girl with dysgerminoma. *Mol. Cancer* **8,** 52.

Goetze, S., Bungenstock, A., Czupalla, C., Eilers, F., Stawowy, P., Kintscher, U., Spencer-Hansch, C., Graf, K., Nurnberg, B., Law, R. E., *et al.* (2002). Leptin induces endothelial cell migration through Akt, which is inhibited by PPARgamma-ligands. *Hypertension* **40,** 748–754.

Greer, E. L., Oskoui, P. R., Banko, M. R., Maniar, J. M., Gygi, M. P., Gygi, S. P., and Brunet, A. (2007). The energy sensor AMP-activated protein kinase directly regulates the mammalian FOXO3 transcription factor. *J. Biol. Chem.* **282,** 30107–30119.

Guerre-Millo, M. (2008). Adiponectin: An update. *Diabetes Metab.* **34,** 12–18.

Hasumi, Y., Baba, M., Ajima, R., Hasumi, H., Valera, V. A., Klein, M. E., Haines, D. C., Merino, M. J., Hong, S. B., Yamaguchi, T. P., *et al.* (2009). Homozygous loss of BHD causes early embryonic lethality and kidney tumor development with activation of mTORC1 and mTORC2. *Proc. Natl. Acad. Sci. USA* **106,** 18722–18727.

Hongo, S., Watanabe, T., Arita, S., Kanome, T., Kageyama, H., Shioda, S., and Miyazaki, A. (2009). Leptin modulates ACAT1 expression and cholesterol efflux from human macrophages. *Am. J. Physiol.* **297,** E474–E482.

Horiguchi, A., Sumitomo, M., Asakuma, J., Asano, T., Zheng, R., Asano, T., Nanus, D. M., and Hayakawa, M. (2006a). Increased serum leptin levels and over expression of leptin receptors are associated with the invasion and progression of renal cell carcinoma. *J. Urol.* **176,** 1631–1635.

Horiguchi, A., Sumitomo, M., Asakuma, J., Asano, T., Zheng, R., Asano, T., Nanus, D. M., and Hayakawa, M. (2006b). Leptin promotes invasiveness of murine renal cancer cells via extracellular signal-regulated kinases and rho dependent pathway. *J. Urol.* **176,** 1636–1641.

Horiguchi, A., Ito, K., Sumitomo, M., Kimura, F., Asano, T., and Hayakawa, M. (2008). Decreased serum adiponectin levels in patients with metastatic renal cell carcinoma. *Jpn J. Clin. Oncol.* **38,** 106–111.

Horton, J. D., Goldstein, J. L., and Brown, M. S. (2002). SREBPs: Activators of the complete program of cholesterol and fatty acid synthesis in the liver. *J. Clin. Invest.* **109,** 1125–1131.

Hua, X., Nohturfft, A., Goldstein, J. L., and Brown, M. S. (1996). Sterol resistance in CHO cells traced to point mutation in SREBP cleavage-activating protein. *Cell* **87,** 415–426.

Hughes, M. D., Kapllani, E., Alexander, A. E., Burk, R. D., and Schoenfeld, A. R. (2007). HIF-2alpha downregulation in the absence of functional VHL is not sufficient for renal cell differentiation. *Cancer Cell Int.* **7,** 13.

Isaacs, J. S., Jung, Y. J., Mole, D. R., Lee, S., Torres-Cabala, C., Chung, Y. L., Merino, M., Trepel, J., Zbar, B., Toro, J., *et al.* (2005). HIF overexpression correlates with biallelic loss of fumarate hydratase in renal cancer: Novel role of fumarate in regulation of HIF stability. *Cancer Cell* **8,** 143–153.

Izem, L., and Morton, R. E. (2007). Possible role for intracellular cholesteryl ester transfer protein in adipocyte lipid metabolism and storage. *J. Biol. Chem.* **282,** 21856–21865.

Jemal, A., Siegel, R., Ward, E., Murray, T., Xu, J., Smigal, C., and Thun, M. J. (2006). Cancer statistics, 2006. *CA Cancer J. Clin.* **56,** 106–130.

Kaelin, W. G., Jr. (2004). The von Hippel-Lindau tumor suppressor gene and kidney cancer. *Clin. Cancer Res.* **10,** 6290S–6295S.

Kaelin, W. G. (2005). The von Hippel-Lindau tumor suppressor protein: Roles in cancer and oxygen sensing. *Cold Spring Harb. Symp. Quant. Biol.* **70,** 159–166.

Kanayama, T., Arito, M., So, K., Hachimura, S., Inoue, J., and Sato, R. (2007). Interaction between sterol regulatory element-binding proteins and liver receptor homolog-1 reciprocally suppresses their transcriptional activities. *J. Biol. Chem.* **282,** 10290–10298.

Kim, C. M., Vocke, C., Torres-Cabala, C., Yang, Y., Schmidt, L., Walther, M., and Linehan, W. M. (2006). Expression of hypoxia inducible factor-1alpha and 2alpha in genetically distinct early renal cortical tumors. *J. Urol.* **175,** 1908–1914.

Kita, A., Yamasaki, H., Kuwahara, H., Moriuchi, A., Fukushima, K., Kobayashi, M., Fukushima, T., Takahashi, R., Abiru, N., Uotani, S., *et al.* (2005). Identification of the promoter region required for human adiponectin gene transcription: Association with CCAAT/enhancer binding protein-beta and tumor necrosis factor-alpha. *Biochem. Biophys. Res. Commun.* **331,** 484–490.

Klinghoffer, Z., Yang, B., Kapoor, A., and Pinthus, J. H. (2009). Obesity and renal cell carcinoma: Epidemiology, underlying mechanisms and management considerations. *Expert Rev. Anticancer Ther.* **9,** 975–987.

Lee, J. N., Song, B., DeBose-Boyd, R. A., and Ye, J. (2006). Sterol-regulated degradation of Insig-1 mediated by the membrane-bound ubiquitin ligase gp78. *J. Biol. Chem.* **281,** 39308–39315.

Lee, J., Brauweiler, A., Rudolph, M., Hooper, J. E., Drabkin, H. A., and Gemmill, R. M. (2010). TRC8 is sterol regulated and interacts with lipid and protein biosynthetic pathways. *Mol. Cancer Res.* **8**(1), 93–106.

Li, L., Zhang, L., Zhang, X., Yan, Q., Minamishima, Y. A., Olumi, A. F., Mao, M., Bartz, S., and Kaelin, W. G., Jr. (2007). Hypoxia-inducible factor linked to differential kidney cancer risk seen with type 2A and type 2B VHL mutations. *Mol. Cell. Biol.* **27,** 5381–5392.

Lim, C. T., Kola, B., and Korbonits, M. (2009). AMPK as a mediator of hormonal signalling. *J. Mol. Endocrinol.* **22,** 22.

Lipworth, L., Tarone, R. E., and McLaughlin, J. K. (2006). The epidemiology of renal cell carcinoma. *J. Urol.* **176,** 2353–2358.

Lorick, K. L., Jensen, J. P., Fang, S., Ong, A. M., Hatakeyama, S., and Weissman, A. M. (1999). RING fingers mediate ubiquitin-conjugating enzyme (E2)-dependent ubiquitination. *Proc. Natl. Acad. Sci. USA* **96,** 11364–11369.

Matsumoto, K., Fujiwara, Y., Nagai, R., Yoshida, M., and Ueda, S. (2008). Expression of two isozymes of acyl-coenzyme A: Cholesterol acyltransferase-1 and -2 in clear cell type renal cell carcinoma. *Int. J. Urol.* **15,** 166–170.

Mukhin, Y. V., Gooz, M., Raymond, J. R., and Garnovskaya, M. N. (2006). Collagenase-2 and -3 mediate epidermal growth factor receptor transactivation by bradykinin B2 receptor in kidney cells. *J. Pharmacol. Exp. Ther.* **318,** 1033–1043.

Nickerson, M. L., Warren, M. B., Toro, J. R., Matrosova, V., Glenn, G., Turner, M. L., Duray, P., Merino, M., Choyke, P., Pavlovich, C. P., *et al.* (2002). Mutations in a novel gene lead to kidney tumors, lung wall defects, and benign tumors of the hair follicle in patients with the Birt-Hogg-Dube syndrome. *Cancer Cell* **2,** 157–164.

Nohturfft, A., Hua, X., Brown, M. S., and Goldstein, J. L. (1996). Recurrent G-to-A substitution in a single codon of SREBP cleavage-activating protein causes sterol resistance in three mutant Chinese hamster ovary cell lines. *Proc. Natl. Acad. Sci. USA* **93,** 13709–13714.

Okuda, H., Saitoh, K., Hirai, S., Iwai, K., Takaki, Y., Baba, M., Minato, N., Ohno, S., and Shuin, T. (2001). The von Hippel-Lindau tumor suppressor protein mediates ubiquitination of activated atypical protein kinase. C. *J. Biol. Chem.* **276,** 43611–43617.

Ouchi, N., Kobayashi, H., Kihara, S., Kumada, M., Sato, K., Inoue, T., Funahashi, T., and Walsh, K. (2004). Adiponectin stimulates angiogenesis by promoting cross-talk between AMP-activated protein kinase and Akt signaling in endothelial cells. *J. Biol. Chem.* **279,** 1304–1309.

Paillasse, M. R., de Medina, P., Amouroux, G., Mhamdi, L., Poirot, M., and Silvente-Poirot, S. (2009). Signaling through cholesterol esterification: A new pathway for the cholecystokinin 2 receptor involved in cell growth and invasion. *J. Lipid Res.* **50,** 2203–2211.

Pascual, D., and Borque, A. (2008). Epidemiology of kidney cancer. *Adv. Urol.* 782381.

Pinthus, J. H., Kleinmann, N., Tisdale, B., Chatterjee, S., Lu, J. P., Gillis, A., Hamlet, T., Singh, G., Farrokhyar, F., and Kapoor, A. (2008). Lower plasma adiponectin levels are associated with larger tumor size and metastasis in clear-cell carcinoma of the kidney. *Eur. Urol.* **54,** 866–873.

Pischon, T., Lahmann, P. H., Boeing, H., Tjonneland, A., Halkjaer, J., Overvad, K., Klipstein-Grobusch, K., Linseisen, J., Becker, N., Trichopoulou, A., *et al.* (2006). Body size and risk of renal cell carcinoma in the European Prospective Investigation into Cancer and Nutrition (EPIC). *Int. J. Cancer* **118,** 728–738.

Poland, K. S., Azim, M., Folsom, M., Goldfarb, R., Naeem, R., Korch, C., Drabkin, H. A., Gemmill, R. M., and Plon, S. E. (2007). A constitutional balanced t(3;8)(p14;q24.1) translocation results in disruption of the TRC8 gene and predisposition to clear cell renal cell carcinoma. *Genes Chromosomes Cancer* **46,** 805–812.

Pollard, P. J., Briere, J. J., Alam, N. A., Barwell, J., Barclay, E., Wortham, N. C., Hunt, T., Mitchell, M., Olpin, S., Moat, S. J., *et al.* (2005). Accumulation of Krebs cycle intermediates and over-expression of HIF1alpha in tumours which result from germline FH and SDH mutations. *Hum. Mol. Genet.* **14,** 2231–2239.

Porstmann, T., Santos, C. R., Griffiths, B., Cully, M., Wu, M., Leevers, S., Griffiths, J. R., Chung, Y. L., and Schulze, A. (2008). SREBP activity is regulated by mTORC1 and contributes to Akt-dependent cell growth. *Cell Metab.* **8**, 224–236.

Rawson, R. B., Zelenski, N. G., Nijhawan, D., Ye, J., Sakai, J., Hasan, M. T., Chang, T. Y., Brown, M. S., and Goldstein, J. L. (1997). Complementation cloning of S2P, a gene encoding a putative metalloprotease required for intramembrane cleavage of SREBPs. *Mol. Cell* **1**, 47–57.

Reeves, G. K., Pirie, K., Beral, V., Green, J., Spencer, E., and Bull, D. (2007). Cancer incidence and mortality in relation to body mass index in the Million Women Study: Cohort study. *BMJ (Clinical Research ed.)* **335**, 1134.

Righi, V., Mucci, A., Schenetti, L., Tosi, M. R., Grigioni, W. F., Corti, B., Bertaccini, A., Franceschelli, A., Sanguedolce, F., Schiavina, R., *et al.* (2007). *Ex vivo* HR-MAS magnetic resonance spectroscopy of normal and malignant human renal tissues. *Anticancer Res.* **27**, 3195–3204.

Roe, J. S., Kim, H., Lee, S. M., Kim, S. T., Cho, E. J., and Youn, H. D. (2006). p53 stabilization and transactivation by a von Hippel-Lindau protein. *Mol. Cell* **22**, 395–405.

Rowley, C. W., Staloch, L. J., Divine, J. K., McCaul, S. P., and Simon, T. C. (2006). Mechanisms of mutual functional interactions between HNF-4alpha and HNF-1alpha revealed by mutations that cause maturity onset diabetes of the young. *Am. J. Physiol. Gastrointest. Liver Physiol.* **290**, G466–G475.

Sakai, J., Rawson, R. B., Espenshade, P. J., Cheng, D., Seegmiller, A. C., Goldstein, J. L., and Brown, M. S. (1998). Molecular identification of the sterol-regulated luminal protease that cleaves SREBPs and controls lipid composition of animal cells. *Mol. Cell* **2**, 505–514.

Samanic, C., Chow, W. H., Gridley, G., Jarvholm, B., and Fraumeni, J. F., Jr. (2006). Relation of body mass index to cancer risk in 362,552 Swedish men. *Cancer Causes Control* **17**, 901–909.

Sato, R., Yang, J., Wang, X., Evans, M. J., Ho, Y. K., Goldstein, J. L., and Brown, M. S. (1994). Assignment of the membrane attachment, DNA binding, and transcriptional activation domains of sterol regulatory element-binding protein-1 (SREBP-1). *J. Biol. Chem.* **269**, 17267–17273.

Schiefelbein, D., Goren, I., Fisslthaler, B., Schmidt, H., Geisslinger, G., Pfeilschifter, J., and Frank, S. (2008). Biphasic regulation of HMG-CoA reductase expression and activity during wound healing and its functional role in the control of keratinocyte angiogenic and proliferative responses. *J. Biol. Chem.* **283**, 15479–15490.

Seo, J. B., Moon, H. M., Noh, M. J., Lee, Y. S., Jeong, H. W., Yoo, E. J., Kim, W. S., Park, J., Youn, B. S., Kim, J. W., *et al.* (2004). Adipocyte determination- and differentiation-dependent factor 1/sterol regulatory element-binding protein 1c regulates mouse adiponectin expression. *J. Biol. Chem.* **279**, 22108–22117.

Setiawan, V. W., Stram, D. O., Nomura, A. M., Kolonel, L. N., and Henderson, B. E. (2007). Risk factors for renal cell cancer: The multiethnic cohort. *Am. J. Epidemiol.* **166**, 932–940.

Shackelford, D. B., and Shaw, R. J. (2009). The LKB1–AMPK pathway: Metabolism and growth control in tumour suppression. *Nat. Rev.* **9**, 563–575.

Sharpe, L. J., and Brown, A. J. (2008). Rapamycin down-regulates LDL-receptor expression independently of SREBP-2. *Biochem. Biophys. Res. Commun.* **373**, 670–674.

Sierra-Honigmann, M. R., Nath, A. K., Murakami, C., Garcia-Cardena, G., Papapetropoulos, A., Sessa, W. C., Madge, L. A., Schechner, J. S., Schwabb, M. B., Polverini, P. J., *et al.* (1998). Biological action of leptin as an angiogenic factor. *Science (New York, NY)* **281**, 1683–1686.

Spyridopoulos, T. N., Petridou, E. T., Dessypris, N., Terzidis, A., Skalkidou, A., Deliveliotis, C., and Chrousos, G. P. (2009). Inverse association of leptin levels with renal cell carcinoma: Results from a case-control study. *Hormones (Athens, Greece)* **8**, 39–46.

Teresi, R. E., and Waite, K. A. (2008). PPARgamma, PTEN, and the Fight against Cancer. *PPAR Res.* **2008,** 932632.
Tomita, K., Tamiya, G., Ando, S., Kitamura, N., Koizumi, H., Kato, S., Horie, Y., Kaneko, T., Azuma, T., Nagata, H., *et al.* (2005). AICAR, an AMPK activator, has protective effects on alcohol-induced fatty liver in rats. *Alcohol. Clin. Exp. Res.* **29,** 240S–245S.
Tugnoli, V., Trinchero, A., and Tosi, M. R. (2004). Evaluation of the lipid composition of human healthy and neoplastic renal tissues. *Ital. J. Biochem.* **53,** 169–182.
Yang, T., Espenshade, P. J., Wright, M. E., Yabe, D., Gong, Y., Aebersold, R., Goldstein, J. L., and Brown, M. S. (2002). Crucial step in cholesterol homeostasis: Sterols promote binding of SCAP to INSIG-1, a membrane protein that facilitates retention of SREBPs in ER. *Cell* **110,** 489–500.
Yang, H., Minamishima, Y. A., Yan, Q., Schlisio, S., Ebert, B. L., Zhang, X., Zhang, L., Kim, W. Y., Olumi, A. F., and Kaelin, W. G., Jr. (2007). pVHL acts as an adaptor to promote the inhibitory phosphorylation of the NF-kappaB agonist Card9 by CK2. *Mol. Cell* **28,** 15–27.
Yao, M., Zhou, R. H., Petreaca, M., Zheng, L., Shyy, J., and Martins-Green, M. (2006). Activation of sterol regulatory element-binding proteins (SREBPs) is critical in IL-8-induced angiogenesis. *J. Leukoc. Biol.* **80,** 608–620.
Yokoyama, C., Wang, X., Briggs, M. R., Admon, A., Wu, J., Hua, X., Goldstein, J. L., and Brown, M. S. (1993). SREBP-1, a basic-helix-loop-helix-leucine zipper protein that controls transcription of the low density lipoprotein receptor gene. *Cell* **75,** 187–197.
Yoon, S., Lee, M. Y., Park, S. W., Moon, J. S., Koh, Y. K., Ahn, Y. H., Park, B. W., and Kim, K. S. (2007). Up-regulation of acetyl-CoA carboxylase alpha and fatty acid synthase by human epidermal growth factor receptor 2 at the translational level in breast cancer cells. *J. Biol. Chem.* **282,** 26122–26131.
Zager, R. A., and Kalhorn, T. F. (2000). Changes in free and esterified cholesterol: Hallmarks of acute renal tubular injury and acquired cytoresistance. *Am. J. Pathol.* **157,** 1007–1016.
Zhou, G., Myers, R., Li, Y., Chen, Y., Shen, X., Fenyk-Melody, J., Wu, M., Ventre, J., Doebber, T., Fujii, N., *et al.* (2001). Role of AMP-activated protein kinase in mechanism of metformin action. *J. Clin. Invest.* **108,** 1167–1174.
Zimmer, M., Doucette, D., Siddiqui, N., and Iliopoulos, O. (2004). Inhibition of hypoxia-inducible factor is sufficient for growth suppression of $VHL^{-/-}$ tumors. *Mol. Cancer Res.* **2,** 89–95.

Regulatory T Cells in Cancer

Dimitrios Mougiakakos, Aniruddha Choudhury, Alvaro Lladser, Rolf Kiessling, and C. Christian Johansson

Department of Oncology and Pathology, Karolinska University Hospital, Cancer Center Karolinska R8:01, Stockholm, Sweden

---

At the present time, regulatory T cells (Tregs) are an integral part of immunology but the route from discovery of "suppressive" lymphocytes in the 1980s to the current established concept of Tregs almost 20 years later has been a rollercoaster ride. Tregs are essential for maintaining self-tolerance as defects in their compartment lead to severe autoimmune diseases. This vitally important function exists alongside the detrimental effects on tumor immunosurveillance and antitumor immunity. Beginning with the identification of $CD4^+CD25^+$ Tregs in 1995, the list of Treg subsets, suppressive mechanisms, and knowledge about their various origins is steadily growing. Increase in Tregs within tumors and circulation of cancer patients, observed in early studies, implied their involvement in pathogenesis and disease progression. Several mechanisms, ranging

Advances in CANCER RESEARCH

0065-230X/10 $35.00
DOI: 10.1016/S0065-230X(10)07003-X

from proliferation to specific trafficking networks, have been identified to account for their systemic and/or local accumulation. Since various immunotherapeutic approaches are being utilized for cancer therapy, various strategies to overcome the antagonistic effects exerted by Tregs are being currently explored. An overview on the biology of Tregs present in cancer patients, their clinical impact, and methods for modulating them is given in this review. Despite the extensive studies on Tregs in cancer many questions still remain unanswered. Even the paradigm that Tregs generally are disadvantageous for the control of malignancies is now under scrutiny. Insight into the specific role of Tregs in different types of neoplasias is the key for targeting them in a way that is beneficial for the clinical outcome.

I. INTRODUCTION

A. Discovery and Fall

The current view on immunology can arguably be thought to begin with the discovery that adaptive immunity is composed of two major types of lymphocytes; the B (bone marrow-derived) and T (thymus-derived) cells (Miller, 1961; Mosier, 1967). Almost concurrently, anecdotal observations were already extending the role of T cells, beyond functioning as effectors and positive regulators, to suppressors of immunological responses. Pioneering studies by Gershon and Kondo in the early 1970s demonstrated for the first time that lymphocytes can suppress T cell responses in an antigen-specific manner (Gershon and Kondo, 1970) and that transfer of antigen-experienced T cells into naïve mice can lead to an antigen-specific tolerance by attenuating T cell activity (Gershon and Kondo, 1971). With great foresight, this cell population was named "suppressor cells" and fit perfectly into the dogma of homeostatic immunoregulation. It was hypothesized that by sustaining quantitatively and qualitatively optimal responses, the immune system facilitated an efficient elimination of pathogens and simultaneously prevented autoimmunity (Penhale *et al.*, 1973). Based on the observations that T cells from tumor bearing hosts were endowed with immunosuppressive capacities preventing the rejection of even highly immunogenic tumors by immunocompetent hosts, potential interconnections between "suppressor cells" and malignancies were presumed (Berendt and North, 1980; Fujimoto *et al.*, 1975). Despite the great significance of these findings, a growing skepticism led to a major loss of momentum and interest for almost 20 years. The main reasons for this were the failure to unequivocally define these cells together with a number of key misleading publications on MHC regions postulated as characteristic for "suppressor cells;" in particular, the illusory I-J locus as well as T-T suppressor hybridomas not transcribing T cell receptor (TCR) genes (Moller, 1988; Simpson, 2008).

B. Renaissance Through Steady Characterization

Finally, in 1995 Sakaguchi and colleagues initiated the renaissance of the "suppressive cells" (Sakaguchi *et al.*, 1995). In very elegant experiments they showed that transfer of thymic CD25-depleted T cells induced autoimmune diseases in athymic nude mice, while addition of a small proportion of $CD4^{+}CD25^{+}$ T cells was sufficient to maintain tolerance. Accordingly, the CD25 molecule was the first promising candidate for a phenotypic definition of "suppressive cells" that were named as thymus-derived naturally occurring regulatory T cells (nTregs). CD25 is the α-chain of the high-affinity receptor for interleukin-2 (IL-2R). Although nTregs do not produce IL-2 (Allan *et al.*, 2005) they are vitally dependent on IL-2 production by their environment. This is markedly illustrated by the development of a lethal lymphoproliferative disease in mice deficient for IL-2 or the IL-2Rβ, which resulted in dysregulated T cell activation and severe alterations within the nTreg compartment (Suzuki *et al.*, 1995). The constitutive expression of the IL-2R on nTregs may reflect this dependence on external IL-2. Several models to date have explored how IL-2 signaling contributes to suppressive function, thymic development, and homeostasis of Tregs (Bayer *et al.*, 2005; Furtado *et al.*, 2002; Setoguchi *et al.*, 2005). Interestingly, IL-2 is one of the primary cytokines secreted by effector T cells upon stimulation (Sojka *et al.*, 2004), and drives proliferation and clonal expansion of T cells (Morgan *et al.*, 1976). In parallel, IL-2 appears to be crucial for mechanisms involved in the termination of T cell responses, thereby forming a sophisticated negative feedback circuit.

Although CD25 was sufficient to characterize and further analyze a relatively homogeneous population of nTregs in mice, the same approach was rather challenging in humans. The reason is the limited specificity provided by CD25, whose intrinsic expression at varying levels can be noted in approximately 30% of the T cells and is further upregulated on effector T cells upon stimulation (Baecher-Allan *et al.*, 2004). Unlike mice kept under pathogen-free conditions, humans are continually exposed to immunogenic stimuli resulting in T cell activation and potential CD25 upregulation. In pathological conditions associated with ongoing inflammation this problem is even more pronounced. Consequently, studying Tregs in autoimmune and malignant diseases is complicated further. It may even be speculated that past studies describing $CD25^{+}$ Tregs as functionally defective may have been influenced by contamination of activated $CD25^{+}$ effector T cells (Dejaco *et al.*, 2006). In the steady effort to define Tregs more accurately, it was demonstrated that up to 5% of human peripheral $CD4^{+}$ T cells that express CD25 at high levels are endowed with strong immunosuppressive capacities. This observation narrowed the phenotype of human Tregs further down to CD4^{+}CD25^{high} T cells (Baecher-Allan *et al.*, 2001). Due to the lack of a standardized methodological cut off point for CD25^{high} expression, comparability between clinical studies remained difficult and elevated levels of CD25 expression on effector T cells under conditions of severe inflammatory activity could not be excluded (Han *et al.*, 2008; Seddiki *et al.*, 2006).

Efforts to identify the genetic defects responsible for the severe autoimmune disorders in patients with the IPEX (Immunodysregulation, Polyendocrinopathy, Enteropathy, X-linked) syndrome led to the discovery of germline mutations resulting in a *FOXP3* gene deletion on the X-chromosome (Bennett *et al.*, 2001; Chatila *et al.*, 2000). The *FOXP3* gene encodes for a transcription factor (TF) of the forkhead-box/winged-helix family. Extensive studies in mice and humans revealed the critical importance of the FOXP3 TF as a master regulator of nTreg development and function. Late double-positive lymphocytes that already express FOXP3 at early thymic developmental stages appear to be destined for the nTreg lineage (Tai *et al.*, 2005; Zhou *et al.*, 2009). Ectopic expression of FOXP3 by retroviral gene transfer in CD4^{+}CD25^{-} T cells has been shown *in vitro* and *in vivo* to result in phenotypic and functional suppressive cells demonstrating the plasticity of lymphocytes and the pivotal role of FOXP3 for nTregs (Fontenot *et al.*, 2003; Hori *et al.*, 2003). Concordant to the CD25 expression-based characterization of Tregs, the majority of CD4^{+}FOXP3^{+} T cells were found to be CD25^{high} (Baecher-Allan *et al.*, 2004; Roncador *et al.*, 2005). FOXP3 dimerizes with the nuclear factor of activated T cells (NF-AT) leading to suppression of IL-2, IL-4, and interferon-γ (IFN-γ) expression, while inducing CD25, cytotoxic T lymphocyte antigen 4 (CTLA-4), and gluco-corticoid-induced TNF receptor family-related gene/protein (GITR) (Lopes *et al.*, 2007; Wu *et al.*, 2006). Like CD25, both CTLA-4 and GITR are also upregulated on effector T cells upon activation (Ermann and Fathman, 2003; Roncador *et al.*, 2005; Tai *et al.*, 2005). Although FOXP3 is presently considered the most reliable (intracellular) phenotypic marker for nTregs, major concerns arose when it became evident that FOXP3 expression could be transiently induced in CD4^{+} and CD8^{+} effector T cells upon stimulation, albeit at lower levels (Gavin *et al.*, 2006; Roncador *et al.*, 2005; Roncarolo and Gregori, 2008; Walker *et al.*, 2003; Ziegler, 2007). Consequently, Zou and colleagues suggested a combination of FOXP3 and intracellular cytokine staining, especially for IL-2, IFN-γ, and tumor necrosis factor-α (TNF-α), as an accurate tool to identify nTregs based on the fact that activated FOXP3^{+} conventional T cells express these polyfunctional cytokines in contrast to nTregs (Kryczek *et al.*, 2009). A promising approach to overcome these impediments can be initiated at the epigenetic level. A major criterion for the lineage commitment of nTregs is the sustained, stable expression of FOXP3 as compared to the transient expression found in $FOXP3^{+}$ effector T cells. A static gene expression can be achieved stably through remodeling of the chromatin structure by epigenetic modifications like DNA methylation. In fact a specific methylation pattern, particularly a demethylated DNA sequence within the FOXP3 locus, associated with stable FOXP3 expression upon *in vitro* expansion, was identified as nTreg-specific and defined as a Treg-specific demethylated region (Baron *et al.*, 2007). This methodology has recently been further optimized allowing enumeration of nTregs in clinical samples such as peripheral blood (PB) and tissues (Wieczorek *et al.*, 2009). Furthermore, two studies have demonstrated that expression of the IL-7R α-chain (CD127) is a useful marker for discriminating between activated conventional T cells and nTregs (Liu *et al.*, 2006b; Seddiki *et al.*, 2006). Suppressive $CD4^{+}$ T cells are negative or weakly positive for CD127, which inversely correlates with the FOXP3 expression, regardless of the CD25 levels. Consequently, the following proposed phenotype of $CD4^{+}CD25^{+}CD127^{low/neg}FOXP3^{+}$ T cells corresponds to the majority of nTregs. Importantly, this phenotype allows a more homogeneous purification of viable $CD4^{+}CD25^{+}CD127^{low/neg}$ nTregs.

The characterization of "suppressive cells" based on CD25 expression heralded a new era of Treg research. More than 10 years later this process is still ongoing and has definitely gained momentum. One of the research areas with the strongest interest in Treg biology has traditionally been cancer research. The biology of human Tregs and their various subtypes, their complex role in cancer and translational approaches in modern cancer therapy are discussed in subsequent sections.

II. REGULATORY T CELL SUBSETS

Several studies have demonstrated that nTregs are primarily formed by high-avidity selection of CD4 single-positive thymocytes through major histocompatibility complex (MHC) class II-dependent TCR interactions (Apostolou *et al.*, 2002; Bensinger *et al.*, 2001; Fontenot *et al.*, 2005b; Jordan *et al.*, 2001; Larkin *et al.*, 2008; Modigliani *et al.*, 1996; Sakaguchi, 2001). However, other contributory mechanisms like selective survival rather than induced differentiation (van Santen *et al.*, 2004) or the expression of the TF AIRE (autoimmune regulator) by medullary thymic epithelial cells are also implicated (Liston *et al.*, 2003). In addition to sustaining self-tolerance, Tregs control a broad spectrum of immune responses including those against tumor cells, allergens, pathogenic microbes as well as allogeneic transplants and the fetus during pregnancy (Baecher-Allan and Anderson, 2006; Battaglia and Roncarolo, 2006;

Chatila, 2005; Mills, 2004; Zenclussen, 2006). Although Tregs could be integrated into an overall T cell population with suppressive properties there is an increasing number of reports on various Treg subsets with distinct development, phenotype and functions (Jiang and Chess, 2006) (summarized in Table 1). It has become apparent that under various conditions, Tregs that are termed adaptive or induced Tregs (iTregs) can be generated extrathymically. Suboptimal antigenic stimulation within specific cytokine milieus, particularly rich in transforming growth factor-β (TGF-β), can result *in vivo* and *in vitro* in the induction of iTregs from conventional T cells (Apostolou and von Boehmer, 2004; Kretschmer *et al.*, 2005; Roncarolo *et al.*, 2006). Physiologically, Treg induction in mesenteric lymph nodes (LNs) and the enteric lamina propria in response to gut flora and food antigens is a major mediator of oral tolerance (Coombes *et al.*, 2007; Mucida *et al.*, 2005; Sun *et al.*, 2007). Furthermore, iTregs are also found in chronically inflamed or transplanted tissues as well as tumors, all of which typically have an altered cytokine milieu (Cobbold *et al.*, 2004; Curotto de Lafaille *et al.*, 2008; Liu *et al.*, 2007). To date several phenotypically and functionally distinct iTreg subsets of both CD4 and CD8 lineage have been described. The most delineated populations include IL-10$^+$ T regulatory 1 (Tr1), TGF-β T helper (Th) 3, CD4$^+$CD25$^+$ nTreg-like, CD8$^+$CD25$^+$, and CD8$^+$CD28$^-$ cells.

A. Naturally Occurring CD4$^+$ Regulatory T cells

As described in the previous sections, most CD4$^+$ nTregs produced by the normal thymus constitutively express CD25 and represent a functionally mature population. Development and function of nTregs depend on the expression of the FOXP3 TF. The *FOXP3* gene contains one AP-1 (Activator Protein-1) and six NF-AT binding sites (Mantel *et al.*, 2006). Previous studies have shown that FOXP3 is a repressor of the *Il2*, *Il4*, and *Ifng* gene transcription through direct interaction with NF-κB and NF-AT. Formation of NF-AT–FOXP3 complexes is essential for the suppressive activity (Bettelli *et al.*, 2005). At the same time this complex is involved in the upregulation of CD25, CTLA-4, and GITR expression (Wu *et al.*, 2006). One hallmark of nTregs is anergy manifested by their inability to proliferate and produce IL-2 upon TCR stimulation. IL-2 is a critically important cytokine for their generation and normal activity *in vivo* (Malek *et al.*, 2002; Suzuki *et al.*, 1995; Wolf *et al.*, 2001). In addition to IL-2, other γ-chain cytokines such as IL-4, IL-7, and IL-15 have also been reported to play a role in the development and suppressive capacity of nTregs (Cupedo *et al.*, 2005; Thornton *et al.*, 2004; Yates *et al.*, 2007). Early studies on TGF-β and TGF-βR

Table I Regulatory T Cell Subsets and Suppressive Mechanisms

Cell type	Origin	Phenotype	Suppressive mechanisms	References
Naturally occurring Tregs	Thymus			
CD4 nTregs		$CD4^{+}CD25^{+}FOXP3^{+}CD127^{-/low}$ $CTLA\text{-}4^{+}LAG\text{-}3^{+}GITR^{+}$	Contact, cytotoxicity, IL-10, TGF-β	Sakaguchi (2004)
CD8 nTregs		$CD8^{+}CD25^{+}FOXP3^{+}CTLA\text{-}4^{+}CD122^{+}$	Contact	Fontenot *et al.* (2005a), Rifa'i *et al.* (2004)
Adaptive/Induced Tregs	Periphery			
CD4 nTreg-like		$CD4^{+}CD25^{+}FOXP3^{+}CTLA\text{-}4^{+}GITR^{+}$	Contact (requires IL-2 and TGF-β)	Apostolou and von Boehmer (2004)
Tr1		$CD4^{+}CD25^{-/low}FOXP3^{-/low}$	IL-10	Groux *et al.* (1997)
Th3		$CD4^{+}CD25^{+}FOXP3^{+}$	TGF-β, IL-10 (to a lesser extent)	Chen *et al.* (1994)
CD8 iTregs		$CD8^{+}CD25^{+}FOXP3^{+}$	IL-10, TGF-β	Chaput *et al.* (2009), Wei *et al.* (2005)
CD8 iTregs		$CD8^{+}CD25^{+}CD28{-}FOXP3^{+}$ $CTLA\text{-}4^{+}GITR^{+}$	Contact, IL-10, ILT3, ILT4	Cortesini *et al.* (2001)

knockout mice did not indicate an involvement of the TGF-β pathway in the development of nTregs; findings were strengthened by recent observations that in the absence of TGF-β signaling IL-2 compensates for its effects (Liu *et al.*, 2008).

With regard to the function of nTregs, it is now established that nTregs suppress activation and expansion of cells from adaptive as well as innate immunity hampering cellular and humoral immune responses. Effector and memory T cells of both $CD4^+$ and $CD8^+$ compartments are efficiently suppressed by $CD4^+CD25^+FOXP3^+$ nTregs with regard to activation, proliferation, and function (Levings *et al.*, 2001; Piccirillo and Shevach, 2001; Takahashi *et al.*, 1998; Thornton and Shevach, 1998). Proliferation, immunoglobulin (Ig) production, and Ig class switch of B cells can be suppressed by nTregs, partly mediated by TGF-β secretion (Lim *et al.*, 2005; Nakamura *et al.*, 2004). Furthermore, nTregs have been shown to inhibit the function of natural killer (NK) cells and NKT cells as well as the function and maturation of dendritic cells (DCs) (Azuma *et al.*, 2003; Ghiringhelli *et al.*, 2005a; Misra *et al.*, 2004). Immature DCs, on the other hand, provide aberrant stimuli to naïve T cells and potentially transform them to iTregs, thereby forming a positive loop. Macrophages that are entering the tissues can switch between proinflammatory M1 and anti-inflammatory M2 phenotypes. A tolerogenic milieu, which is typically found in tumors, skews macrophages toward an M2 phenotype. In experiments performed *in vitro* nTregs induced an analogous immunosuppressive M2-like alternative activation phenotype in macrophages (Tiemessen *et al.*, 2007).

B. Induced (Adaptive) $CD4^+$ Regulatory T Cells

While nTregs play a critical role in regulating self-tolerance, iTregs are thought to be responsible for governing the immune response to a wide variety of microbial and tissue antigens. They develop in peripheral lymphoid tissues from naive T cells normally at very low frequencies in a steady state and endow the immune system with an extraordinary environmental adaptability. The physiological processes and environmental conditions driving their development are as yet incompletely determined. Up till now, tumor-induced Tregs are phenotypically indistinguishable from other iTregs and often also from nTregs. However, it remains to be further investigated whether tumor-associated iTregs acquire specific characteristics contributed by the tumor environment. A prerequisite for iTreg development is TCR triggering of naïve T cells by antigenic stimulation under conditions not optimal for the generation of effector T cells. The circumstances under which iTregs are induced are wideranging and may include among others the presence of certain cytokines most notably high levels of IL-2, IL-10, or TGF-β, low dose of antigens and antigen presenting cells (APCs) exhibiting alterations in maturation and function (Curotto de Lafaille and Lafaille, 2009; Lohr *et al.*, 2006). It is obvious that the local microenvironment is the key to the generation of iTregs. Tumor cells can directly initiate the induction of Tregs through several factors including CD70, cyclooxygenase-2 (COX-2), indoleamine 2,3-dioxygenase (IDO), IL-10, Galectin-1, and TGF-β (Bergmann *et al.*, 2007; Curti *et al.*, 2007; Juszczynski *et al.*, 2007; Li *et al.*, 2007; Liu *et al.*, 2007; Yang *et al.*, 2007). In addition, neoplastic cells can modulate recruited or local APCs to become tolerogenic, which thereby strongly contribute to the induction of Tregs within the microenvironment or the local LNs.

Several subsets of $CD4^+$ iTregs have been described, which differ but also overlap with regard to their phenotype, function, and mechanisms of suppression. Well-established subsets of $CD4^+$ iTregs are the Th3, Tr1, and $CD25^+FOXP3^+$ nTreg-like cells. Th3 cells are defined by their production of large amounts of TGF-β that they utilize for direct suppression and the creation of a tolerogenic milieu and to a lesser extent IL-4 and IL-10 (Chen *et al.*, 1994). This subset is one of the earliest regulatory populations described *in vivo* following oral tolerance toward myelin basic protein (MBP) and suppressing the induction of MBP-specific experimental autoimmune encephalitis (Chen *et al.*, 1994). Th3 generation appears to be triggered in an antigen-dependent fashion but suppression is antigen-independent, leading to the term "bystander suppression". Tr1 cells were initially observed to develop *in vitro* in the presence of high dose of IL-10 and chronic antigenic stimulation. They produce high levels of IL-10 and negligible amounts of IL-2 and IL-4, if any (Groux *et al.*, 1997). In accordance to the *in vitro* results Tr1 cells could also be generated *in vivo* by multiple rounds of stimulation with immature DCs in presence of IL-10 (Levings *et al.*, 2005). In contrast to a minor proportion of Th3 cells, nTreg-like cells and nTregs, Tr1 cells express no or low levels of FOXP3 and CD25 (Bacchetta *et al.*, 2005; Foussat *et al.*, 2003; Levings *et al.*, 2002). Like Th3 cells, Tr1 cells require TCR ligation in order to acquire suppressive activity, and once activated Tr1 cells can mediate bystander suppression. Tr1 cells and their supernatants containing IL-10 directly suppress T cells but can also reduce the capacity of DCs to induce alloantigen-specific T cell responses. Cancer is often associated with complement activation. Stimulation via the CD46 molecule, which is a receptor for the complement factors CD3b and CD4b and widely expressed on lymphocytes can lead to the generation of $IL\text{-}10^+$ Tr1 cells when combined with TCR triggering (Kemper *et al.*, 2003). The highly suppressive $FOXP3^+$ iTregs called nTreg-like cells express CD25, CTLA-4, and GITR and to date several settings leading to their generation from naïve T cells have been described.

Antigenic stimulation in the presence of TGF-β or IL-2 can lead to the induction of this suppressive phenotype in naive T cells (Apostolou and von Boehmer, 2004). Studies in mice have suggested that conversion of $CD4^+CD25^-$ T cells to nTreg-like cells *in vivo* requires costimulation via B7 (CD80 and CD86) molecules (Liang *et al.*, 2005). Another rather antagonistic key cytokine is IL-6, which abolishes the conversion to suppressive iTregs and at the same time promotes the generation of Th17 cells. Cumulatively, the observations emphasize the role of soluble factors and cytokines in determining cell differentiation from tolerogenic to responsive subtypes and vice versa (Korn *et al.*, 2008).

C. Naturally Occurring and Induced $CD8^+$ Regulatory T Cells

Although, $CD4^+$ Tregs have been the focus of Treg research, $CD8^+$ Tregs are increasingly emerging as crucial components in the negative control of immune responses. Interestingly, $CD8^+$ suppressor cells were already described together with their $CD4^+$ counterparts in the early 1970s (Gershon and Kondo, 1970). Similar to $CD4^+$ Tregs, Tregs from the $CD8^+$ lineage may develop intrathymically as well as in peripheral tissue. $CD8^+CD25^+FOXP3^+CTLA\text{-}4^+$ nTregs have been identified in several studies in rodents and humans and act mainly in a cell-to-cell contact-dependent fashion (Cosmi *et al.*, 2003, 2004; Fontenot *et al.*, 2005a; Rifa'i *et al.*, 2004; Xystrakis *et al.*, 2004a,b). Peripherally induced $CD8^+$ iTregs are generated from naïve $CD8^+CD25^-$ T cells upon antigenic stimulation (Mills, 2004). $CD8^+$ Tregs described in humans with mycobacterial infections expressed lymphocyte-activation gene 3 (LAG-3) and suppressed T cell activation by CC chemokine ligand 4 secretion, which interferes with TCR signaling (Joosten *et al.*, 2007) whereas $CD8^+$ Tregs in systemic lupus erythematodes patients produced significant amounts of TGF-β (Zhang *et al.*, 2009). Recent reports also describe $CD8^+$ Tregs in cancer patients. In prostate cancer patients, $CD8^+$ Tregs were described to be $CD25^+CD122^+FOXP3^+$ and partly $GITR^+$. Their suppressive activity was mediated via cell-to-cell contact as well as through yet unidentified soluble factors other than IL-10 or TGF-β (Kiniwa *et al.*, 2007). $CD8^+CD25^+FOXP3^+$ Tregs in colorectal cancer were positive for TGF-β (Chaput *et al.*, 2009). Tumor plasmacytoid DCs (pDCs) from ovarian cancer patients induced $CD8^+$ iTregs *in vitro* which corroborates with the *ex vivo* data showing an accumulation of $CD8^+$ Tregs in ascites, draining LNs and PB of the patients (Wei *et al.*, 2005). In this particular setting suppression was mainly mediated by secreted IL-10 underlining the plasticity of the suppressive phenotype as well as its dependence on the shaping milieu. The proposed model of induction and activation of the $CD8^+$ Tregs at the tumor site is analogous to $CD4^+$ Tregs. $CD8^+$ Tregs accumulate in tumor tissues (Chaput *et al.*, 2009; Kiniwa *et al.*, 2007; Wei *et al.*, 2005) and can be activated in a peptide-specific manner as recently shown in various types of tumors (Andersen *et al.*, 2009). Another type of $CD8^+$ iTregs is $CD8^+CD28^-$ iTregs, which was first described in the allogeneic setting induced through MHC class I peptide stimulation, but is also found in cancer patients (Cortesini *et al.*, 2001; Filaci and Suciu-Foca, 2002; Suciu-Foca *et al.*, 2005). $CD8^+CD28^-$ iTregs have been shown to be suppressive via contact-dependent mechanisms, IL-10 secretion as well as upregulation of inhibitory immunoglobulin-like transcript (ILT) receptors ILT3 and ILT4 on APCs (Filaci *et al.*, 2007; Suciu-Foca and Cortesini, 2007). Characterization and understanding of $CD8^+$ Tregs is at its inception and consequently subclassification and function is relatively tentative and will surely be modified and expanded in the future.

III. MECHANISMS MEDIATING THE SUPPRESSIVE FUNCTION

In the past decade extensive studies have been performed to further explore the underlying cellular and molecular mechanisms of Treg-mediated immunomodulation (summarized in Fig. 1), which has led to significant improvement in our understanding.

Proliferation and cytokine production of conventional T cells can be inhibited upon TCR activation of Tregs (Takahashi *et al.*, 1998; Thornton and Shevach, 1998). This process is cell-to-cell contact dependent and leads to an inhibition of IL-2 production. Functional activity can be rescued by the administration of IL-2 and activating anti-CD28 antibodies, which implies a disruption of costimulatory signaling being involved. Furthermore, CTLA-4 and LAG-3 surface molecules constitutively expressed on nTregs contribute to the cell-to-cell-dependent suppressive mechanisms via interactions with CD80 and CD86 on APCs (Huang *et al.*, 2004; Sakaguchi, 2004). CTLA-4 is a ligand for CD80 and CD86, possessing a higher affinity than the CD28, thereby directly competing with the costimulatory signal transduction. Blockage of CTLA-4 *in vivo* results in the development of organ-specific autoimmune diseases (Sakaguchi, 2004). Another role of CTLA-4 could be that it directly exerts suppressive activity through induction of the enzyme IDO in DCs via interaction with their CD80 and CD86 (Fallarino *et al.*, 2006). IDO catalyzes the conversion of tryptophan into kynurenine, leading to (A) tryptophan depletion and (B) generation of immunosuppressive metabolites, both of which attenuate T cell function (Fallarino *et al.*, 2006). It has also been proposed that binding of CTLA-4 to CD80 and CD86 mediates their downregulation on DCs in a negative feedback manner (Misra *et al.*,

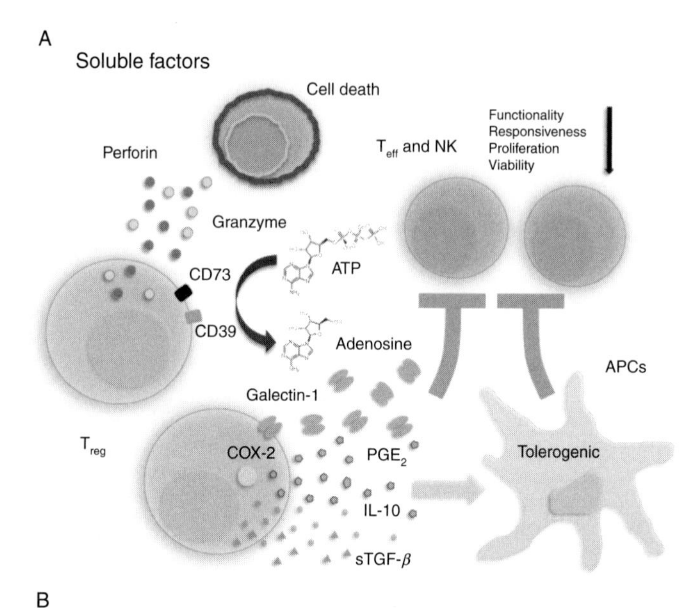

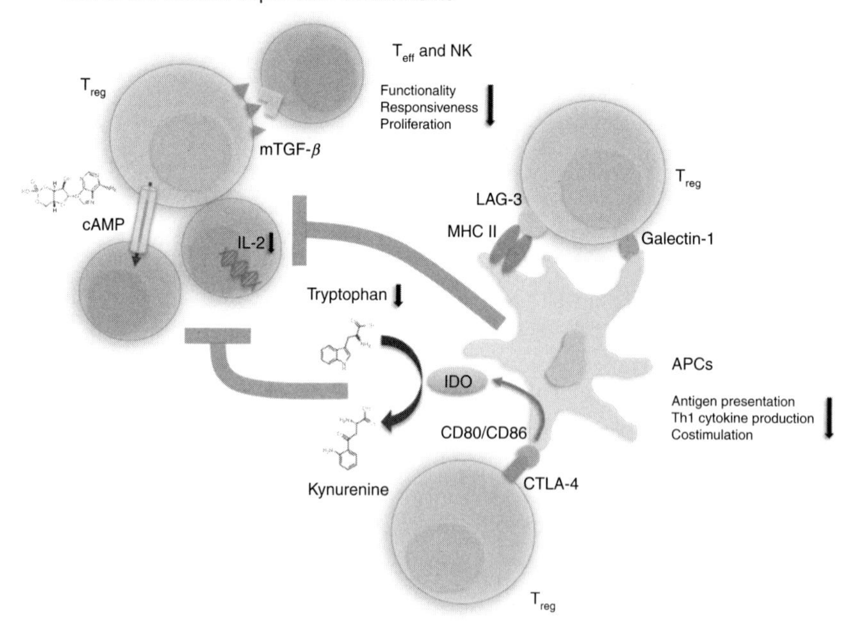

Fig. 1 (Continued)

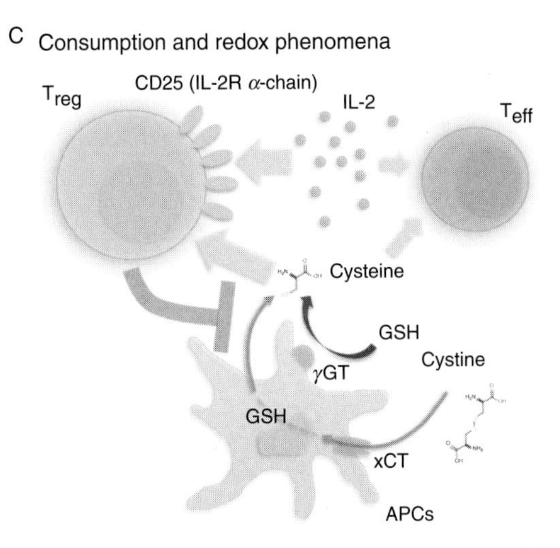

Fig. 1 Regulatory T cell-mediated immunosuppression. (A) Several soluble factors released by regulatory T cells (Tregs) (e.g., Galectin-1, Prostaglandin E_2 [PGE_2]) may directly suppress or induce cell death (e.g., Perforin, Granzyme) of effector T (T_{eff}) and NK cells. Ectoenzymes located on the cell membrane of Tregs (e.g., CD39, CD73) mediate the metabolization of ATP to Adenosine, a potential suppressant of T cells. Antigen presenting cells (APCs) are functionally modulated (e.g., by IL-10, soluble TGF-β [sTGF-β]) contributing to a tolerogenic tumor milieu. (B) Cell-to-cell contact between Tregs and immune cells is obligatory for certain direct and indirect suppressive pathways. Tregs weaken T_{eff} and NK cell responses by membrane-bound TGF-β (mTGF-β) as well as cAMP "injections." Close interaction with APCs (e.g., via LAG-3, Galectin) reduces their immunostimulatory capacity through attenuation of costimulation and antigen presentation, while increases their tolerizing potential, especially by a CTLA-4-mediated upregulation of the enzyme Indoleamine 2,3-dioxygenase (IDO). IDO activity leads to a depletion of tryptophan accompanied by an accumulation of kynurenine, both with a negative impact on T cells. (C) Proper function of Tregs depends on IL-2 produced by other cells. Tregs express high levels of CD25, a component of the IL-2 high-affinity receptor, enabling them to withdraw IL-2 from their local environment. Tregs alter the redox balance of T cells by inhibition of their supply of thiols provided by APCs mainly in form of cysteines, which are obligatory for an efficient activation. (See Page 1 in Color Section at the back of the book.)

2004). Consequently, further activation of T cells by the DCs is abrogated which leads to aberrant stimulation and generation of iTregs. LAG-3 is a CD4 homologue expressed on nTregs upon activation and on certain $CD8^+$ Tregs (Joosten *et al.*, 2007). The LAG-3 binds to MHC class II molecules expressed on several types of APCs and is required for maximal suppressive activity (Huang *et al.*, 2004). Unlike CTLA-4, mice deficient for LAG-3 do not develop severe autoimmunity. Recent studies suggest the involvement of LAG-3 in suppressing DC maturation and immunostimulatory capacity by recruitment of SH2-domain-containing protein tyrosine phosphatase 1 (Liang *et al.*, 2008). Gene expression analyses have shown that *GITR* transcription is under the control of the FOXP3 TF and is thus highly, but not exclusively, expressed in Tregs (McHugh *et al.*, 2002; Shimizu *et al.*, 2002). Studies on T cells from GITR-deficient mice have revealed that ligation of GITR on naïve $CD4^+CD25^-$ T cells is involved in the inhibition of the Treg-mediated suppression (Stephens *et al.*, 2004). In addition to GITR there are several molecules found in Tregs that contribute to the control of Treg-mediated suppression including toll-like receptors (TLRs) like TLR2 and TLR8 (Peng *et al.*, 2005; Sutmuller *et al.*, 2006).

In contrast to the requirement for cell-to-cell contact for suppression by Tregs *in vitro*, there are numerous reports that indicate the need for soluble factors such as IL-10 and TGF-β for suppression *in vivo*. Several studies on rodents especially in models of autoimmune diseases, like colitis or asthma, have demonstrated the importance of IL-10 for Treg-mediated immunosuppression (Annacker *et al.*, 2001, 2003; Hawrylowicz and O'Garra, 2005; Tang *et al.*, 2004). However, *in vitro* experiments with human nTreg-clones did not show secretion of IL-10, but only of TGF-β (Roncarolo *et al.*, 2006). Similarly, IL-10 and TGF-β are rarely detectable in the supernatants from suppression assays with nTregs *in vitro* (Sakaguchi, 2004). In contrast, adaptive Tr1 cells and selected $CD8^+$ T cells produce and secrete substantial amounts of IL-10. Interestingly, membrane-bound TGF-β can be found on Tregs and is implicated in mediating nTreg suppression of T and NK cells in a cell-to-cell contact-dependent manner (Chen *et al.*, 2005; Ghiringhelli *et al.*, 2005a). In patients with gastrointestinal stromal tumors (GIST) an inverse correlation between NK cell activation and Treg expansion was observed. Subsequent analyses revealed that Tregs utilized membrane-bound TGF-β to attenuate the cytotoxic function of NK cells and downregulate the expression of the activating NKG2D receptor (Ghiringhelli *et al.*, 2005a). The controversy regarding the role of IL-10 and TGF-β for nTreg-mediated suppression is ongoing and inferences appear to strongly depend on the model studied. Another newly identified inhibitory cytokine belonging to the IL-12 heterodimeric family is IL-35, which is found in murine Tregs. IL-35 may contribute to the function of Tregs but is not constitutively expressed in human Tregs and warrants further investigation (Bardel *et al.*, 2008; Collison *et al.*, 2007). Galectin-1, a member of a highly conserved family of β-galactoside-binding proteins is preferentially expressed on human Tregs and upregulated upon TCR activation (Garin *et al.*, 2007). It is secreted as a homodimer and binds glycoproteins such as CD45, CD43, and CD7 leading to growth arrest, apoptosis as well as abrogation of proinflammatory cytokine production in activated T cells. Blocking galectin-1 clearly reduces the maximal intrinsic inhibitory efficacy of both mouse and human Tregs. However, it is still not clear whether galectin-1 works *in vivo* mainly as a soluble factor or exerts its suppressive effect via cell-to-cell contact. Induced Tregs secrete T cell suppressive prostaglandin (PG) E_2, which is generated by COX-2 (Mahic *et al.*, 2006) and COX-2^+

iTregs were noted in colorectal cancer patients in whom T cell function could be restored by the COX inhibitor indomethacin (Yaqub *et al.*, 2008).

As previously described, Tregs require IL-2 for a proper function, which they do not produce themselves and need conventional T cells as their main source *in vivo*. Accordingly, the nTregs express increased levels of the high-affinity heterotrimeric receptor for IL-2 composed of CD25, CD122, and CD132. Competitive depletion of available IL-2 by Tregs and the resultant starvation of activated, dividing T cells has been proposed as a minor suppressive mechanism at minimum within the tumor microenvironment (Pandiyan *et al.*, 2007; von Boehmer, 2005). Exhaustion of free thiol groups by a process similar to cytokine depletion can also produce a negative effect on activated T cells. Conventional T cells require thiols for efficient activation. Activated T cells need cysteine as they lack transporters for its oxidized form, cystine. It has been shown that DCs create a cysteine-rich milieu by intra- and extracellular redox reactions thereby providing cysteine to the T cells (Angelini *et al.*, 2002). Tregs interfere with this process with one very likely mechanism being competitive consumption of thiols including cysteine, as Tregs exhibit increased levels of intra- and extracellular thiols (Mougiakakos *et al.*, 2009; Yan *et al.*, 2009).

The perforin/granzyme pathway classically mediates cytolytic effects of $CD8^{+}$ T and NK cells. Perforins traffic granzymes into target cells, whereas granzyme A and B induce apoptosis by cleaving important substrates. Tregs utilize this system to initiate cytolysis of monocytes, B and T cells as well as DCs (Gondek *et al.*, 2005; Grossman *et al.*, 2004; Zhao *et al.*, 2006). Granzyme A expression by human Tregs has been established; however, the expression of granzyme B remains equivocal (Grossman *et al.*, 2004). One study demonstrated in a mouse tumor model that up to 30% of Tregs located at the tumor site utilize the perforin/granzyme B pathway to suppress antitumor responses suggesting a tumor-driven induction of cytolytic Tregs (Cao *et al.*, 2007). In another recent study, Wilms Tumor 1 (WT1)-specific Treg clones from leukemia patients, upregulated granzyme B upon peptide stimulation. These cells had an nTreg-like $CD4^{+}CD25^{+}CD127^{neg}$ $FOXP3^{+}GITR^{+}$ phenotype and induced cytolysis of APCs (Lehe *et al.*, 2008). Nevertheless, the role of cytolysis as a major suppressive mechanism *in vivo* remains unresolved and is further expanded by the addition of TRAIL/DR5 and galectin pathways as potential cytolytic mechanisms (Ren *et al.*, 2007; Toscano *et al.*, 2007). As described previously nTregs are anergic. In this context it has been observed that elevated cyclic adenosine monophosphate (cAMP) levels in Tregs contribute to their anergic state. Formation of gap junctions between Tregs and effector T cells permits diffusion of cAMP following the concentration gradient into effector T cells inducing suppression through the cAMP–protein kinase A type I–C-terminal Src kinase inhibitory pathway (Bopp *et al.*, 2007). Additionally, it has recently been reported that Tregs express ectoenzymes like CD73 that cleave extracellular adenosine triphosphate (ATP) generating adenosine, which inhibits T cell function through the adenosine receptor 2A (Vignali *et al.*, 2008).

IV. REGULATORY T CELLS IN CANCER

The role of the immune system in cancerogenesis and tumor progression has been the subject of much controversy since the 1950s when Burnet and Thomas formulated their concept of "tumor immunosurveillance"; a process through which the immune system recognizes and (ideally) eliminates self-cells that have undergone malignant transformation (Burnet, 1957). Numerous observations in clinical and experimental settings have fortified this concept that was further advanced by the model of "immune editing." According to this theory, multiple factors generated by the oncogenic process counteract the immune system cumulatively hampering an efficient immune response and facilitating the "tumor escape" (Dunn *et al.*, 2002). Tregs as regulatory elements have the ability to actively suppress immune responses and represent a predominant tolerance-inducing modality (Sakaguchi *et al.*, 2008). Already in the early days of the discovery of the suppressor cells, observations from tumor mouse models indicated a central (negative) role of Tregs in immunosurveillance; namely hindering an efficient tumor eradication. Tumor cells, in particular methylcholanthrene-induced fibrosarcomas, elicited measureable T cell responses that were not sufficient to eradicate the tumors due to the development of tumor-induced suppressor T cell activity within the $CD4^+$ T cell population (Berendt and North, 1980; Dye and North, 1981). In the following part of the review, we have focused mainly on the impact of Tregs in patients with solid tumors and hematological malignancies. The underlying biological mechanisms and targeted therapeutic interventions are discussed.

A. Regulatory T Cells in Solid Malignancies

The vast majority of the studies on Tregs in cancer are performed on patients with solid malignancies. It is obligatory to take into consideration that virtually all of these studies were carried out during the period when the phenotype of Tregs was being refined thereby complicating direct comparisons between studies. Shortly after the publication on the existence of $CD4^+CD25^{high}$ Tregs in the PB of healthy individuals (Baecher-Allan *et al.*, 2001) the group of Carl June was the first to provide direct evidence that patients with epithelial malignancies, in particular ovarian and non-small-cell lung cancer (NSCLC) displayed increased levels of CD4^{+}CD25^{high} Tregs in the circulation and within the tumor infiltrating lymphocytes (TILs). These cells constitutively expressed CTLA-4 and exhibited suppressive effects by inhibiting the proliferation of conventional T cells and IFN-γ production. The suppressive activity was partly mediated by TGF-β (Woo *et al.*, 2001, 2002). In patients with pancreatic and breast cancer, increased levels of cells with similar phenotype were found in the PB, LNs, and tumor tissue. These cells were positive for IL-10, TGF-β, and CTLA-4 (Liyanage *et al.*, 2002). Furthermore, results from these initial studies strongly indicated a tropism of Tregs toward tumor sites as their proportion in draining LNs and TILs was higher than that expected theoretically, based on their frequencies in PB. In addition, the first Treg cell lines derived from autologous cocultures of tumor cells and lymphocytes from colorectal cancer patients were generated. These cells displayed tumor-dependent expansion and suppressed both allogeneic and autologous T cell responses independent of cell-to-cell contact via TGF-β (Somasundaram *et al.*, 2002). One of the first proposed mechanisms underlying the activation and induction of Tregs was heavy-chain Ferritin (H-Ferritin), which is produced in large amounts by melanoma cells. Melanoma patients exhibited a significant positive correlation between serum levels of H-Ferritin and increased Treg frequencies and activation (Gray *et al.*, 2003; Javia and Rosenberg, 2003; Viguier *et al.*, 2004). Several studies on gastro-esophageal cancers also reported that increased frequencies of IL-10-producing CD4^{+}CD25^{high} Tregs can be found in PB, TILs, draining LNs, and ascites fluid, which were strongly associated to disease stage (Ichihara *et al.*, 2003; Kawaida *et al.*, 2005; Kono *et al.*, 2006; Sasada *et al.*, 2003). Importantly, the proportion of Tregs was significantly reduced in patients to almost physiological levels upon curative surgery. Furthermore, the level of Tregs rebounded at the timepoint of postoperative recurrent disease, strongly indicating an interconnection between tumor burden and Treg accumulation (Kono *et al.*, 2006). It has been shown that CD4^{+}CD25^{+} Tregs are capable of suppressing NK cell-mediated cytotoxicity in patients with various types of epithelial tumors including lung, breast, and colorectal cancer (Wolf *et al.*, 2003). Upon identification of FOXP3 as a more reliable marker for Tregs and potentially as a surrogate measure for their suppressive function, an increasing number of subsequent studies included FOXP3 in their staining panels such as the pivotal work carried out by Tyler J. Curiel and colleagues on ovarian cancer patients (Curiel *et al.*, 2004). In this comprehensive study it was convincingly demonstrated that CD4^{+}CD25^{+}FOXP3^{+} Tregs were present in PB, malignant ascites, tumoral tissue, and draining LNs. Interestingly, Treg levels in tumor-draining LNs were lower as compared to control LNs and tonsils and decreased with increasing disease stage. One of the proposed mechanisms underlying this phenomenon was the presence of the chemokine CCL22. Secreted by ovarian cancer cells and tumor-associated macrophages (TAMs), a concentration gradient of CCL22, which binds to CCR4 expressed on Tregs, is generated and thereby mediates migration of Tregs away from the draining LNs toward the CCL22-rich tumor microenvironment. It is worth mentioning that physiologically CCL22 facilitates the encounter between DCs and activated antigen-specific T cells suggesting that tumors elegantly capture this process in order to efficiently suppress activated effector cells (Tang and Cyster, 1999). Similar findings regarding Treg trafficking and redistribution have been largely made in various types of malignancies (Gobert *et al.*, 2009; Haas *et al.*, 2008; Olkhanud *et al.*, 2009; Qin *et al.*, 2009; Shevach, 2004), pointing toward the need for examining the distribution of Tregs in multiple tissue compartments since quantification of Tregs in PB alone may not accurately portray Treg frequency or trafficking.

Analysis of subset frequency for effector cells such as NK and T cells together with Tregs revealed that a shift of the Treg/effector T cell ratio was often linked to the tumor burden and disease course (Gao *et al.*, 2007; Leffers *et al.*, 2009; Sato *et al.*, 2005). The global interest in Tregs resulted in several analogous studies on Treg (-subsets) in different types of malignancies including melanoma (Viguier *et al.*, 2004), hepato-cellular carcinoma (HCC) (Kobayashi *et al.*, 2007; Ormandy *et al.*, 2005), Ewing sarcoma (Brinkrolf *et al.*, 2009), head-and-neck (Schaefer *et al.*, 2005), prostate (Kiniwa *et al.*, 2007; Miller *et al.*, 2006), ovarian (Kryczek *et al.*, 2005; Wolf *et al.*, 2005), breast (Leong *et al.*, 2006), colorectal (Chaput *et al.*, 2009; Ling *et al.*, 2007), and pancreatic cancer (Liyanage *et al.*, 2002). Despite the fact that the preponderance of results indicated a negative impact of Tregs in carcinogenesis and disease progression, some findings raised doubts with regard to this "simplification". The presence of Tregs was in fact correlated to positive prognosis in head-and-neck as well as gastric cancer (Badoual *et al.*, 2006; Haas *et al.*, 2009). These *prima facie* contradictory findings gained further credibility from studies in animal models of colorectal and gastric cancer providing further evidence for the plasticity of Tregs and their rather complex role in immunoregulation (Erdman *et al.*, 2003, 2005, 2009; Gounaris *et al.*, 2009). It must be emphasized that these anecdotal exceptions do not negate the perception that Tregs hamper "immune surveillance" but rather they present a more holistic view of their functional repertoire. Tregs are *per se* associated with immunosuppression and anti-inflammatory activity. Consequently, by counteracting inflammatory processes Tregs may mediate an anticarcinogenic effect given that inflammation-initiated carcinogenesis and tumor progression is a well-established model (Colotta *et al.*, 2009; Marshall *et al.*, 2004). Under certain proinflammatory conditions characterized by elevated levels of IL-6, IL-1β, IL-23, and lactic acid, Tregs can convert from anti- to proinflammatory,

IL-17^+ cells. Thus, Treg populations with contradictory functions can coexist at elevated levels in tumor tissue. One speculation is that functionally reversed Tregs may contribute at an early stage to the escalation of cancer-associated inflammation and subsequently during the course of disease inhibitory Tregs suppress tumor-specific responses as implied by most studies.

B. Regulatory T Cells in Hematologic Malignancies

Various studies on the role of Tregs in hematologic diseases have been reported providing a more complex mosaic of diverse observations. In Hodgkin's lymphoma (HL), the draining LNs, rich in infiltrating B and T cells as well as macrophages, showed the presence of Tregs, which suppressed T cells via CTLA-4 and IL-10, thus contributing to an ineffective clearance of Hodgkin's disease-associated Sternberg Reed cells (Marshall *et al.*, 2004). Results from studies on immune effector cells indicated that a more immunoreactive environment is associated with a worse outcome in HL. In accordance, the presence of FOXP3^+ Tregs cells appeared to have a positive impact on event-free and disease-free survival in HL, especially when noted together with low infiltration of cytotoxic TIA-1^+CD8^+ T cells (Alvaro *et al.*, 2005). In chronic lymphocytic leukemia (CLL), increased levels of circulating CD4^+CD25^{high} Tregs have been observed and mediate T cell suppression through CTLA-4 (Beyer *et al.*, 2005; Motta *et al.*, 2005). Interestingly, CLL, a chronic B cell-derived leukemia, is associated with hypoglobulinemia that has been found to inversely correlate with the Treg frequency. This observation indicates a direct suppressive effect of Tregs on Ig production; an observation that has been further bolstered by basic studies on the suppressive effects of Tregs on B cells (Lim *et al.*, 2005). In addition, patients with CLL treated with the nucleoside analogue Fludarabine showed a selective reduction of Tregs (Beyer *et al.*, 2005). In B cell-derived non-Hodgkin lymphomas (B-NHLs) as well as acute myeloid leukemia (AML), Tregs were also overrepresented (Wang *et al.*, 2005b; Yang *et al.*, 2006a,b). In AML, the proportion of apoptotic (7-AAD$^+$) and proliferating (Ki67^+) cells among Tregs was higher in patients as compared to healthy controls. It was later demonstrated in independent studies that Tregs can have a rapid turnover rate and may be generated from rapidly dividing, highly differentiated memory CD4^+ T cells. They are also relatively susceptible to apoptotic stimuli partly due to critically short telomeres and reduced telomerase activity (Vukmanovic-Stejic *et al.*, 2006). The cumulative evidence indicates that accumulation of Tregs associated with malignancies may result from the proliferation of a preexisting pool, rather than blockade in senescence. Myelodysplastic syndrome (MDS) is often regarded as the antecedent condition for AML. Parallel to AML, MDS patients exhibit increased Treg frequencies and a skewed $CD8^+$ T cell/Treg ratio toward Tregs. Furthermore, high-risk subgroups of MDS and disease progression to more aggressive MDS subtypes were accompanied by an increase of Treg levels, suggesting a direct role of Tregs in progression and malignant transformation (Hamdi *et al.*, 2009; Kordasti *et al.*, 2007). Some hematologic malignancies display quantitative and functional deficits of the Treg compartment, for example, cutaneous T cell lymphoma (Tiemessen *et al.*, 2006) and multiple myeloma (Prabhala *et al.*, 2006). There is an ongoing discussion how the inflammatory component of the disease, manifested for example by high levels of IL-6 in multiple myeloma, may impact the Treg compartment and whether functional Tregs may have a direct suppressive effect on malignant clones.

C. Regulatory T Cells as Biomarkers

As it became increasingly evident that levels of Tregs often correlate with tumor burden and disease progression, their role as predictors of disease prognosis was explored. In gastric cancer, patients with higher frequencies of circulating Tregs had a worse survival (Kono *et al.*, 2006; Sasada *et al.*, 2003). Interestingly, an evaluation of primary gastric cancer material revealed that merely increased presence of Tregs did not strongly correlate with prognosis but in fact the pattern of localization predicted the outcome. In particular, a diffuse intratumoral distribution predicted a shortened survival as compared to a peritumoral pattern (Mizukami *et al.*, 2008b). A persistent Treg infiltration in tumors that were radically resected was also associated with a worse prognosis (Perrone *et al.*, 2008). The significance of the topological distribution of Tregs at the tumor site was also observed by our group in patients with uveal melanoma, where only intratumoral localization of Tregs was an independent negative prognostic factor in contrast to peritumoral formation (Mougiakakos *et al.*, in press). An increased number of circulating Tregs is associated with high mortality and reduced survival in patients with HCC (Fu *et al.*, 2007). However, only Tregs in the center of advanced HCC and not at the noncancerous margins were of negative impact (Gao *et al.*, 2007; Kobayashi *et al.*, 2007). Obviously, the evidence is far from conclusive since Treg localization has been assessed in only a minority of reported studies. In patients with ovarian cancer, reduced survival correlated with increasing Treg numbers (Curiel *et al.*, 2004). Immunohistochemical (IHC) analysis of tumor specimens from 117 patients with epithelial ovarian cancer demonstrated that a skewing of the $CD8^+$ T cell/Treg ratio toward Tregs correlated with a poor prognosis (Sato *et al.*, 2005). A similar study in cervical cancer evaluated the $CD8^+$ T cell/Treg ratio as well as the MHC class I expression (Jordanova *et al.*, 2008). Other studies in NSCLC examined the $CD3^+$ T cell/Treg ratio (Petersen *et al.*, 2006) while in

HCC ratio of activated Granzyme B^+ $CD8^+$ T cell/Treg was measured (Gao *et al.*, 2007). Thus, the relative proportion of negative regulators like Tregs to effector T cells in the tumor infiltrate may be of greater significance for prognosis than absolute numbers of Tregs in itself. Consistent with these findings, results from breast cancer patients suggest that Tregs negatively affect overall and relapse-free survival (Bates *et al.*, 2006; Gobert *et al.*, 2009). Increased levels of tumor infiltrating Tregs define a new high-risk subgroup within the cohort of breast cancer patients positive for estrogen receptors, serving as a predictive marker for late relapse (Bates *et al.*, 2006). In order to better understand and define the role of infiltrating Tregs in breast cancer, Tregs were assessed in two different locations: within the tumor tissue and the surrounding lymphoid aggregates. The Tregs within the lymphoid infiltrates were identified as the ones with the leading negative impact on disease course and outcome, suggesting that at this site they counteract the recruited effector lymphocytes by abrogating their reactivation (Gobert *et al.*, 2009). In ovarian cancer, a prominent colocalization of Tregs and $CD8^+$ T cells within the tumor tissue has also been observed (Curiel *et al.*, 2004). Patients with breast cancer who show complete responses to chemotherapy have a persistence of $CD8^+$ TILs and a total disappearance of Tregs, indicating that immune responses released from negative regulation may cofacilitate chemotherapy-mediated complete regression of tumor cells (Ladoire *et al.*, 2008). Studies linking the presence of Tregs to a worse outcome have been performed in various other malignancies as well including colorectal, pancreatic, and renal cancer (Griffiths *et al.*, 2007; Hiraoka *et al.*, 2006; Ling *et al.*, 2007).

Although most studies link Tregs to a poor disease course and outcome, data from other investigations show the opposite. In patients with follicular, Hodgkin's, and cutaneous T cell lymphoma, head-and-neck as well as colorectal cancer high numbers of intratumoral Tregs are associated with longer disease-free and event-free survival (Alvaro *et al.*, 2005; Badoual *et al.*, 2006; Carreras *et al.*, 2006; Gjerdrum *et al.*, 2007; Klemke *et al.*, 2006; Lee *et al.*, 2008; Salama *et al.*, 2009; Tiemessen *et al.*, 2006; Tzankov *et al.*, 2008). The role of Tregs in cancer is complex as it is not identical for all types of cancers and even differs at distinct phases of disease course for the same type of malignancy. This can clearly be exemplified by observations in ovarian cancer, where presence of Tregs is an unfavorable predictor for an unselected group of patients (Curiel *et al.*, 2004) but a positive factor for overall survival in a subgroup of patients with advanced disease (Leffers *et al.*, 2009). It has been shown in murine tumor models that elimination of Tregs before tumor establishment was beneficial for the survival in contrast to established tumors as Tregs dominated multiple immune evasion mechanisms early on but not during late phases of tumor development (Elpek *et al.*, 2007). Compelling studies showing that Tregs can have anticancerous effects through their anti-inflammatory role have also been described (Erdman *et al.*, 2003, 2005,

2009; Gounaris *et al.*, 2009). Malignancies characterized by massive infiltration of proinflammatory cells that drive the neoplastic process, may actually benefit from Treg infiltration. It has been demonstrated that Tregs can exert an anti-inflammatory effect not only on cells of the adaptive immunity but also on the innate immunity, which is strongly involved in the inflammatory responses (Tiemessen *et al.*, 2007; Venet *et al.*, 2006). A possible scenario in hematological malignancies may be that Tregs directly suppress the malignant clone and may thereby have antineoplastic effects. For instance, it has been shown that Tregs can kill B cells and potentially malignant B cell clones too may be targeted (Lim *et al.*, 2005; Zhao *et al.*, 2006). The same applies to T cell and myeloid-derived malignancies, where nonmalignant counterparts are known to be under the control of Tregs.

V. ACCUMULATION OF REGULATORY T CELLS

A. Compartmental Redistribution

Increasing evidence confirms the hypothesis that Tregs selectively migrate to the site where regulation is required (Fig. 2A). This system, relying on interactions between chemokines/chemokine-receptors and integrins/integrin-receptors (Wei *et al.*, 2006), is often usurped by tumors. Curiel and colleagues were the first to show in ovarian cancer a CCL22-orchestrated migration of CCR4-expressing Tregs toward tumor tissue and malignant ascites (Curiel *et al.*, 2004). In addition to tumor cells, bystander cells especially of myeloid origin including TAMs are sources of CCL22. Expression of CCL22 can be upregulated in myeloid cells *in vitro* upon addition of tumor cells and/or tumor supernatant. To date, a CCL22-mediated Treg attraction has been observed in several types of neoplastic diseases including breast, prostate cancer, and B-NHLs (Gobert *et al.*, 2009; Miller *et al.*, 2006; Qin *et al.*, 2009; Yang *et al.*, 2006a). Decreased expression of CD62L (L-selectin) and CCR7 on infiltrating Tregs as compared to circulating counterparts substantiates active recruitment of these cells to the site of action. In regional LNs, the majority of the Tregs express CD62L (80%) and CCR7 (50%) (Huehn and Hamann, 2005). Tregs internalize CCR4 upon binding of CCL22 which accounts for the varying levels of CCR4 on Tregs found in the circulation, draining LNs, and tumor microenvironment (Gobert *et al.*, 2009). CCL17 is another ligand for CCR4 and has been shown to be involved in Treg trafficking in gastric cancer and HL (Ishida *et al.*, 2006; Mizukami *et al.*, 2008a). Supporting these observations, major CCL17 and CCL22 sources like tolerogenic DCs, immature myeloid cells, and TAMs can be found in different tumor microenvironments (Penna *et al.*, 2002). In pancreatic

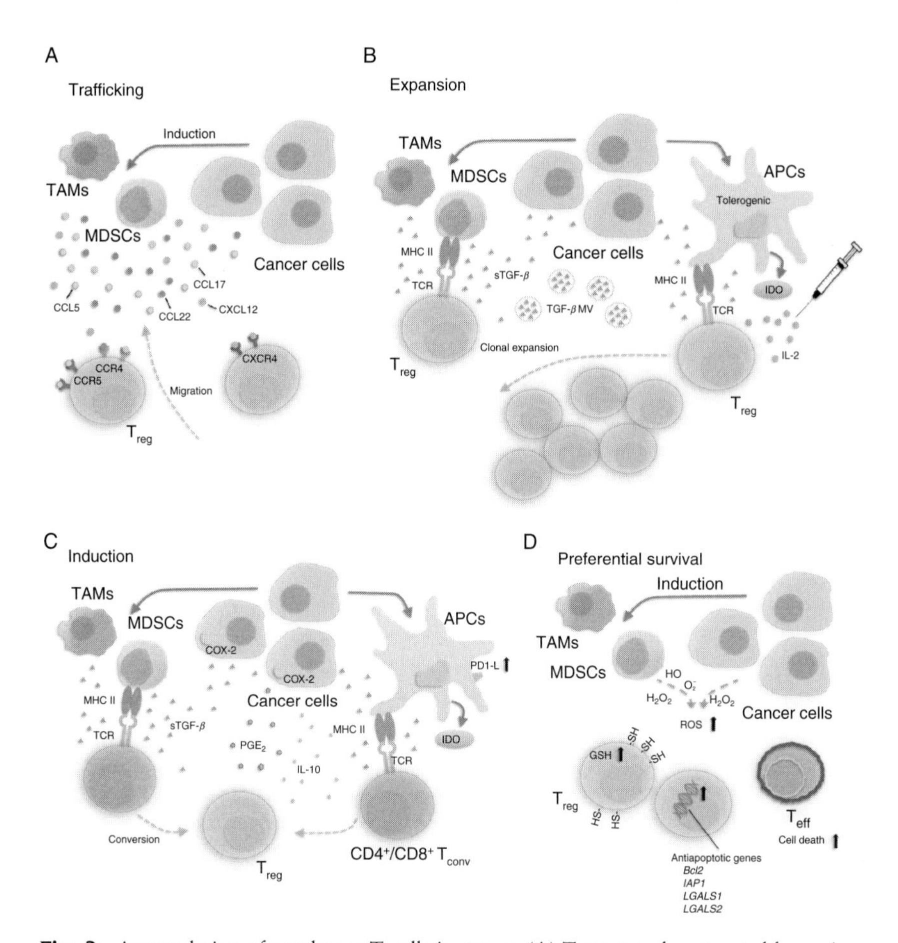

Fig. 2 Accumulation of regulatory T cells in cancer. (A) Tregs may be attracted by various chemokines (CCL5, CCL17, CCL22, CXCL12) to the tumor site. Cancerous cells and/or bystanding tumor-associated macrophages (TAMs) and myeloid-derived suppressor cells (MDSCs) secrete these chemokines of which Tregs possess the corresponding receptors (CCR4, CCR5, CXCR4). (B) Preexisting Tregs expand upon (suboptimal) antigen stimulation provided by APCs, TAMs, and MDSCs within an overall tolerizing environment. TGF-β directly secreted or carried in microvesicles (MV) as well as IDO play a central role in this process. Administration of IL-2 as a component of therapeutic schemes in malignancies may drive such a Treg expansion. (C) Tregs can be generated *de novo* from conventional $CD4^+$ and $CD8^+$ T cells (T_{conv}). Several factors, among others TGF-β, Prostaglandin E_2 (PGE_2), IL-10, and IDO in conjunction with (suboptimal) T cell activation have been identified to favor this induction of Tregs. (D) In the tumor microenvironment, reactive oxygen species (ROS) produced mainly by cancer and myeloid cells (e.g., TAMs, MDSCs) are responsible for high levels of oxidative stress, which is harmful for immune cells. Tregs depict a better protection against oxidative stress as compared to conventional effector T cells (T_{eff}), as they possess higher amounts of intracellular glutathione (GSH) and surface thiols (–SH). Furthermore, Tregs in cancer patients show a higher expression of antiapoptotic genes (*Bcl2, IAP1, LGALS1, LGALS2*) as compared to their counterparts from healthy donors indicating an increased resistance toward apoptotic stimuli. (See Page 2 in Color Section at the back of the book.)

adenocarcinoma, the migration of Tregs is partly driven by CCR5 chemotaxis. Tregs from these patients express CCR5 and tumors secrete the cognate ligand CCL5. In a murine tumor model of pancreatic cancer, tumor growth was significantly inhibited by reducing CCL5 production by tumor cells or by systemic administration of a CCR5 antagonist (Tan *et al.*, 2009). Interestingly IL-2, which is utilized as an immunologic adjuvant in cancer therapies, modifies Treg trafficking. IL-2 can lead to an upregulation of CCR4 expression on Tregs and thereby potentially drive migration toward the tumor site. In addition, Tregs have been observed to exhibit increased CXCR4 levels upon IL-2 treatment in patients with ovarian carcinoma (Wei *et al.*, 2007). CXCR4 is the receptor for CXCL12, also known as stromal-derived-factor (SDF-1), which is strongly associated with the regulation of organ-specific metastases in various cancers (Kryczek *et al.*, 2005). A recent report on cervical cancer showed that expression of CXCL12 in the tumor tissue positively correlates with the tumor infiltration of $FOXP3^+$ Tregs and cancer progression (Jaafar *et al.*, 2009). Dependent on activation status and tissue localization, Tregs can express a plethora of chemokine receptors including CCR2, CCR4, CCR5, CCR7, CCR8, CXCR4, and CXCR5 and thus are responsive to a variety of ligands. An interesting aspect is the role of the cancer-related inflammatory component for Treg recruitment. Indeed, it has been shown that Tregs migrate toward sites of inflammation. This process is mediated partly by the integrin CD103 ($\alpha_E\beta_7$), which interacts with E-cadherin, and CCR2 (Wei *et al.*, 2006), though it remains to be elucidated whether it contributes to Treg migration in cancer patients. A comprehensive analysis of the cytokine pattern in tumors combined with a characterization of chemokine receptors expressed on tumor infiltrating Tregs may help to address some of the unanswered questions.

B. Expansion

Much evidence directly or indirectly suggest that cancers not only attract but also facilitate proliferation of different Treg subsets as they appear to be highly activated and underwent proliferation when investigated in tumor patients (Fig. 2B). Physiologically, Tregs have been observed to exhibit high turnover rates (Vukmanovic-Stejic *et al.*, 2006). Tregs isolated from cancer patients depict a decreased content of TCR excision circles (TRECs) as compared to Tregs from healthy donors, which points toward proliferation rather than a mere redistribution (Wolf *et al.*, 2006). An increased proportion of proliferating $Ki67^+$ Tregs has also been shown in various types of cancers including breast cancer and AML (Gobert *et al.*, 2009; Wang *et al.*, 2005b). TGF-β, an autonomous regulator of tumor initiation, progression, immune escape, and metastasis in epithelial cells has been observed to play a central role for peripheral expansion of Tregs (Huber *et al.*, 2004; Yamagiwa *et al.*, 2001). Tumor cells are capable of producing TGF-β, and in addition can modulate myeloid-derived suppressor cells (MDSCs) (Filipazzi *et al.*, 2007), and immature DCs (Ghiringhelli *et al.*, 2005b) to become major sources of TGF-β. Several studies have also shown that Tregs, especially the Th3 Treg subtype, produce TGF-β in its membrane-bound or secreted form, which besides mediating suppression may also act as an autocrine pathway of stimulating self-expansion (Nakamura *et al.*, 2004). Both MDSCs and immature DCs express MHC II and costimulatory molecules at low levels, which may be sufficient to elicit Treg but not effector T cell responses since weak or diminished TCR signaling (e.g., by rapamycin) can favor Treg expansion (Battaglia *et al.*, 2006). Self- and non-self antigens can drive Treg activation and proliferation manifested by a skewed TCR repertoire and further implicated by the importance of APC presence at the site of inflammation and/or cancer in such a process (Belkaid and Oldenhove, 2008; Kumar, 2004). Several studies in mouse models have provided evidence to support these observations (Walker, 2004). Mature APCs are now also being implicated in the expansion of Tregs, in contrast to earlier thought that only immature or aberrant APCs promote Treg expansion (Lundqvist *et al.*, 2005). IDO is a key immunomodulatory enzyme found in the tumor tissue or in APCs of the draining LNs and is linked to tumor-associated immunosuppression and tumor-induced tolerance (Munn and Mellor, 2007). It was recently shown that IDO expressed by APCs could directly activate Tregs and promote their proliferation (Baban *et al.*, 2009; Chung *et al.*, 2009). Ligation of CD80 and CD86 by CTLA-4, constitutively expressed on Tregs increases the functional activity of IDO forming a positive feedback loop (Fallarino *et al.*, 2003). TLRs have been increasingly demonstrated to have roles beyond mere antimicrobial surveillance to multiple physiologic functions as they are also regulated by several intrinsic ligands. TLRs can be found in Tregs and are of significance to their function (van Maren *et al.*, 2008). Activation of TLR2, in particular by heat shock protein 60 (Hsp60), leads to proliferation of Tregs and an increased production of IL-10 and TGF-β (Caramalho *et al.*, 2003; Liu *et al.*, 2006a). Members of the Hsp-family released by (dying) tumor cells within the tumor microenvironment can serve either as immunostimulatory signals or be immunosuppressive as in the case of Hsp60. TLR4 and TLR5 stimulation by lipopolysaccharide and flagellin, respectively, can lead to Treg activation and proliferation, although their exact role warrants further investigation (Caramalho *et al.*, 2003). Cumulatively, these findings suggest that activation of certain TLRs by proinflammatory bacterial by-products can promote Treg proliferation in the absence of APCs. Tregs are found to express higher levels of TLRs as compared to conventional T cells which is suggesting a greater degree of environmental control. Tumor-derived microvesicles (MVs) constitute a potent mechanism by which malignancies transform the host microenvironment. Tumor cells actively release these endosome-derived 50–100 nm organelles (exosomes) that systemically exert protumorigenic effects as they can be found in virtually all body fluids. MVs carrying membrane-bound TGF-β, which skews $CD4^+$ T cell responses in favor of Tregs and deter cytotoxic cells have been identified in tumor patients (Clayton *et al.*, 2007). Recently, MVs isolated directly from patient's sera were shown to induce Treg proliferation (Wieckowski *et al.*, 2009). As described previously, IL-2 plays a major role *in vivo* for Treg maintenance and expansion via STAT-dependent mechanisms (Zorn *et al.*, 2006). STAT3 and STAT5 bind to a highly conserved binding site located in the first intron of the *FOXP3* gene. Consequently, patients with a STAT5b deficiency have been observed to have decreased numbers of $CD4^+CD25^{high}$ T cells, which display low FOXP3 levels and diminished suppressive function (Cohen *et al.*, 2006). Treatment with IL-2 commonly used for patients with renal cancer and melanoma may result in an increase of Treg frequency and suppressive activity in patients; IL-2-based therapy of cancer thus requires a more judicious appraisal, an outlook supported by recent reports on melanoma and renal cell carcinoma patients treated with IL-2. Discussions about the substitution of IL-2 with other immunostimulatory cytokines sharing the γc receptor such as IL-7, IL-15, and IL-21 are currently ongoing (Ahmadzadeh and Rosenberg, 2006; Jensen *et al.*, 2009; van der Vliet *et al.*, 2007).

C. *De Novo* Generation

Tregs can amass at tumor sites by *de novo* generation from naïve and memory $CD4^+$ and $CD8^+$ T cells as recently shown in B-NHL (Ai *et al.*, 2009) (Fig. 2C). Intensive efforts have been undertaken to determine exactly the tumor-derived factors promoting such a Treg *de novo* generation in order to explore avenues of potential intervention. It is apparent that malignant cells as well as other cells of the tumor microenvironment are involved in this process utilizing various mechanisms. In contrast to the intrathymic Treg generation, TGF-β holds a crucial role in peripheral development of induced Tregs. Antigen-mediated stimulation of the TCR in the presence of TGF-β induces Tregs; a mechanism that has been explored in multiple models (Chen *et al.*, 2003; Liu *et al.*, 2007; Yamagiwa *et al.*, 2001). Is should be pointed out that the promoter region of the *FOXP3* gene in these iTregs depicts more methylated nucleotides as compared to nTregs, indicating a less stable suppressive phenotype (Zhou *et al.*, 2009). Activin A, a member of the TGF-β family induced by inflammatory signals, was recently found to promote peripheral Treg conversion, suggesting a redundancy within the members of the TGF-β family (Huber *et al.*, 2009). The fact that TGF-β is associated with diverse cancer types emphasizes the significance of this pathway. Tumor cells not only produce and secrete significant amounts of TGF-β, but also modulate cells of the tumor microenvironment, especially APCs, turning them into additional sources of soluble or even membrane-bound TGF-β (Filipazzi *et al.*, 2007; Ghiringhelli *et al.*, 2005b). Interestingly, TGF-β fuels an autoreactive loop by upregulating FOXP3, which downregulates SMAD7 and thereby leads to an increased TGF-β expression (Fantini *et al.*, 2004). Akin to TGF-β, IL-10 is the second most prominent cytokine involved in Treg induction. IL-10 is also associated with various types of cancers. Early on during tumor growth, antigenic stimulation of conventional T cells in the presence of IL-10 led to the generation of Tr1 cells in a B16 melanoma model (Seo *et al.*, 2001). Hemeoxygenase (HO)-1, inducible by inflammation and oxidative stress, may be involved in this process as it maintains DCs in an immature stage and promotes IL-10 production (Chauveau *et al.*, 2005). APCs are the interface of innate and adaptive immunity orchestrating numerous immunological responses. The net direction of adaptive immunity toward anergy or reactivity strongly depends on APCs; their developmental stage, activation, and costimulatory potential. Malignancies regularly suppress APC differentiation in the tumor microenvironment and thereby potentially drive Treg conversion. Minute antigen presentation in combination with weak costimulation, also termed subimmunogenic conditions, can convert conventional T cells to Tregs even in the absence of TGF-β (Kretschmer *et al.*, 2005). Observations from single injection of immature DCs pulsed with influenza matrix peptide and keyhole limpet hemocyanin in two healthy individuals provides evidence for this pathway of Treg induction (Dhodapkar *et al.*, 2001). Tregs that arose in this manner were capable of responding subsequently to optimal antigen presentation and expanding without losing their suppressive functions. This observation partly explains how functionally mature DCs that typically stimulate effector T cells can facilitate the expansion of available Tregs (Banerjee *et al.*, 2006; Lundqvist *et al.*, 2005). MDSCs are a new emerging population of suppressive cells that have yet not been thoroughly characterized. MDSCs are increased in cancer patients and potentially can induce Tregs. Hoechst and colleagues demonstrated that MDSCs from patients with HCC, characterized as CD14^{+}HLA-DR^{-} cells, induced two suppressive populations including nTreg-like CD4^{+}CD25^{+}FOXP3^{+} and IL10^{+} Tr1-like cells (Hoechst *et al.*, 2008). In ovarian cancer patients pDCs can directly induce CD8^{+}IL-10^{+} Tregs (Wei *et al.*, 2005). Subsequent investigations revealed that IDO is essential for this pDC-mediated Treg induction (Chen *et al.*, 2008), and appears to be strongly involved in cancer-related Treg conversion (Baban *et al.*, 2009; Liu *et al.*, 2007; Munn *et al.*, 2004). In melanoma patients, increased levels of H-Ferritin have been associated to increased levels of CD4^{+}CD25^{+} iTregs. This Tr1 induction was mediated by a modulation of DCs by means of increased expression of CD86 and programmed death 1 ligand (PD1-L) (Gray *et al.*, 2003). Upregulation of CD86 and PD1-L have also been observed upon combined administration of vaccines and TLR3 agonists leading to attenuated $CD8^+$ T cell responses (Pulko *et al.*, 2009). Coinhibitory signaling by PD1-L is important for TGF-β-mediated Treg induction (Wang *et al.*, 2008) and is significant for the suppression noted in T-lymphoproliferative diseases, promoting Treg induction among other effects (Wilcox *et al.*, 2009). A profound expression of COX-2, mediating the production of PGE_2, can be found in numerous inflammatory and malignant processes. PGE_2 can directly induce and expand Tr1 as shown in glioma, head-and-neck and lung cancer (Akasaki *et al.*, 2004; Bergmann *et al.*, 2007; Sharma *et al.*, 2005). Additionally, PGE_2 can indirectly increase immunosuppression by facilitating the generation of aberrant or immature myeloid cells. Like TGF-β, a positive feedback loop seems to be present as COX-2 utilized by iTregs for suppressive activity may concurrently drive their own generation (Mahic *et al.*, 2006). Recent clinical studies on HCC (Gao *et al.*, 2009), uveal melanoma (Mougiakakos *et al.*, in press), and renal cancer (Li *et al.*, 2009) have linked COX-2 expression to Treg infiltration and clinical prognosis. Additional cross-talk between cancer-related APCs and Tregs involving the inhibitory molecules B7.H3 and B7.H4 is under current investigation (Kryczek *et al.*, 2007; Mahnke *et al.*, 2007a). ICOS is an activation marker, which binds to the stimulatory molecule B7.H2 on APCs, and is expressed on Tregs in breast cancer and melanoma patients. The subset of $ICOS^{high}$ Tregs represents a hyperactivated population with increased suppressive properties and the ability to induce surrounding clusters of Tr1 cells (Gobert *et al.*, 2009; Strauss *et al.*, 2008). Of course several counteracting mechanisms do exist, explaining how there can even exist a paucity of Treg conversion in inflammatory milieus as exemplified by IL-6 possessing a prominent role by abolishing Treg induction and generating Th17 effector cells instead (Korn *et al.*, 2008). The balance of these factors consequently determines the extent of Treg induction and expansion.

D. Preferential Survival

In addition to redistribution, expansion, and conversion, a fourth mechanism may contribute to the accumulation of Tregs in cancer patients (Fig. 2D). We have demonstrated that nTregs are more resistant toward oxidative stress-mediated cell death compared to conventional $CD4^+$ T cells from healthy individuals (Mougiakakos *et al.*, 2009) as well as advanced melanoma patients (unpublished data). Moreover, nTregs maintained their suppressive properties at hydrogen peroxide levels that were lethal for 50% (LD_{50}) of conventional $CD4^+$ T cells. Increase in cell surface thiol groups (–SH) and intracellular glutathione content (the main thiol-containing redox buffer) appears to be the major mediators of these protective effects of nTregs (unpublished data). Oxidative stress is known to be increased in several tumor types and can negatively affect cellular immunity (Mehrotra *et al.*, 2009). Both tumor cells and bystander myeloid cells contribute to increased oxidative stress. The expression of the enzyme HO-1, which has anti-inflammatory and antioxidative function adds to the suppressive function of nTregs and may partly contribute to the observed resistance toward oxidative stress (Brusko *et al.*, 2005). Furthermore, a recent study suggests that nTregs themselves contribute to a pro-oxidative local milieu, potentially by consumption of free thiols as a part of their suppressive repertoire (Yan *et al.*, 2009). Cumulatively, these results sustain recent observations that in comparison to healthy individuals, Tregs from patients with several types of epithelial cancers were less affected by apoptosis-inducing stimuli than other lymphocyte subsets examined (Stanzer *et al.*, 2008). In CLL, it has been shown that nTregs express higher levels of the antiapoptotic *Bcl2* and *IAP1* genes as compared to Tregs from healthy donors, indicating a switch toward increased survival in tumor-associated Tregs, reflected by *in vitro* assays depicting a reduced sensitivity toward CD95 ligation and p53-dependent (Fludarabine) and -independent (Roscovitine) apoptosis (Jak *et al.*, 2009). Similarly, gene expression analysis on Tregs from renal cancer patients revealed 49 genes to be differentially expressed. The most prominent genes observed to be overexpressed were *LGALS1* and *LGALS3*; both are galectin genes involved in control of apoptosis as well as implicated in the downregulation of the proapoptotic genes *BAX* and *TNFRSF25* leading to a shifted balance toward survival and fitness of Tregs (Jeron *et al.*, 2009).

VI. ANTIGEN SPECIFICITY OF TREGS IN CANCER

As nTregs, like conventional T cells are educated in the thymus, possess somatically rearranged TCRs and recognize self-Ags they should in theory be able to recognize tumor-associated antigens (TAAs). Mouse studies show that antigen-specific Tregs may be more suppressive compared to nonspecific Tregs. Wang and colleagues were the first ones to generate Treg clones specific for the LAGE1 cancer testis antigen from TILs of melanoma patients. These Treg clones required antigen-specific activation for an efficient suppressive activity (Wang *et al.*, 2004). The same group identified Tregs in melanoma patients specific for the tumor-specific ARTC-1 (Antigen Recognized by T Cells 1) antigen (Wang *et al.*, 2005a). The Tregs specific for LAGE1 and ARTC-1 were similar to thymic-derived nTregs in terms of FOXP3, GITR, CTLA-4, and CD25 expression as well as cytokine production. Circulating IL-10^{+} Tregs, reactive against gp100, TRP1 (melanoma tissue differentiation antigens), NY-ESO-1 (cancer/testis antigen), and survivin (member of the inhibitor of apoptosis protein family) have subsequently been identified in metastatic melanoma patients (Vence *et al.*, 2007). A common feature of suppression exerted by these cells detected in melanoma patients was the need for cell-to-cell contact. In addition to the findings in melanoma, Tregs specific for WT1, an antigen overexpressed by several human leukemias were identified in AML patients. These cells displayed an nTreg-like phenotype and suppressed T cell responses *in vitro* independent of cell-to-cell contact (Lehe *et al.*, 2008). Tregs specific for telomerase, CEA, EGFR, Mucin-1, and HER2/neu have been detected in colorectal cancer patients, suggesting that these Tregs control TAA-specific effector cell responses in an antigen-selective manner (Bonertz *et al.*, 2009). Human papilloma virus (HPV) is the major risk factor for cervical cancers as it is directly involved in the process of carcinogenesis. High Treg frequency in the PB correlates with persistence of premalignant lesions caused by HPV infection (Molling *et al.*, 2007). Tumor cells express the HPV-encoded oncoproteins E6 and E7. In malignant tissue as well as draining LNs, E6- and E7-specific Tregs can be detected, which links viral antigen-specific Tregs to local immunosuppression in patients without a generalized immunodeficiency (van der Burg *et al.*, 2007). Theoretically, a vaccination against HPV may lead to an induction, activation, or expansion of such preexisting viral antigen-specific Tregs. In two out of six patients who received vaccination with E6/E7 long peptides, an expansion of suppressive antigen-specific Tregs that reached levels as high as those observed for effector $CD4^+$ T cells was reported (Welters *et al.*, 2008). Taken together, these results convincingly demonstrate that Tregs specific for self as well as foreign peptides expressed by tumor cells do exist. Furthermore, antigen-specific Tregs are not restricted to the $CD4^+$ T cell compartment as patients with melanoma, renal and breast cancer show significantly increased frequencies of circulating suppressive $CD8^+$ Tregs specific for HO-1 (Andersen *et al.*, 2009). These HO-1-specific $CD8^+$ Tregs exhibited a stronger suppressive function than $CD4^+CD25^+$ nTregs and not only hampered effector T cells directly but also protected directly the tumor target cells from an efficient CTL recognition by yet unidentified mechanisms. HO-1 is a late phase anti-inflammatory enzyme that promotes tolerogenic DCs producing IL-10. Cancer-mediated inflammation can theoretically lead to an increased HO-1 production in the tumor microenvironment, and thereby result in activation and expansion of HO-1-specific Tregs. These findings together with results from mouse models (Zhou *et al.*, 2006) raise concerns regarding the potential adverse effects of vaccination strategies that utilize self or foreign proteins expressed by tumors in order to elicit efficient $CD4^+$ and $CD8^+$ T cell responses. It seems possible or even likely that immunization with certain antigens may expand suppressive Tregs antagonizing the positive effects on effector T cells.

VII. CANCER VACCINES AND REGULATORY T CELLS

Abundant evidence exists that clinical responses to cancer vaccines are influenced by the disease stage at the time of vaccination. Tumor burden and Treg levels typically tend to go hand-in-hand. For example, patients with advanced melanoma have significantly higher circulating Tregs than those with minimal residual disease (Nicholaou *et al.*, 2009). Tregs may be induced or expanded by cancer vaccines as illustrated in studies with melanoma patients, where immunological and clinical responses pre- and/or postvaccination with either NY-ESO-1 protein or DCs pulsed with allogeneic cell lysate (TRIMEL) were reported to be associated with the presence of $CD4^+CD25^+FOXP3^+CD127^-$ and TGF-β-producing Th3 cells (Lopez *et al.*, 2009; Nicholaou *et al.*, 2009). Patients vaccinated with a NY-ESO-1 DNA vaccine had measurable T cell responses, which were clearly suppressed by $CD4^+CD25^+FOXP3^+$ Tregs (Gnjatic *et al.*, 2009). B-CLL patients who received autologous DCs loaded with tumor lysates had specific $CD8^+$ T cell responses against the TAAs RHAMM or fibromodulin, which correlated positively with levels of IL-12 in serum and inversely with $CD4^+CD25^+FOXP3^+$ Treg frequency (Hus *et al.*, 2008). As Tregs obviously represent a major obstacle for efficient cancer immunotherapies, Treg depletion has emerged as an adjuvant therapy that effectively synergizes with different cancer vaccine approaches in animal models. However, the effect of Treg depletion in these models was greatest when done immediately before or after tumor inoculation (Knutson, 2006; Onizuka, 1999; Sutmuller, 2001). Depletion of Tregs after the establishment of tumors often fails to significantly improve the therapeutic outcome (Elpek, 2007) implying a tumor stage-dependent impact of Tregs in cancer control. Several ongoing clinical studies in cancer are aimed at exploring Treg depletion in combination with different immunotherapeutic approaches. In renal cancer patients, elimination of Tregs using an immunotoxin, followed by vaccination with tumor RNA-transfected DCs significantly improved the induction of tumor-specific T cell responses (Dannull *et al.*, 2005). Similar results could be obtained with DC or peptide-based vaccination strategies combined with Treg depletion in colorectal and breast cancer as well as melanoma patients (Mahnke *et al.*, 2007b; Morse *et al.*, 2008; Rech and Vonderheide, 2009). Data from animal models support the concept that tumor-associated Tregs can be expanded in response to therapeutic vaccination and suppress the concomitantly generated effector T cells (Zhou *et al.*, 2006). In melanoma patients vaccinated either with specific peptides or APCs loaded with tumor lysates a significant increase of IL-10-producing Tr1 cells was noted in PB postvaccination (Chakraborty *et al.*, 2004). Melanoma patients who received MAGE-A3 peptide vaccines had an increased specific $CD4^+$ T cell response as detected by HLA class II tetramers.

After flow cytometric enrichment of MAGE-A3-specific $CD4^+$ T cells a substantial proportion of the subsequently generated clones showed phenotypic and functional characteristics of nTregs and Th3 cells, which exhibited suppressive activity *in vitro* upon peptide stimulation (Francois *et al.*, 2009). Thus therapeutic vaccination in cancer may potentially be a double-edged sword and expansion of Tregs and resultant immune suppression may ensue rather than boosting of effector T cell activity. Consequently, strategies to disarm Tregs should be considered as essential components of immunotherapeutical approaches and are a feature of virtually all prominent vaccine trials in recent times. In this scenario, a combination of vaccines with agents modulating Tregs is one additional option already under clinical evaluation. The microenvironment where T cells encounter the antigen and get primed by the local APCs plays a major role for the balance between tolerogenesis and immunogenesis. Features such as antigen availability (Turner *et al.*, 2009) and a Th1-biasing cytokine milieu (Nishikawa *et al.*, 2005) are considered as critical variables determining the resulting polarization of T cells. Vaccine adjuvants modulating this milieu by, for example, increasing the production of type-1 interferons may be a promising strategy to interfere with the induction of antigen-specific Tregs. The studies discussed in this section raise major concerns regarding the design of cancer vaccines and can explain at least partially the low objective responses observed in many clinical studies. In addition, these observations strongly suggest that monitoring of the Treg compartment is as important as the evaluation of the effector cell arm in patients receiving immunotherapies.

VIII. TARGETING REGULATORY T CELLS IN CANCER THERAPY

Taken together Tregs regardless origin, impede tumor surveillance and appear in many cases to be directly linked to the disease pathogenesis. In studies dating back to the 1980s performed by Robert North and colleagues, Treg depletion was shown to be an elegant approach for increasing immune reactivity against cancer. Especially to date, where various forms of immunotherapies find their way into cancer treatment it appears inevitable to counteract the suppressive effects of Tregs. Nevertheless, the impact of modulating Tregs is not trivial as it may result in unwanted side effects most notably autoimmunological phenomena. Furthermore, targeting of Tregs has to be restricted to malignancies, where Tregs have been shown to be undoubtedly linked to deleterious effects. Different strategies aimed to deplete Tregs or to functionally inactivate Tregs are currently under development or in clinical evaluation (selected studies are summarized in Table 2).

Type(s) of malignancy	No.	Depletion regimen	Treatment responses	References
Metastatic melanoma	13	CPM (60 mg/kg/d 2d) + Flu (25 mg/m^2/d 5d) prior ACT	Objective responses in 6 pts, AID in 5 pts	Dudley *et al.* (2002)
Metastatic melanoma	35	CPM (60 mg/kg/d 2d) + Flu (25 mg/m^2/d 5d) prior ACT	Objective responses in 18 pts, AID in 13 pts	Dudley *et al.* (2005)
Metastatic melanoma	93	CPM (60 mg/kg/d 2d) + Flu (25 mg/m^2/d 3d) + TBI (2 Gy or 12 Gy) prior ACT	Objective responses in 50–70% of pts, 4 CRs	Dudley *et al.* (2008)
Various types of metastatic solid tumors	9	Metronomic CPM (50 mg p.o., 2d/1 w)	PD in 3 pts, SD (2–3 months) in 4 pts	Ghiringhelli *et al.* (2007)
Chronic lymphocytic leukemia	73	Fludarabine-containing therapies	Reduced Treg frequency/function	Beyer *et al.* (2005)
Metastatic breast cancer (ongoing study)	3	Daclizumab (1 mg/m^2; single dose) 1 w prior peptide vaccination	Improved responses to vaccination in all pts	Rech and Vonderheide (2009)
Metastatic melanoma and renal carcinoma	13	ONTAK (9 or 18 μg/kg; successive doses)	No objective responses, no Treg depletion	Attia *et al.* (2005)
Metastatic renal cell carcinoma	10	ONTAK (18 μg/kg; single dose) + tumor RNA-transfected DC vaccine	Improved CTL responses, reduced Treg levels	Dannull *et al.* (2005)
Metastatic melanoma	7	ONTAK (5 or 18 μg/kg; successive doses) prior peptide vaccination	Peptide-specific CTLs in 5/6 pts, PD in 5 pts	Mahnke *et al.* (2007b)
Metastatic melanoma, renal cell cancer	15	Ipilumimab (1–9 mg/kg; successive doses)	Objective responses in 8 pts, AID in 5 pts	Maker *et al.* (2005)
B cell non-Hodgkin lymphoma	18	Ipilumimab (1–3 mg/kg; successive doses)	1 CR and 1 PR	Ansell *et al.* (2009)
Metastatic melanoma	14	Ipilumimab (3 mg/kg; successive doses) + peptide vaccination	2 CRs and 1 PR, AID in 6 pts	Phan *et al.* (2003)
Metastatic melanoma, ovarian cancer	20	Ipilumimab (3 mg/kg; successive doses) upon tumor cell vaccination (GVAX)	SD in 8 pts, PR in 4 pts, PD in 8 pts	Hodi *et al.* (2008)

Abbreviations: n, number of patients; CPM, cyclophosphamide; Flu, fludarabine; ACT, adoptive cell transfer; pts, patients; AID, autoimmune disease; TBI, total body irradiation; Gy, gray; PD, progressive disease; SD, stable disease; CTL, cytotoxic T lymphocyte; CR, complete remission; PR, partial remission.

Notes: Daclizumab, humanized anti-CD25 antibody (Zenapax); ONTAK, diphtheria toxin-interleukin-2 fusion protein (Denileukin diftitox); Ipilumimab, human anti-CTLA-4 antibody (MDX-010).

A. Depletion of Regulatory T Cells

The concept of "suppressing the suppressors" goes back to the 1980s beginning with the revolutionary studies by Robert North, who hypothesized that the antitumor effect of cyclophosphamide (CPM) in murine experimental cancer models was due to the depletion of by that time unidentified suppressor T cells (North, 1982). CPM, a DNA alkylating drug, is a standard chemotherapeutic agent utilized in numerous chemotherapy regimens since the 1950s. Observations in patients with autoimmune and malignant diseases treated with CPM revealed that while the drug was immunosuppressive at high dosages, low-dose CPM had an immunostimulatory effect. Studies in mice indicated that the immunostimulatory effects of low-dose CPM were due to selected depletion of Tregs (Ercolini *et al.*, 2005; Lutsiak *et al.*, 2005). Low-dose CPM has been shown to significantly reduce $CD4^{+}CD25^{+}$ Tregs but not the total T cell population (Lutsiak *et al.*, 2005). In a Her2/neu transgenic breast cancer mouse model, combination of peptide vaccination with CPM led to a decreased number of circulating Tregs and a parallel boost in tumor-specific, high-avidity $CD8^{+}$ T cells increasing tumor protection (Ercolini *et al.*, 2005). Potential mechanisms of action include induction of apoptosis, decrease of homeostatic proliferation as well as attenuation of suppressive function (Taieb *et al.*, 2006). Dudley and colleagues have performed clinical trials on patients with therapy-refractory metastatic melanoma by adoptively transferring autologous T cells after preconditioning with the Treg depleting agents CPM and fludarabine. Objective clinical responses were noted in an astonishing 50–70% of the patients (Dudley *et al.*, 2002, 2005, 2008). Low-dose "metronomic" CPM administration in end-stage cancer patients selectively depletes $CD4^{+}CD25^{+}FOXP3^{+}$ Tregs and restores function of T cells and NK cells (Ghiringhelli *et al.*, 2007). However, Treg depletion with low dose of CPM is short-lived, lasting only for 5–6 days. As mentioned previously, fludarabine a cytotoxic purine analog used in hematologic malignancies has been shown to decrease Treg frequencies and abolish their suppressive activity in CLL patients (Beyer *et al.*, 2005).

CD25, the high-affinity IL-2Rα, is constitutively expressed on major subsets of Tregs, especially nTregs. It is also transiently expressed on effector T cells initially during their activation, complicating a CD25-based Treg targeting strategy. Nevertheless, the anti-CD25 monoclonal antibody PC61, originally identified as a monoclonal antibody against the murine IL-2R, has been shown to abrogate suppressive function of $CD4^{+}CD25^{+}$ Tregs enhancing tumor rejection in mouse cancer models (Onizuka *et al.*, 1999; Shimizu *et al.*, 1999; Tanaka *et al.*, 2002). Whether PC61 in fact depletes Tregs or rather inactivates Tregs is still unclear (Kohm *et al.*, 2006). Two different antihuman CD25 antibodies, basiliximab (Zenapax) and daclizumab (Simulect) have been approved for transplantation, autoimmune diseases, and cancer. In metastatic breast cancer patients, treatment with Simulect resulted in a marked depletion of circulating Tregs and peptide vaccination against TAAs generated an effective CTL response (Rech and Vonderheide, 2009). The development of the recombinant IL-2 diphtheria toxin conjugate called denileukin diftitox DAB_{389}IL-2 (ONTAK) was considered a breakthrough for some types of malignancies. It is Food and Drug Administration (FDA) approved for treatment of cutaneous T cell leukemia/lymphoma (Olsen *et al.*, 2001). ONTAK has a short half-life of 60 min and is designed to target cells expressing the high-affinity IL-2R. Upon internalization via endocytosis diphtheria toxin inhibits protein synthesis leading to apoptotic cell death (Figgitt *et al.*, 2000). Based on promising results in initial studies, ONTAK has been used in combination with other therapies in the treatment of diseases like B- and T-NHLs (Dang *et al.*, 2004; Foss *et al.*, 2005; Frankel *et al.*, 2006). There are conflicting reports whether ONTAK depletes Tregs or rather inhibits their function (Attia *et al.*, 2005; Dannull *et al.*, 2005; Vaclavkova *et al.*, 2006). In a study on melanoma patients it was shown *in vitro* and *in vivo* that ONTAK treatment resulted in both decreased numbers and a reduced function of Tregs (Mahnke *et al.*, 2007b). However, several facts need consideration when incorporating ONTAK into therapeutic regimens. In addition to Tregs, ONTAK may also target $CD25^+$ effector T cells. Moreover, Treg homeostasis is very robust and Treg levels recover rapidly following depletion to pretreatment levels or even exceed them. In order to achieve an optimal treatment efficacy, different application schemes and dosage protocols have to be carefully evaluated aiming for an ideal balance between depletion of Tregs and enhancement of effector T cell response.

B. Targeting Function of Regulatory T Cells

Another target molecule on Tregs is CTLA-4, which is involved in mediating suppression as described previously. Like CD25, CTLA-4 can also be expressed on activated $CD4^+$ and $CD8^+$ T cells (Egen *et al.*, 2002). This potential blocking of CTLA-4 function on many levels, including Tregs as well as effector T cells may be responsible for a superior efficacy (Egen *et al.*, 2002). However, it remains still to be elucidated, which is the predominant mechanism mediating the observed anti-CTLA-4 effects. Currently, two humanized anti-CTLA-4 antibodies, Ipilimumab (MDX-010) and Tremelimumab (CP-675206), have been used in phase I/II clinical trials. Results from a study on patients with advanced stage metastatic melanoma and renal cancer imply that antitumor effects are due to a direct enhancement of $CD4^+$ and $CD8^+$ T cell activity rather than inhibition or depletion of Tregs

(Maker *et al.*, 2005). In a mouse model expressing human instead of mouse CTLA-4 it was elegantly demonstrated that CTLA-4 blockade of Tregs alone failed to enhance antitumor responses (Peggs *et al.*, 2009). In contrast, concomitant blockade on both effector T cells and Tregs leads to a synergistic effect with maximal antitumor activity. In several phase I trials including mostly melanoma patients, but also ovarian and prostate cancer as well as B-NHL, blockage of CTLA-4 resulted in tumor regression, but in some cases it also generated severe autoimmune adverse effects (Ansell *et al.*, 2009; Dranoff, 2005; Maker *et al.*, 2005; Phan *et al.*, 2003). Autoimmunity may be minimized by altering the schedule of administration, dose and nature of the therapeutic antibody as well as the concomitant treatment, such as vaccines against certain TAAs. In a recent study on patients with metastatic melanoma and ovarian cancer, periodic infusion of anti-CTLA-4 antibodies after vaccination with autologous tumor cells secreting GM-CSF generated clinical antitumor immunity, and importantly, did not induce any grade 3 or grade 4 toxicity (Hodi *et al.*, 2008). Therapy-induced tumor necrosis correlated with intratumoral $CD8^+$ effector T cell/Treg ratio detected in post-treatment biopsies.

GITR, a molecule constitutively expressed on nTregs but also at lower levels on activated conventional T cells has also been considered a target for Treg depletion and functional inhibition (Nocentini and Riccardi, 2005; Shimizu *et al.*, 2002). *In vitro* stimulation of GITR in murine Tregs resulted in reduced suppressive activity, but this could not be reproduced in human Tregs (Kanamaru *et al.*, 2004; Levings *et al.*, 2002; Shimizu *et al.*, 2002). Tumor-bearing mice treated with the agonistic anti-GITR antibody DTA-1, or a GITR ligand showed decreased intratumoral Treg recruitment together with the generation of a potent specific antitumor response (Ko *et al.*, 2005; Levings *et al.*, 2002). Future studies are obligatory in order to evaluate the feasibility of such an approach for the treatment of cancer patients.

TLRs are widely expressed on multiple human cells and represent the first line of immunological defense through recognition of various pathogen-associated molecular patterns. TLRs are involved in DC maturation and activation of TLR pathways in DCs has been shown to prevent conversion of conventional T cells into Tregs (Iwasaki and Medzhitov, 2004). As described previously, Tregs express various TLRs, and thereby TLR ligands may have direct (positive or negative) effects on Tregs. Activation of TLR8 by natural or synthetic ligands independently of presence of DCs has been shown to reverse Treg function and augments *in vivo* tumor immunity in mouse models (Peng *et al.*, 2005; Sutmuller *et al.*, 2006). Stimulation of TLR signaling may be of particular importance for vaccination strategies, since appropriate TLR stimulation may overturn Treg-mediated tolerance (Yang *et al.*, 2004). Specific adjuvants providing vaccines with such properties are currently under investigation.

C. Disrupting Intratumoral Homing of Regulatory T Cells

As described in previous sections, chemokine/chemokine receptor interactions are vital to the migration of Tregs into the tumor microenvironment. One of the most important interplay is the one between CCL22 secreted by tumor and tumor conditioned myeloid cells, and CCR4, which is highly expressed on Tregs. Blocking of CCL22 significantly reduces the migration of Tregs into ovarian tumors as demonstrated in a preclinical murine xenograft model (Curiel *et al.*, 2004). In addition, CCL5–CCR5 interaction is crucial for Treg attraction in pancreatic adenocarcinoma (Tan *et al.*, 2009). Disrupting the CCL5–CCR5 signaling reduces Treg migration into the tumor bed also leading to significant tumor reduction. Both CCL5 and CCL22 are also involved in trafficking of effector T cells; a fact that needs to be taken into account during development of potential targeting strategies. Altogether, interfering with Treg trafficking represents a promising and very elegant potential approach in the treatment of cancer. However, it needs to be determined to what extent blocking of chemokine–chemokine receptor signaling will affect other cell types obligatory for an efficient immune response.

D. Modulation of Regulatory T Cell Proliferation/Conversion

As described in previous sections, DCs, regardless of maturation status, are involved in activation and induction of Tregs. One central molecule in that process is the enzyme IDO, which is highly expressed in tolerogenic myeloid and pDCs (Chen *et al.*, 2008; Chung *et al.*, 2009). Binding of CTLA-4 on CD80 and/or CD86 triggers IDO activity in DCs (Fallarino *et al.*, 2003), thus aforementioned anti-CTLA-4 treatment may interfere with the IDO pathway. Phase I clinical trials treating patients with relapsed or refractory solid tumor with the IDO inhibitor 1-methyl-D-tryptophan (D-1MT) are currently ongoing. In addition, animal studies have demonstrated that IDO-mediated immunosuppression can be reversed by celecoxib treatment (Lee *et al.*, 2009). Celecoxib is a specific inhibitor of the PGE_2-producing enzyme COX-2. The production of PGE_2 directly stimulates Treg expansion (Akasaki *et al.*, 2004) or indirectly facilitates Treg recruitment by promoting tolerogenic APCs (Bergmann *et al.*, 2007). Furthermore, Tregs themselves can suppress immune responses through PGE_2 secretion (Mahic *et al.*, 2006; Yaqub *et al.*, 2008) which further supports the evaluation of COX-2 inhibitors in the treatment of malignancies known to show high COX-2 and Treg levels such as HCC and renal cancer (Gao *et al.*, 2009;

Li *et al.*, 2009). PGE_2 stimulates expression of the enzyme aromatase through a cAMP-dependent pathway. Aromatase inhibitors, in particular letrozole (Femara) used to treat breast cancer patients have been shown to reduce the number of circulating Tregs, potentially by disrupting the PGE_2–aromatase pathway (Generali *et al.*, 2009).

In ovarian cancer patients, pDCs directly induce IL-10-producing $CD8^+$ Tregs (Wei *et al.*, 2005; Zou *et al.*, 2001). Tumor cells can produce CXCL12 and thereby attract pDCs expressing the specific receptor CXCR4. Blocking of the CXCL12–CXCR4 interaction induces apoptosis of tumor-related pDCs and abrogates their chemotaxis (Zou *et al.*, 2001). Furthermore, Tregs may upregulate CXCR4 upon IL-2 treatment (Wei *et al.*, 2006) or hypoxic conditions (Schioppa *et al.*, 2003), often noted in cancer. Therefore, agents like AMD-3100 used in HIV patients that antagonize the CXCR4 function may also be useful in the treatment of cancer.

Coinhibitory signaling through PD1-L is involved in the induction of Tregs (Gray *et al.*, 2003; Krupnick *et al.*, 2005; Wang *et al.*, 2008). Blockade of PD1-L on Tregs (Wang *et al.*, 2009) augments human tumor-specific T cell responses (Curiel *et al.*, 2003). An anti-PD-1 monoclonal IgG4 antibody, MDX-1106 (Ono-4538) is currently in a phase II trial for various types of cancer including melanoma, colon and lung cancer. Impact on clinical course, toxicities, and T cell subsets remains to be seen (Brahmer *et al.*, 2009).

Antiangiogenic treatment of colorectal cancer patients with the humanized anti-VEGF antibody bevacizumab (Avastin) induced a decrease in the levels of Tregs. The observations correlate with animal studies demonstrating a direct and positive correlation between VEGF expression and Treg levels (Li *et al.*, 2006). Expression of VEGF receptor-2 (VEGFR-2) within the T cell compartment is restricted to Tregs (Suzuki *et al.*, 2009). However, it is presently unclear whether the observed effects result directly from inhibiting VEGFR-2 or via an unknown intermediary mechanism (Wada *et al.*, 2009).

Two main cytokines involved in Treg induction and function are IL-10 and TGF-β. Disrupting their pathways may be useful for reducing the frequency and function of Tregs. Inhibitors of TGF-β for clinical use are currently under development and include anti-TGF-β antibodies, soluble TGF-β receptors as well as the antisense oligonucleotide, AP-12009, which blocks TGF-β expression and is currently being tested in phase I/II clinical trials.

E. Targeting the Antioxidative Capacity of Regulatory T Cells

As described previously, malignant diseases result in increased levels of oxidative stress mediated by reactive oxygen species (ROS) (Kusmartsev *et al.*, 2004; Szatrowski and Nathan, 1991). The detrimental effect of ROS on effector cells of the immune system is well established and described in malignant and chronic inflammatory diseases (Gringhuis *et al.*, 2000; Li *et al.*, 2008; Malmberg *et al.*, 2001; Schmielau and Finn, 2001). Paradoxically, Treg levels are often increased in this hostile (for lymphocytes) milieu as described recently (Mougiakakos *et al.*, 2009). The mechanism underlying the increased resistance of Tregs toward oxidative stress is currently unclear but appears to be linked to the increased intracellular and surface thiol content. Nevertheless, the identification of this mechanistic pathway could provide yet another means for targeting Tregs in order to restore a "balance of power" between Tregs and conventional T cells as regards to oxidative stress susceptibility.

IX. CONCLUDING REMARKS

Tregs efficiently suppress innate and adaptive immunity. Despite the extensive research that has been carried out, many aspects of Treg biology in cancer remain to be explored. Vast majority of preclinical and clinical studies have linked the presence of Tregs to an increased risk for development as well as progression of cancer. This paradigm is currently under scrutiny as it has been convincingly shown that Tregs can act in a beneficial fashion in inflammatory driven malignancies, explaining controversial reports on some types of cancers, where Tregs were actually associated with a better disease course and outcome. In the context of controversial data regarding the impact of Tregs in cancer, it is important to point out the lack of comparability between distinct studies as differences in methodologies, enumeration strategies, Treg characterization as well as inclusion criteria for selection of patient groups have been substantial. The identification of Tregs specific for self and non-self TAAs is already leading to a major reevaluation of vaccine designs. Vaccination with tumor-specific peptides comprises the risk of boosting and/or inducing peptide-specific Tregs, which could thereby hamper the potential antitumor response. Strategies to incorporate adjuvants counteracting this process such as local induction of high IL-6 levels at site of antigen encountering or triggering of Treg-inhibiting TLRs are currently undertaken and evaluated in preclinical models. Altogether, the Treg population in cancer patients constitutes a very dynamic system as regards to subsets, origins, modes of suppression, and mechanisms leading to their accumulation. Interestingly, Tregs appear to generate a self-amplifying system by the production of cytokines that act in a positive feedback fashion and indirectly by promoting tolerogenic APCs. This complex system is at the same time a boon and a bane. On the one hand, it demands extensive efforts in order to decrypt all its building blocks, and on the other hand in-depth insight will allow us more specific and elegant interventions into this web of tumor-associated immunosuppression. New technological achievements, like nanoparticles used as vehicles for a loco-regional delivery of Treg-targeting molecules may be very useful in our attempts to modulate Tregs at the site of their action in order to strengthen host surveillance and/or promote vaccine-induced immunity whenever it is considered beneficial for the clinical course of the particular type of cancer in question.

ACKNOWLEDGMENTS

This work was supported by grants from the Swedish Cancer Society, the Cancer Society of Stockholm, the Swedish Medical Research Council, an ALF-Project grant from the Stockholm City Council, and the German Research Foundation (DFG).

REFERENCES

Ahmadzadeh, M., and Rosenberg, S. A. (2006). IL-2 administration increases CD4+ CD25(hi) Foxp3+ regulatory T cells in cancer patients. *Blood* **107**, 2409–2414.

Ai, W. Z., Hou, J. Z., Zeiser, R., Czerwinski, D., Negrin, R. S., and Levy, R. (2009). Follicular lymphoma B cells induce the conversion of conventional CD4+ T cells to T-regulatory cells. *Int. J. Cancer* **124**, 239–244.

Akasaki, Y., Liu, G., Chung, N. H., Ehtesham, M., Black, K. L., and Yu, J. S. (2004). Induction of a CD4+ T regulatory type 1 response by cyclooxygenase-2-overexpressing glioma. *J. Immunol.* **173**, 4352–4359.

Allan, S. E., Passerini, L., Bacchetta, R., Crellin, N., Dai, M., Orban, P. C., Ziegler, S. F., Roncarolo, M. G., and Levings, M. K. (2005). The role of 2 FOXP3 isoforms in the generation of human CD4+ Tregs. *J. Clin. Invest.* **115**, 3276–3284.

Alvaro, T., Lejeune, M., Salvado, M. T., Bosch, R., Garcia, J. F., Jaen, J., Banham, A. H., Roncador, G., Montalban, C., and Piris, M. A. (2005). Outcome in Hodgkin's lymphoma can be predicted from the presence of accompanying cytotoxic and regulatory T cells. *Clin. Cancer Res.* **11**, 1467–1473.

Andersen, M. H., Sorensen, R. B., Brimnes, M. K., Svane, I. M., Becker, J. C., and thor Straten, P. (2009). Identification of heme oxygenase-1-specific regulatory CD8+ T cells in cancer patients. *J. Clin. Invest.* **119**, 2245–2256.

Angelini, G., Gardella, S., Ardy, M., Ciriolo, M. R., Filomeni, G., Di Trapani, G., Clarke, F., Sitia, R., and Rubartelli, A. (2002). Antigen-presenting dendritic cells provide the reducing extracellular microenvironment required for T lymphocyte activation. *Proc. Natl. Acad. Sci. USA* **99**, 1491–1496.

Annacker, O., Pimenta-Araujo, R., Burlen-Defranoux, O., Barbosa, T. C., Cumano, A., and Bandeira, A. (2001). CD25+ CD4+ T cells regulate the expansion of peripheral CD4 T cells through the production of IL-10. *J. Immunol.* **166**, 3008–3018.

Annacker, O., Asseman, C., Read, S., and Powrie, F. (2003). Interleukin-10 in the regulation of T cell-induced colitis. *J. Autoimmun.* **20**, 277–279.

Ansell, S. M., Hurvitz, S. A., Koenig, P. A., LaPlant, B. R., Kabat, B. F., Fernando, D., Habermann, T. M., Inwards, D. J., Verma, M., Yamada, R., Erlichman, C., Lowy, I., *et al.* (2009). Phase I study of ipilimumab, an anti-CTLA-4 monoclonal antibody, in patients with relapsed and refractory B-cell non-Hodgkin lymphoma. *Clin. Cancer Res.* **15**, 6446–6453.

Apostolou, I., and von Boehmer, H. (2004). In vivo instruction of suppressor commitment in naive T cells. *J. Exp. Med.* **199**, 1401–1408.

Apostolou, I., Sarukhan, A., Klein, L., and von Boehmer, H. (2002). Origin of regulatory T cells with known specificity for antigen. *Nat. Immunol.* **3**, 756–763.

Attia, P., Maker, A. V., Haworth, L. R., Rogers-Freezer, L., and Rosenberg, S. A. (2005). Inability of a fusion protein of IL-2 and diphtheria toxin (Denileukin Diftitox, DAB389IL-2, ONTAK) to eliminate regulatory T lymphocytes in patients with melanoma. *J. Immunother.* **28**, 582–592.

Azuma, T., Takahashi, T., Kunisato, A., Kitamura, T., and Hirai, H. (2003). Human CD4+ CD25+ regulatory T cells suppress NKT cell functions. *Cancer Res.* **63**, 4516–4520.

Baban, B., Chandler, P. R., Sharma, M. D., Pihkala, J., Koni, P. A., Munn, D. H., and Mellor, A. L. (2009). IDO activates regulatory T cells and blocks their conversion into Th17-like T cells. *J. Immunol.* **183**, 2475–2483.

Bacchetta, R., Gregori, S., and Roncarolo, M. G. (2005). CD4+ regulatory T cells: mechanisms of induction and effector function. *Autoimmun. Rev.* **4**, 491–496.

Badoual, C., Hans, S., Rodriguez, J., Peyrard, S., Klein, C., Agueznay Nel, H., Mosseri, V., Laccourreye, O., Bruneval, P., Fridman, W. H., Brasnu, D. F., and Tartour, E. (2006). Prognostic value of tumor-infiltrating CD4+ T-cell subpopulations in head and neck cancers. *Clin. Cancer Res.* **12**, 465–472.

Baecher-Allan, C., and Anderson, D. E. (2006). Regulatory cells and human cancer. *Semin. Cancer Biol.* **16**, 98–105.

Baecher-Allan, C., Brown, J. A., Freeman, G. J., and Hafler, D. A. (2001). CD4+CD25high regulatory cells in human peripheral blood. *J. Immunol.* **167**, 1245–1253.

Baecher-Allan, C., Viglietta, V., and Hafler, D. A. (2004). Human CD4+CD25+ regulatory T cells. *Semin. Immunol.* **16**, 89–98.

Banerjee, D. K., Dhodapkar, M. V., Matayeva, E., Steinman, R. M., and Dhodapkar, K. M. (2006). Expansion of FOXP3high regulatory T cells by human dendritic cells (DCs) in vitro and after injection of cytokine-matured DCs in myeloma patients. *Blood* **108**, 2655–2661.

Bardel, E., Larousserie, F., Charlot-Rabiega, P., Coulomb-L'Hermine, A., and Devergne, O. (2008). Human CD4+ CD25+ Foxp3+ regulatory T cells do not constitutively express IL-35. *J. Immunol.* **181**, 6898–6905.

Baron, U., Floess, S., Wieczorek, G., Baumann, K., Grutzkau, A., Dong, J., Thiel, A., Boeld, T. J., Hoffmann, P., Edinger, M., Turbachova, I., Hamann, A., *et al.* (2007). DNA demethylation in the human FOXP3 locus discriminates regulatory T cells from activated FOXP3(+) conventional T cells. *Eur. J. Immunol.* **37**, 2378–2389.

Bates, G. J., Fox, S. B., Han, C., Leek, R. D., Garcia, J. F., Harris, A. L., and Banham, A. H. (2006). Quantification of regulatory T cells enables the identification of high-risk breast cancer patients and those at risk of late relapse. *J. Clin. Oncol.* **24**, 5373–5380.

Battaglia, M., and Roncarolo, M. G. (2006). Induction of transplantation tolerance via regulatory T cells. *Inflamm. Allergy Drug Targets* **5**, 157–165.

Battaglia, M., Stabilini, A., Migliavacca, B., Horejs-Hoeck, J., Kaupper, T., and Roncarolo, M. G. (2006). Rapamycin promotes expansion of functional CD4+CD25+FOXP3+ regulatory T cells of both healthy subjects and type 1 diabetic patients. *J. Immunol.* **177**, 8338–8347.

Bayer, A. L., Yu, A., Adeegbe, D., and Malek, T. R. (2005). Essential role for interleukin-2 for CD4(+)CD25(+) T regulatory cell development during the neonatal period. *J. Exp. Med.* **201**, 769–777.

Belkaid, Y., and Oldenhove, G. (2008). Tuning microenvironments: induction of regulatory T cells by dendritic cells. *Immunity* **29**, 362–371.

Bennett, C. L., Christie, J., Ramsdell, F., Brunkow, M. E., Ferguson, P. J., Whitesell, L., Kelly, T. E., Saulsbury, F. T., Chance, P. F., and Ochs, H. D. (2001). The immune dysregulation, polyendocrinopathy, enteropathy, X-linked syndrome (IPEX) is caused by mutations of FOXP3. *Nat. Genet.* **27,** 20–21.

Bensinger, S. J., Bandeira, A., Jordan, M. S., Caton, A. J., and Laufer, T. M. (2001). Major histocompatibility complex class II-positive cortical epithelium mediates the selection of CD4 (+)25(+) immunoregulatory T cells. *J. Exp. Med.* **194,** 427–438.

Berendt, M. J., and North, R. J. (1980). T-cell-mediated suppression of anti-tumor immunity. An explanation for progressive growth of an immunogenic tumor. *J. Exp. Med.* **151,** 69–80.

Bergmann, C., Strauss, L., Zeidler, R., Lang, S., and Whiteside, T. L. (2007). Expansion of human T regulatory type 1 cells in the microenvironment of cyclooxygenase 2 overexpressing head and neck squamous cell carcinoma. *Cancer Res.* **67,** 8865–8873.

Bettelli, E., Dastrange, M., and Oukka, M. (2005). Foxp3 interacts with nuclear factor of activated T cells and NF-kappa B to repress cytokine gene expression and effector functions of T helper cells. *Proc. Natl. Acad. Sci. USA* **102,** 5138–5143.

Beyer, M., Kochanek, M., Darabi, K., Popov, A., Jensen, M., Endl, E., Knolle, P. A., Thomas, R. K., von Bergwelt-Baildon, M., Debey, S., Hallek, M., and Schultze, J. L. (2005). Reduced frequencies and suppressive function of CD4+CD25hi regulatory T cells in patients with chronic lymphocytic leukemia after therapy with fludarabine. *Blood* **106,** 2018–2025.

Bonertz, A., Weitz, J., Pietsch, D. H., Rahbari, N. N., Schlude, C., Ge, Y., Juenger, S., Vlodavsky, I., Khazaie, K., Jaeger, D., Reissfelder, C., Antolovic, D., *et al.* (2009). Antigen-specific Tregs control T cell responses against a limited repertoire of tumor antigens in patients with colorectal carcinoma. *J. Clin. Invest.* **119,** 3311–3321.

Bopp, T., Becker, C., Klein, M., Klein-Hessling, S., Palmetshofer, A., Serfling, E., Heib, V., Becker, M., Kubach, J., Schmitt, S., Stoll, S., Schild, H., *et al.* (2007). Cyclic adenosine monophosphate is a key component of regulatory T cell-mediated suppression. *J. Exp. Med.* **204,** 1303–1310.

Brahmer, J. R., Topalian, S. L., Powderly, J., Wollner, I., Picus, J., Drake, C. G., Stankevich, E., Korman, A., Pardoll, A., and Lowy, I. (2009). Phase II experience with MDX-1106 (Ono-4538), an anti-PD-1 monoclonal antibody, in patients with selected refractory or relapsed malignancies. *J. Clin. Oncol.* ASCO Annual Meeting Proceedings 27, 3018.

Brinkrolf, P., Landmeier, S., Altvater, B., Chen, C., Pscherer, S., Rosemann, A., Ranft, A., Dirksen, U., Juergens, H., and Rossig, C. (2009). A high proportion of bone marrow T cells with regulatory phenotype (CD4+CD25hiFoxP3+) in Ewing sarcoma patients is associated with metastatic disease. *Int. J. Cancer* **125,** 879–886.

Brusko, T. M., Wasserfall, C. H., Agarwal, A., Kapturczak, M. H., and Atkins, M. A. (2005). An integral role for heme oxygenase-1 and carbon monoxide in maintaining peripheral tolerance by CD4+CD25+ regulatory T cells. *J. Immunol.* **174,** 5181–5186.

Burnet, M. (1957). Cancer; a biological approach. I. The processes of control. *Br. Med. J.* **1,** 779–786.

Cao, X., Cai, S. F., Fehniger, T. A., Song, J., Collins, L. I., Piwnica-Worms, D. R., and Ley, T. J. (2007). Granzyme B and perforin are important for regulatory T cell-mediated suppression of tumor clearance. *Immunity* **27,** 635–646.

Caramalho, I., Lopes-Carvalho, T., Ostler, D., Zelenay, S., Haury, M., and Demengeot, J. (2003). Regulatory T cells selectively express toll-like receptors and are activated by lipopolysaccharide. *J. Exp. Med.* **197,** 403–411.

Carreras, J., Lopez-Guillermo, A., Fox, B. C., Colomo, L., Martinez, A., Roncador, G., Montserrat, E., Campo, E., and Banham, A. H. (2006). High numbers of tumor-infiltrating FOXP3-positive regulatory T cells are associated with improved overall survival in follicular lymphoma. *Blood* **108,** 2957–2964.

Chakraborty, N. G., Chattopadhyay, S., Mehrotra, S., Chhabra, A., and Mukherji, B. (2004). Regulatory T-cell response and tumor vaccine-induced cytotoxic T lymphocytes in human melanoma. *Hum. Immunol.* **65**, 794–802.

Chaput, N., Louafi, S., Bardier, A., Charlotte, F., Vaillant, J. C., Menegaux, F., Rosenzwajg, M., Lemoine, F., Klatzmann, D., and Taieb, J. (2009). Identification of CD8+CD25+Foxp3+ suppressive T cells in colorectal cancer tissue. *Gut* **58**, 520–529.

Chatila, T. A. (2005). Role of regulatory T cells in human diseases. *J. Allergy Clin. Immunol.* **116**, 949–959; quiz 960.

Chatila, T. A., Blaeser, F., Ho, N., Lederman, H. M., Voulgaropoulos, C., Helms, C., and Bowcock, A. M. (2000). JM2, encoding a fork head-related protein, is mutated in X-linked autoimmunity-allergic disregulation syndrome. *J. Clin. Invest.* **106**, R75–R81.

Chauveau, C., Remy, S., Royer, P. J., Hill, M., Tanguy-Royer, S., Hubert, F. X., Tesson, L., Brion, R., Beriou, G., Gregoire, M., Josien, R., Cuturi, M. C., *et al.* (2005). Heme oxygenase-1 expression inhibits dendritic cell maturation and proinflammatory function but conserves IL-10 expression. *Blood* **106**, 1694–1702.

Chen, Y., Kuchroo, V. K., Inobe, J., Hafler, D. A., and Weiner, H. L. (1994). Regulatory T cell clones induced by oral tolerance: suppression of autoimmune encephalomyelitis. *Science* **265**, 1237–1240.

Chen, W., Jin, W., Hardegen, N., Lei, K. J., Li, L., Marinos, N., McGrady, G., and Wahl, S. M. (2003). Conversion of peripheral CD4+CD25- naive T cells to CD4+CD25+ regulatory T cells by TGF-beta induction of transcription factor Foxp3. *J. Exp. Med.* **198**, 1875–1886.

Chen, M. L., Pittet, M. J., Gorelik, L., Flavell, R. A., Weissleder, R., von Boehmer, H., and Khazaie, K. (2005). Regulatory T cells suppress tumor-specific CD8 T cell cytotoxicity through TGF-beta signals in vivo. *Proc. Natl. Acad. Sci. USA* **102**, 419–424.

Chen, W., Liang, X., Peterson, A. J., Munn, D. H., and Blazar, B. R. (2008). The indoleamine 2,3-dioxygenase pathway is essential for human plasmacytoid dendritic cell-induced adaptive T regulatory cell generation. *J. Immunol.* **181**, 5396–5404.

Chung, D. J., Rossi, M., Romano, E., Ghith, J., Yuan, J., Munn, D. H., and Young, J. W. (2009). Indoleamine 2,3-dioxygenase-expressing mature human monocyte-derived dendritic cells expand potent autologous regulatory T cells. *Blood* **114**, 555–563.

Clayton, A., Mitchell, J. P., Court, J., Mason, M. D., and Tabi, Z. (2007). Human tumor-derived exosomes selectively impair lymphocyte responses to interleukin-2. *Cancer Res.* **67**, 7458–7466.

Cobbold, S. P., Castejon, R., Adams, E., Zelenika, D., Graca, L., Humm, S., and Waldmann, H. (2004). Induction of foxP3+ regulatory T cells in the periphery of T cell receptor transgenic mice tolerized to transplants. *J. Immunol.* **172**, 6003–6010.

Cohen, A. C., Nadeau, K. C., Tu, W., Hwa, V., Dionis, K., Bezrodnik, L., Teper, A., Gaillard, M., Heinrich, J., Krensky, A. M., Rosenfeld, R. G., and Lewis, D. B. (2006). Cutting edge: Decreased accumulation and regulatory function of CD4+ CD25(high) T cells in human STAT5b deficiency. *J. Immunol.* **177**, 2770–2774.

Collison, L. W., Workman, C. J., Kuo, T. T., Boyd, K., Wang, Y., Vignali, K. M., Cross, R., Sehy, D., Blumberg, R. S., and Vignali, D. A. (2007). The inhibitory cytokine IL-35 contributes to regulatory T-cell function. *Nature* **450**, 566–569.

Colotta, F., Allavena, P., Sica, A., Garlanda, C., and Mantovani, A. (2009). Cancer-related inflammation, the seventh hallmark of cancer: links to genetic instability. *Carcinogenesis* **30**, 1073–1081.

Coombes, J. L., Siddiqui, K. R., Arancibia-Carcamo, C. V., Hall, J., Sun, C. M., Belkaid, Y., and Powrie, F. (2007). A functionally specialized population of mucosal CD103+ DCs induces Foxp3+ regulatory T cells via a TGF-beta and retinoic acid-dependent mechanism. *J. Exp. Med.* **204**, 1757–1764.

Cortesini, R., LeMaoult, J., Ciubotariu, R., and Cortesini, N. S. (2001). CD8+CD28- T suppressor cells and the induction of antigen-specific, antigen-presenting cell-mediated suppression of Th reactivity. *Immunol. Rev.* **182**, 201–206.

Cosmi, L., Liotta, F., Lazzeri, E., Francalanci, M., Angeli, R., Mazzinghi, B., Santarlasci, V., Manetti, R., Vanini, V., Romagnani, P., Maggi, E., Romagnani, S., *et al.* (2003). Human CD8+CD25+ thymocytes share phenotypic and functional features with CD4+CD25+ regulatory thymocytes. *Blood* **102,** 4107–4114.

Cosmi, L., Liotta, F., Angeli, R., Mazzinghi, B., Santarlasci, V., Manetti, R., Lasagni, L., Vanini, V., Romagnani, P., Maggi, E., Annunziato, F., and Romagnani, S. (2004). Th2 cells are less susceptible than Th1 cells to the suppressive activity of CD25+ regulatory thymocytes because of their responsiveness to different cytokines. *Blood* **103,** 3117–3121.

Cupedo, T., Nagasawa, M., Weijer, K., Blom, B., and Spits, H. (2005). Development and activation of regulatory T cells in the human fetus. *Eur. J. Immunol.* **35,** 383–390.

Curiel, T. J., Wei, S., Dong, H., Alvarez, X., Cheng, P., Mottram, P., Krzysiek, R., Knutson, K. L., Daniel, B., Zimmermann, M. C., David, O., Burow, M., *et al.* (2003). Blockade of B7-H1 improves myeloid dendritic cell-mediated antitumor immunity. *Nat. Med.* **9,** 562–567.

Curiel, T. J., Coukos, G., Zou, L., Alvarez, X., Cheng, P., Mottram, P., Evdemon-Hogan, M., Conejo-Garcia, J. R., Zhang, L., Burow, M., Zhu, Y., Wei, S., *et al.* (2004). Specific recruitment of regulatory T cells in ovarian carcinoma fosters immune privilege and predicts reduced survival. *Nat. Med.* **10,** 942–949.

Curotto de Lafaille, M. A., and Lafaille, J. J. (2009). Natural and adaptive foxp3+ regulatory T cells: more of the same or a division of labor? *Immunity* **30,** 626–635.

Curotto de Lafaille, M. A., Kutchukhidze, N., Shen, S., Ding, Y., Yee, H., and Lafaille, J. J. (2008). Adaptive Foxp3+ regulatory T cell-dependent and -independent control of allergic inflammation. *Immunity* **29,** 114–126.

Curti, A., Pandolfi, S., Valzasina, B., Aluigi, M., Isidori, A., Ferri, E., Salvestrini, V., Bonanno, G., Rutella, S., Durelli, I., Horenstein, A. L., Fiore, F., *et al.* (2007). Modulation of tryptophan catabolism by human leukemic cells results in the conversion of CD25- into CD25+ T regulatory cells. *Blood* **109,** 2871–2877.

Dang, N. H., Hagemeister, F. B., Pro, B., McLaughlin, P., Romaguera, J. E., Jones, D., Samuels, B., Samaniego, F., Younes, A., Wang, M., Goy, A., Rodriguez, M. A., *et al.* (2004). Phase II study of denileukin diftitox for relapsed/refractory B-Cell non-Hodgkin's lymphoma. *J. Clin. Oncol.* **22,** 4095–4102.

Dannull, J., Su, Z., Rizzieri, D., Yang, B. K., Coleman, D., Yancey, D., Zhang, A., Dahm, P., Chao, N., Gilboa, E., and Vieweg, J. (2005). Enhancement of vaccine-mediated antitumor immunity in cancer patients after depletion of regulatory T cells. *J. Clin. Invest.* **115,** 3623–3633.

Dejaco, C., Duftner, C., Grubeck-Loebenstein, B., and Schirmer, M. (2006). Imbalance of regulatory T cells in human autoimmune diseases. *Immunology* **117,** 289–300.

Dhodapkar, M. V., Steinman, R. M., Krasovsky, J., Munz, C., and Bhardwaj, N. (2001). Antigen-specific inhibition of effector T cell function in humans after injection of immature dendritic cells. *J. Exp. Med.* **193,** 233–238.

Dranoff, G. (2005). CTLA-4 blockade: unveiling immune regulation. *J. Clin. Oncol.* **23,** 662–664.

Dudley, M. E., Wunderlich, J. R., Robbins, P. F., Yang, J. C., Hwu, P., Schwartzentruber, D. J., Topalian, S. L., Sherry, R., Restifo, N. P., Hubicki, A. M., Robinson, M. R., Raffeld, M., *et al.* (2002). Cancer regression and autoimmunity in patients after clonal repopulation with antitumor lymphocytes. *Science* **298,** 850–854.

Dudley, M. E., Wunderlich, J. R., Yang, J. C., Sherry, R. M., Topalian, S. L., Restifo, N. P., Royal, R. E., Kammula, U., White, D. E., Mavroukakis, S. A., Rogers, L. J., Gracia, G. J., *et al.* (2005). Adoptive cell transfer therapy following non-myeloablative but lymphodepleting chemotherapy for the treatment of patients with refractory metastatic melanoma. *J. Clin. Oncol.* **23,** 2346–2357.

Dudley, M. E., Yang, J. C., Sherry, R., Hughes, M. S., Royal, R., Kammula, U., Robbins, P. F., Huang, J., Citrin, D. E., Leitman, S. F., Wunderlich, J., Restifo, N. P., *et al.* (2008). Adoptive cell therapy for patients with metastatic melanoma: evaluation of intensive myeloablative chemoradiation preparative regimens. *J. Clin. Oncol.* **26,** 5233–5239.

Dunn, G. P., Bruce, A. T., Ikeda, H., Old, L. J., and Schreiber, R. D. (2002). Cancer immunoediting: from immunosurveillance to tumor escape. *Nat. Immunol.* **3,** 991–998.

Dye, E. S., and North, R. J. (1981). T cell-mediated immunosuppression as an obstacle to adoptive immunotherapy of the P815 mastocytoma and its metastases. *J. Exp. Med.* **154,** 1033–1042.

Egen, J. G., Kuhns, M. S., and Allison, J. P. (2002). CTLA-4: new insights into its biological function and use in tumor immunotherapy. *Nat. Immunol.* **3,** 611–618.

Elpek, K. G., Lacelle, C., Singh, N. P., Yolcu, E. S., and Shirwan, H. (2007). CD4+CD25+ T regulatory cells dominate multiple immune evasion mechanisms in early but not late phases of tumor development in a B cell lymphoma model. *J. Immunol.* **178,** 6840–6848.

Ercolini, A. M., Ladle, B. H., Manning, E. A., Pfannenstiel, L. W., Armstrong, T. D., Machiels, J. P., Bieler, J. G., Emens, L. A., Reilly, R. T., and Jaffee, E. M. (2005). Recruitment of latent pools of high-avidity CD8(+) T cells to the antitumor immune response. *J. Exp. Med.* **201,** 1591–1602.

Erdman, S. E., Poutahidis, T., Tomczak, M., Rogers, A. B., Cormier, K., Plank, B., Horwitz, B. H., and Fox, J. G. (2003). CD4+ CD25+ regulatory T lymphocytes inhibit microbially induced colon cancer in Rag2-deficient mice. *Am. J. Pathol.* **162,** 691–702.

Erdman, S. E., Sohn, J. J., Rao, V. P., Nambiar, P. R., Ge, Z., Fox, J. G., and Schauer, D. B. (2005). CD4+CD25+ regulatory lymphocytes induce regression of intestinal tumors in ApcMin/+ mice. *Cancer Res.* **65,** 3998–4004.

Erdman, S. E., Rao, V. P., Olipitz, W., Taylor, C. L., Jackson, E. A., Levkovich, T., Lee, C. W., Horwitz, B. H., Fox, J. G., Ge, Z., and Poutahidis, T. (2009). Unifying roles for regulatory T cells and inflammation in cancer. *Int. J. Cancer* **126,** 1651–1665.

Ermann, J., and Fathman, C. G. (2003). Costimulatory signals controlling regulatory T cells. *Proc. Natl. Acad. Sci. USA* **100,** 15292–15293.

Fallarino, F., Grohmann, U., Hwang, K. W., Orabona, C., Vacca, C., Bianchi, R., Belladonna, M. L., Fioretti, M. C., Alegre, M. L., and Puccetti, P. (2003). Modulation of tryptophan catabolism by regulatory T cells. *Nat. Immunol.* **4,** 1206–1212.

Fallarino, F., Grohmann, U., You, S., McGrath, B. C., Cavener, D. R., Vacca, C., Orabona, C., Bianchi, R., Belladonna, M. L., Volpi, C., Santamaria, P., Fioretti, M. C., *et al.* (2006). The combined effects of tryptophan starvation and tryptophan catabolites down-regulate T cell receptor zeta-chain and induce a regulatory phenotype in naive T cells. *J. Immunol.* **176,** 6752–6761.

Fantini, M. C., Becker, C., Monteleone, G., Pallone, F., Galle, P. R., and Neurath, M. F. (2004). Cutting edge: TGF-beta induces a regulatory phenotype in CD4+CD25- T cells through Foxp3 induction and down-regulation of Smad7. *J. Immunol.* **172,** 5149–5153.

Figgitt, D. P., Lamb, H. M., and Goa, K. L. (2000). Denileukin diftitox. *Am. J. Clin. Dermatol.* **1,** 67–72; discussion 73.

Filaci, G., and Suciu-Foca, N. (2002). CD8+ T suppressor cells are back to the game: are they players in autoimmunity? *Autoimmun. Rev.* **1,** 279–283.

Filaci, G., Fenoglio, D., Fravega, M., Ansaldo, G., Borgonovo, G., Traverso, P., Villaggio, B., Ferrera, A., Kunkl, A., Rizzi, M., Ferrera, F., Balestra, P., *et al.* (2007). CD8+ CD28- T regulatory lymphocytes inhibiting T cell proliferative and cytotoxic functions infiltrate human cancers. *J. Immunol.* **179,** 4323–4334.

Filipazzi, P., Valenti, R., Huber, V., Pilla, L., Canese, P., Iero, M., Castelli, C., Mariani, L., Parmiani, G., and Rivoltini, L. (2007). Identification of a new subset of myeloid suppressor cells in peripheral blood of melanoma patients with modulation by a granulocyte-macrophage colony-stimulation factor-based antitumor vaccine. *J. Clin. Oncol.* **25,** 2546–2553.

Fontenot, J. D., Gavin, M. A., and Rudensky, A. Y. (2003). Foxp3 programs the development and function of CD4+CD25+ regulatory T cells. *Nat. Immunol.* **4,** 330–336.

Fontenot, J. D., Dooley, J. L., Farr, A. G., and Rudensky, A. Y. (2005a). Developmental regulation of Foxp3 expression during ontogeny. *J. Exp. Med.* **202,** 901–906.

Fontenot, J. D., Rasmussen, J. P., Williams, L. M., Dooley, J. L., Farr, A. G., and Rudensky, A. Y. (2005b). Regulatory T cell lineage specification by the forkhead transcription factor foxp3. *Immunity* **22,** 329–341.

Foss, F., Demierre, M. F., and DiVenuti, G. (2005). A phase-1 trial of bexarotene and denileukin diftitox in patients with relapsed or refractory cutaneous T-cell lymphoma. *Blood* **106,** 454–457.

Foussat, A., Cottrez, F., Brun, V., Fournier, N., Breittmayer, J. P., and Groux, H. (2003). A comparative study between T regulatory type 1 and CD4+CD25+ T cells in the control of inflammation. *J. Immunol.* **171,** 5018–5026.

Francois, V., Ottaviani, S., Renkvist, N., Stockis, J., Schuler, G., Thielemans, K., Colau, D., Marchand, M., Boon, T., Lucas, S., and van der Bruggen, P. (2009). The CD4(+) T-cell response of melanoma patients to a MAGE-A3 peptide vaccine involves potential regulatory T cells. *Cancer Res.* **69,** 4335–4345.

Frankel, A. E., Surendranathan, A., Black, J. H., White, A., Ganjoo, K., and Cripe, L. D. (2006). Phase II clinical studies of denileukin diftitox diphtheria toxin fusion protein in patients with previously treated chronic lymphocytic leukemia. *Cancer* **106,** 2158–2164.

Fu, J., Xu, D., Liu, Z., Shi, M., Zhao, P., Fu, B., Zhang, Z., Yang, H., Zhang, H., Zhou, C., Yao, J., Jin, L., *et al.* (2007). Increased regulatory T cells correlate with CD8 T-cell impairment and poor survival in hepatocellular carcinoma patients. *Gastroenterology* **132,** 2328–2339.

Fujimoto, S., Greene, M., and Sehon, A. H. (1975). Immunosuppressor T cells in tumor bearing host. *Immunol. Commun.* **4,** 201–217.

Furtado, G. C., Curotto de Lafaille, M. A., Kutchukhidze, N., and Lafaille, J. J. (2002). Interleukin 2 signaling is required for CD4(+) regulatory T cell function. *J. Exp. Med.* **196,** 851–857.

Gao, Q., Qiu, S. J., Fan, J., Zhou, J., Wang, X. Y., Xiao, Y. S., Xu, Y., Li, Y. W., and Tang, Z. Y. (2007). Intratumoral balance of regulatory and cytotoxic T cells is associated with prognosis of hepatocellular carcinoma after resection. *J. Clin. Oncol.* **25,** 2586–2593.

Gao, Y. W., Chen, Y. X., Wang, Z. M., Jin, J., Li, X. Y., Zhou le, D., Huo, Z., Zhou, J. H., and Chen, W. (2009). Increased expression of cyclooxygenase-2 and increased infiltration of regulatory T cells in tumors of patients with hepatocellular carcinoma. *Digestion* **79,** 169–176.

Garin, M. I., Chu, C. C., Golshayan, D., Cernuda-Morollon, E., Wait, R., and Lechler, R. I. (2007). Galectin-1: a key effector of regulation mediated by CD4+CD25+ T cells. *Blood* **109,** 2058–2065.

Gavin, M. A., Torgerson, T. R., Houston, E., DeRoos, P., Ho, W. Y., Stray-Pedersen, A., Ocheltree, E. L., Greenberg, P. D., Ochs, H. D., and Rudensky, A. Y. (2006). Single-cell analysis of normal and FOXP3-mutant human T cells: FOXP3 expression without regulatory T cell development. *Proc. Natl. Acad. Sci. USA* **103,** 6659–6664.

Generali, D., Bates, G., Berruti, A., Brizzi, M. P., Campo, L., Bonardi, S., Bersiga, A., Allevi, G., Milani, M., Aguggini, S., Dogliotti, L., Banham, A. H., *et al.* (2009). Immunomodulation of FOXP3+ regulatory T cells by the aromatase inhibitor letrozole in breast cancer patients. *Clin. Cancer Res.* **15,** 1046–1051.

Gershon, R. K., and Kondo, K. (1970). Cell interactions in the induction of tolerance: the role of thymic lymphocytes. *Immunology* **18,** 723–737.

Gershon, R. K., and Kondo, K. (1971). Infectious immunological tolerance. *Immunology* **21,** 903–914.

Ghiringhelli, F., Menard, C., Terme, M., Flament, C., Taieb, J., Chaput, N., Puig, P. E., Novault, S., Escudier, B., Vivier, E., Lecesne, A., Robert, C., *et al.* (2005a). CD4+CD25+ regulatory T cells inhibit natural killer cell functions in a transforming growth factor-beta-dependent manner. *J. Exp. Med.* **202,** 1075–1085.

Ghiringhelli, F., Puig, P. E., Roux, S., Parcellier, A., Schmitt, E., Solary, E., Kroemer, G., Martin, F., Chauffert, B., and Zitvogel, L. (2005b). Tumor cells convert immature myeloid dendritic cells into TGF-beta-secreting cells inducing CD4+CD25+ regulatory T cell proliferation. *J. Exp. Med.* **202,** 919–929.

Ghiringhelli, F., Menard, C., Puig, P. E., Ladoire, S., Roux, S., Martin, F., Solary, E., Le Cesne, A., Zitvogel, L., and Chauffert, B. (2007). Metronomic cyclophosphamide regimen selectively depletes CD4+CD25+ regulatory T cells and restores T and NK effector functions in end stage cancer patients. *Cancer Immunol. Immunother.* **56,** 641–648.

Gjerdrum, L. M., Woetmann, A., Odum, N., Burton, C. M., Rossen, K., Skovgaard, G. L., Ryder, L. P., and Ralfkiaer, E. (2007). FOXP3+ regulatory T cells in cutaneous T-cell lymphomas: association with disease stage and survival. *Leukemia* **21,** 2512–2518.

Gnjatic, S., Altorki, N. K., Tang, D. N., Tu, S. M., Kundra, V., Ritter, G., Old, L. J., Logothetis, C. J., and Sharma, P. (2009). NY-ESO-1 DNA vaccine induces T-cell responses that are suppressed by regulatory T cells. *Clin. Cancer Res.* **15,** 2130–2139.

Gobert, M., Treilleux, I., Bendriss-Vermare, N., Bachelot, T., Goddard-Leon, S., Arfi, V., Biota, C., Doffin, A. C., Durand, I., Olive, D., Perez, S., Pasqual, N., *et al.* (2009). Regulatory T cells recruited through CCL22/CCR4 are selectively activated in lymphoid infiltrates surrounding primary breast tumors and lead to an adverse clinical outcome. *Cancer Res.* **69,** 2000–2009.

Gondek, D. C., Lu, L. F., Quezada, S. A., Sakaguchi, S., and Noelle, R. J. (2005). Cutting edge: contact-mediated suppression by CD4+CD25+ regulatory cells involves a granzyme B-dependent, perforin-independent mechanism. *J. Immunol.* **174,** 1783–1786.

Gounaris, E., Blatner, N. R., Dennis, K., Magnusson, F., Gurish, M. F., Strom, T. B., Beckhove, P., Gounari, F., and Khazaie, K. (2009). T-regulatory cells shift from a protective anti-inflammatory to a cancer-promoting proinflammatory phenotype in polyposis. *Cancer Res.* **69,** 5490–5497.

Gray, C. P., Arosio, P., and Hersey, P. (2003). Association of increased levels of heavy-chain ferritin with increased CD4+ CD25+ regulatory T-cell levels in patients with melanoma. *Clin. Cancer Res.* **9,** 2551–2559.

Griffiths, R. W., Elkord, E., Gilham, D. E., Ramani, V., Clarke, N., Stern, P. L., and Hawkins, R. E. (2007). Frequency of regulatory T cells in renal cell carcinoma patients and investigation of correlation with survival. *Cancer Immunol. Immunother.* **56,** 1743–1753.

Gringhuis, S. I., Leow, A., Papendrecht-Van Der Voort, E. A., Remans, P. H., Breedveld, F. C., and Verweij, C. L. (2000). Displacement of linker for activation of T cells from the plasma membrane due to redox balance alterations results in hyporesponsiveness of synovial fluid T lymphocytes in rheumatoid arthritis. *J. Immunol.* **164,** 2170–2179.

Grossman, W. J., Verbsky, J. W., Barchet, W., Colonna, M., Atkinson, J. P., and Ley, T. J. (2004). Human T regulatory cells can use the perforin pathway to cause autologous target cell death. *Immunity* **21,** 589–601.

Groux, H., O'Garra, A., Bigler, M., Rouleau, M., Antonenko, S., de Vries, J. E., and Roncarolo, M. G. (1997). A CD4+ T-cell subset inhibits antigen-specific T-cell responses and prevents colitis. *Nature* **389,** 737–742.

Haas, J., Schopp, L., Storch-Hagenlocher, B., Fritzsching, B., Jacobi, C., Milkova, L., Fritz, B., Schwarz, A., Suri-Payer, E., Hensel, M., and Wildemann, B. (2008). Specific recruitment of regulatory T cells into the CSF in lymphomatous and carcinomatous meningitis. *Blood* **111,** 761–766.

Haas, M., Dimmler, A., Hohenberger, W., Grabenbauer, G. G., Niedobitek, G., and Distel, L. V. (2009). Stromal regulatory T-cells are associated with a favourable prognosis in gastric cancer of the cardia. *BMC Gastroenterol.* **9,** 65.

Hamdi, W., Ogawara, H., Handa, H., Tsukamoto, N., Nojima, Y., and Murakami, H. (2009). Clinical significance of regulatory T cells in patients with myelodysplastic syndrome. *Eur. J. Haematol.* **82,** 201–207.

Han, G. M., O'Neil-Andersen, N. J., Zurier, R. B., and Lawrence, D. A. (2008). CD4+CD25high T cell numbers are enriched in the peripheral blood of patients with rheumatoid arthritis. *Cell. Immunol.* **253,** 92–101.

Hawrylowicz, C. M., and O'Garra, A. (2005). Potential role of interleukin-10-secreting regulatory T cells in allergy and asthma. *Nat. Rev. Immunol.* **5,** 271–283.

Hiraoka, N., Onozato, K., Kosuge, T., and Hirohashi, S. (2006). Prevalence of FOXP3+ regulatory T cells increases during the progression of pancreatic ductal adenocarcinoma and its premalignant lesions. *Clin. Cancer Res.* **12,** 5423–5434.

Hodi, F. S., Butler, M., Oble, D. A., Seiden, M. V., Haluska, F. G., Kruse, A., Macrae, S., Nelson, M., Canning, C., Lowy, I., Korman, A., Lautz, D., *et al.* (2008). Immunologic and clinical effects of antibody blockade of cytotoxic T lymphocyte-associated antigen 4 in previously vaccinated cancer patients. *Proc. Natl. Acad. Sci. USA* **105,** 3005–3010.

Hoechst, B., Ormandy, L. A., Ballmaier, M., Lehner, F., Kruger, C., Manns, M. P., Greten, T. F., and Korangy, F. (2008). A new population of myeloid-derived suppressor cells in hepatocellular carcinoma patients induces CD4(+)CD25(+)Foxp3(+) T cells. *Gastroenterology* **135,** 234–243.

Hori, S., Nomura, T., and Sakaguchi, S. (2003). Control of regulatory T cell development by the transcription factor Foxp3. *Science* **299,** 1057–1061.

Huang, C. T., Workman, C. J., Flies, D., Pan, X., Marson, A. L., Zhou, G., Hipkiss, E. L., Ravi, S., Kowalski, J., Levitsky, H. I., Powell, J. D., Pardoll, D. M., *et al.* (2004). Role of LAG-3 in regulatory T cells. *Immunity* **21,** 503–513.

Huber, S., Schramm, C., Lehr, H. A., Mann, A., Schmitt, S., Becker, C., Protschka, M., Galle, P. R., Neurath, M. F., and Blessing, M. (2004). Cutting edge: TGF-beta signaling is required for the in vivo expansion and immunosuppressive capacity of regulatory CD4+CD25+ T cells. *J. Immunol.* **173,** 6526–6531.

Huber, S., Stahl, F. R., Schrader, J., Luth, S., Presser, K., Carambia, A., Flavell, R. A., Werner, S., Blessing, M., Herkel, J., and Schramm, C. (2009). Activin a promotes the TGF-beta-induced conversion of CD4+CD25- T cells into Foxp3+ induced regulatory T cells. *J. Immunol.* **182,** 4633–4640.

Huehn, J., and Hamann, A. (2005). Homing to suppress: address codes for Treg migration. *Trends Immunol.* **26,** 632–636.

Hus, I., Schmitt, M., Tabarkiewicz, J., Radej, S., Wojas, K., Bojarska-Junak, A., Schmitt, A., Giannopoulos, K., Dmoszynska, A., and Rolinski, J. (2008). Vaccination of B-CLL patients with autologous dendritic cells can change the frequency of leukemia antigen-specific CD8+ T cells as well as CD4+CD25+FoxP3+ regulatory T cells toward an antileukemia response. *Leukemia* **22,** 1007–1017.

Ichihara, F., Kono, K., Takahashi, A., Kawaida, H., Sugai, H., and Fujii, H. (2003). Increased populations of regulatory T cells in peripheral blood and tumor-infiltrating lymphocytes in patients with gastric and esophageal cancers. *Clin. Cancer Res.* **9,** 4404–4408.

Ishida, T., Ishii, T., Inagaki, A., Yano, H., Komatsu, H., Iida, S., Inagaki, H., and Ueda, R. (2006). Specific recruitment of CC chemokine receptor 4-positive regulatory T cells in Hodgkin lymphoma fosters immune privilege. *Cancer Res.* **66,** 5716–5722.

Iwasaki, A., and Medzhitov, R. (2004). Toll-like receptor control of the adaptive immune responses. *Nat. Immunol.* **5,** 987–995.

Jaafar, F., Righi, E., Lindstrom, V., Linton, C., Nohadani, M., Van Noorden, S., Lloyd, T., Poznansky, J., Stamp, G., Dina, R., Coleman, D. V., and Poznansky, M. C. (2009). Correlation of CXCL12 expression and FoxP3+ cell infiltration with human papillomavirus infection and clinicopathological progression of cervical cancer. *Am. J. Pathol.* **175,** 1525–1535.

Jak, M., Mous, R., Remmerswaal, E. B., Spijker, R., Jaspers, A., Yague, A., Eldering, E., Van Lier, R. A., and Van Oers, M. H. (2009). Enhanced formation and survival of CD4+ CD25hi Foxp3+ T-cells in chronic lymphocytic leukemia. *Leuk. Lymphoma* **50,** 788–801.

Javia, L. R., and Rosenberg, S. A. (2003). CD4+CD25+ suppressor lymphocytes in the circulation of patients immunized against melanoma antigens. *J. Immunother.* **26,** 85–93.

Jensen, H. K., Donskov, F., Nordsmark, M., Marcussen, N., and von der Maase, H. (2009). Increased intratumoral FOXP3-positive regulatory immune cells during interleukin-2 treatment in metastatic renal cell carcinoma. *Clin. Cancer Res.* **15,** 1052–1058.

Jeron, A., Pfoertner, S., Bruder, D., Geffers, R., Hammerer, P., Hofmann, R., Buer, J., and Schrader, A. J. (2009). Frequency and gene expression profile of regulatory T cells in renal cell carcinoma. *Tumour Biol.* **30,** 160–170.

Jiang, H., and Chess, L. (2006). Regulation of immune responses by T cells. *N. Engl. J. Med.* **354,** 1166–1176.

Joosten, S. A., van Meijgaarden, K. E., Savage, N. D., de Boer, T., Triebel, F., van der Wal, A., de Heer, E., Klein, M. R., Geluk, A., and Ottenhoff, T. H. (2007). Identification of a human CD8+ regulatory T cell subset that mediates suppression through the chemokine CC chemokine ligand 4. *Proc. Natl. Acad. Sci. USA* **104,** 8029–8034.

Jordan, M. S., Boesteanu, A., Reed, A. J., Petrone, A. L., Holenbeck, A. E., Lerman, M. A., Naji, A., and Caton, A. J. (2001). Thymic selection of CD4+CD25+ regulatory T cells induced by an agonist self-peptide. *Nat. Immunol.* **2,** 301–306.

Jordanova, E. S., Gorter, A., Ayachi, O., Prins, F., Durrant, L. G., Kenter, G. G., van der Burg, S. H., and Fleuren, G. J. (2008). Human leukocyte antigen class I, MHC class I chain-related molecule A, and CD8+/regulatory T-cell ratio: which variable determines survival of cervical cancer patients? *Clin. Cancer Res.* **14,** 2028–2035.

Juszczynski, P., Ouyang, J., Monti, S., Rodig, S. J., Takeyama, K., Abramson, J., Chen, W., Kutok, J. L., Rabinovich, G. A., and Shipp, M. A. (2007). The AP1-dependent secretion of galectin-1 by Reed Sternberg cells fosters immune privilege in classical Hodgkin lymphoma. *Proc. Natl. Acad. Sci. USA* **104,** 13134–13139.

Kanamaru, F., Youngnak, P., Hashiguchi, M., Nishioka, T., Takahashi, T., Sakaguchi, S., Ishikawa, I., and Azuma, M. (2004). Costimulation via glucocorticoid-induced TNF receptor in both conventional and CD25+ regulatory CD4+ T cells. *J. Immunol.* **172,** 7306–7314.

Kawaida, H., Kono, K., Takahashi, A., Sugai, H., Mimura, K., Miyagawa, N., Omata, H., Ooi, A., and Fujii, H. (2005). Distribution of CD4+CD25high regulatory T-cells in tumor-draining lymph nodes in patients with gastric cancer. *J. Surg. Res.* **124,** 151–157.

Kemper, C., Chan, A. C., Green, J. M., Brett, K. A., Murphy, K. M., and Atkinson, J. P. (2003). Activation of human CD4+ cells with CD3 and CD46 induces a T-regulatory cell 1 phenotype. *Nature* **421,** 388–392.

Kiniwa, Y., Miyahara, Y., Wang, H. Y., Peng, W., Peng, G., Wheeler, T. M., Thompson, T. C., Old, L. J., and Wang, R. F. (2007). CD8+ Foxp3+ regulatory T cells mediate immunosuppression in prostate cancer. *Clin. Cancer Res.* **13,** 6947–6958.

Klemke, C. D., Fritzsching, B., Franz, B., Kleinmann, E. V., Oberle, N., Poenitz, N., Sykora, J., Banham, A. H., Roncador, G., Kuhn, A., Goerdt, S., Krammer, P. H., *et al.* (2006). Paucity of FOXP3+ cells in skin and peripheral blood distinguishes Sezary syndrome from other cutaneous T-cell lymphomas. *Leukemia* **20,** 1123–1129.

Knutson, K. L., Dang, Y., Lu, H., Lukas, J., Almand, B., Gad, E., Azeke, E., and Disis, M. L. (2006). IL-2 immunotoxin therapy modulates tumor-associated regulatory T cells and leads to lasting immune-mediated rejection of breast cancers in neu-transgenic mice. *J. Immunol.* **177,** 84–91.

Ko, K., Yamazaki, S., Nakamura, K., Nishioka, T., Hirota, K., Yamaguchi, T., Shimizu, J., Nomura, T., Chiba, T., and Sakaguchi, S. (2005). Treatment of advanced tumors with agonistic anti-GITR mAb and its effects on tumor-infiltrating Foxp3+CD25+CD4+ regulatory T cells. *J. Exp. Med.* **202,** 885–891.

Kobayashi, N., Hiraoka, N., Yamagami, W., Ojima, H., Kanai, Y., Kosuge, T., Nakajima, A., and Hirohashi, S. (2007). FOXP3+ regulatory T cells affect the development and progression of hepatocarcinogenesis. *Clin. Cancer Res.* **13,** 902–911.

Kohm, A. P., McMahon, J. S., Podojil, J. R., Begolka, W. S., DeGutes, M., Kasprowicz, D. J., Ziegler, S. F., and Miller, S. D. (2006). Cutting Edge: Anti-CD25 monoclonal antibody injection results in the functional inactivation, not depletion, of CD4+CD25+ T regulatory cells. *J. Immunol.* **176,** 3301–3305.

Kono, K., Kawaida, H., Takahashi, A., Sugai, H., Mimura, K., Miyagawa, N., Omata, H., and Fujii, H. (2006). CD4(+)CD25high regulatory T cells increase with tumor stage in patients with gastric and esophageal cancers. *Cancer Immunol. Immunother.* **55,** 1064–1071.

Kordasti, S. Y., Ingram, W., Hayden, J., Darling, D., Barber, L., Afzali, B., Lombardi, G., Wlodarski, M. W., Maciejewski, J. P., Farzaneh, F., and Mufti, G. J. (2007). CD4+CD25high Foxp3+ regulatory T cells in myelodysplastic syndrome (MDS). *Blood* **110,** 847–850.

Korn, T., Mitsdoerffer, M., Croxford, A. L., Awasthi, A., Dardalhon, V. A., Galileos, G., Vollmar, P., Stritesky, G. L., Kaplan, M. H., Waisman, A., Kuchroo, V. K., and Oukka, M. (2008). IL-6 controls Th17 immunity in vivo by inhibiting the conversion of conventional T cells into Foxp3+ regulatory T cells. *Proc. Natl. Acad. Sci. USA* **105,** 18460–18465.

Kretschmer, K., Apostolou, I., Hawiger, D., Khazaie, K., Nussenzweig, M. C., and von Boehmer, H. (2005). Inducing and expanding regulatory T cell populations by foreign antigen. *Nat. Immunol.* **6,** 1219–1227.

Krupnick, A. S., Gelman, A. E., Barchet, W., Richardson, S., Kreisel, F. H., Turka, L. A., Colonna, M., Patterson, G. A., and Kreisel, D. (2005). Murine vascular endothelium activates and induces the generation of allogeneic CD4+25+Foxp3+ regulatory T cells. *J. Immunol.* **175,** 6265–6270.

Kryczek, I., Lange, A., Mottram, P., Alvarez, X., Cheng, P., Hogan, M., Moons, L., Wei, S., Zou, L., Machelon, V., Emilie, D., Terrassa, M., *et al.* (2005). CXCL12 and vascular endothelial growth factor synergistically induce neoangiogenesis in human ovarian cancers. *Cancer Res.* **65,** 465–472.

Kryczek, I., Wei, S., Zhu, G., Myers, L., Mottram, P., Cheng, P., Chen, L., Coukos, G., and Zou, W. (2007). Relationship between B7-H4, regulatory T cells, and patient outcome in human ovarian carcinoma. *Cancer Res.* **67,** 8900–8905.

Kryczek, I., Liu, R., Wang, G., Wu, K., Shu, X., Szeliga, W., Vatan, L., Finlayson, E., Huang, E., Simeone, D., Redman, B., Welling, T. H., *et al.* (2009). FOXP3 defines regulatory T cells in human tumor and autoimmune disease. *Cancer Res.* **69,** 3995–4000.

Kumar, V. (2004). Homeostatic control of immunity by TCR peptide-specific Tregs. *J. Clin. Invest.* **114,** 1222–1226.

Kusmartsev, S., Nefedova, Y., Yoder, D., and Gabrilovich, D. I. (2004). Antigen-specific inhibition of CD8+ T cell response by immature myeloid cells in cancer is mediated by reactive oxygen species. *J. Immunol.* **172,** 989–999.

Ladoire, S., Arnould, L., Apetoh, L., Coudert, B., Martin, F., Chauffert, B., Fumoleau, P., and Ghiringhelli, F. (2008). Pathologic complete response to neoadjuvant chemotherapy of breast carcinoma is associated with the disappearance of tumor-infiltrating foxp3+ regulatory T cells. *Clin. Cancer Res.* **14,** 2413–2420.

Larkin, J., III, Rankin, A. L., Picca, C. C., Riley, M. P., Jenks, S. A., Sant, A. J., and Caton, A. J. (2008). CD4+CD25+ regulatory T cell repertoire formation shaped by differential presentation of peptides from a self-antigen. *J. Immunol.* **180,** 2149–2157.

Lee, N. R., Song, E. K., Jang, K. Y., Choi, H. N., Moon, W. S., Kwon, K., Lee, J. H., Yim, C. Y., and Kwak, J. Y. (2008). Prognostic impact of tumor infiltrating FOXP3 positive regulatory T cells in diffuse large B-cell lymphoma at diagnosis. *Leuk. Lymphoma* **49,** 247–256.

Lee, S. Y., Choi, H. K., Lee, K. J., Jung, J. Y., Hur, G. Y., Jung, K. H., Kim, J. H., Shin, C., Shim, J. J., In, K. H., Kang, K. H., and Yoo, S. H. (2009). The immune tolerance of cancer is mediated by IDO that is inhibited by COX-2 inhibitors through regulatory T cells. *J. Immunother.* **32,** 22–28.

Leffers, N., Gooden, M. J., de Jong, R. A., Hoogeboom, B. N., ten Hoor, K. A., Hollema, H., Boezen, H. M., van der Zee, A. G., Daemen, T., and Nijman, H. W. (2009). Prognostic significance of tumor-infiltrating T-lymphocytes in primary and metastatic lesions of advanced stage ovarian cancer. *Cancer Immunol. Immunother.* **58,** 449–459.

Lehe, C., Ghebeh, H., Al-Sulaiman, A., Al Qudaihi, G., Al-Hussein, K., Almohareb, F., Chaudhri, N., Alsharif, F., Al-Zahrani, H., Tbakhi, A., Aljurf, M., and Dermime, S. (2008). The Wilms' tumor antigen is a novel target for human CD4+ regulatory T cells: implications for immunotherapy. *Cancer Res.* **68,** 6350–6359.

Leong, P. P., Mohammad, R., Ibrahim, N., Ithnin, H., Abdullah, M., Davis, W. C., and Seow, H. F. (2006). Phenotyping of lymphocytes expressing regulatory and effector markers in infiltrating ductal carcinoma of the breast. *Immunol. Lett.* **102,** 229–236.

Levings, M. K., Sangregorio, R., and Roncarolo, M. G. (2001). Human cd25(+)cd4(+) t regulatory cells suppress naive and memory T cell proliferation and can be expanded in vitro without loss of function. *J. Exp. Med.* **193,** 1295–1302.

Levings, M. K., Sangregorio, R., Sartirana, C., Moschin, A. L., Battaglia, M., Orban, P. C., and Roncarolo, M. G. (2002). Human CD25+CD4+ T suppressor cell clones produce transforming growth factor beta, but not interleukin 10, and are distinct from type 1 T regulatory cells. *J. Exp. Med.* **196,** 1335–1346.

Levings, M. K., Gregori, S., Tresoldi, E., Cazzaniga, S., Bonini, C., and Roncarolo, M. G. (2005). Differentiation of Tr1 cells by immature dendritic cells requires IL-10 but not CD25+CD4+ Tr cells. *Blood* **105,** 1162–1169.

Li, B., Lalani, A. S., Harding, T. C., Luan, B., Koprivnikar, K., Huan Tu, G., Prell, R., VanRoey, M. J., Simmons, A. D., and Jooss, K. (2006). Vascular endothelial growth factor blockade reduces intratumoral regulatory T cells and enhances the efficacy of a GM-CSF-secreting cancer immunotherapy. *Clin. Cancer Res.* **12,** 6808–6816.

Li, X., Ye, F., Chen, H., Lu, W., Wan, X., and Xie, X. (2007). Human ovarian carcinoma cells generate CD4(+)CD25(+) regulatory T cells from peripheral CD4(+)CD25(-) T cells through secreting TGF-beta. *Cancer Lett.* **253,** 144–153.

Li, W., Lidebjer, C., Yuan, X. M., Szymanowski, A., Backteman, K., Ernerudh, J., Leanderson, P., Nilsson, L., Swahn, E., and Jonasson, L. (2008). NK cell apoptosis in coronary artery disease: relation to oxidative stress. *Atherosclerosis* **199,** 65–72.

Li, J. F., Chu, Y. W., Wang, G. M., Zhu, T. Y., Rong, R. M., Hou, J., and Xu, M. (2009). The prognostic value of peritumoral regulatory T cells and its correlation with intratumoral cyclooxygenase-2 expression in clear cell renal cell carcinoma. *BJU Int.* **103,** 399–405.

Liang, S., Alard, P., Zhao, Y., Parnell, S., Clark, S. L., and Kosiewicz, M. M. (2005). Conversion of CD4+ CD25- cells into CD4+ CD25+ regulatory T cells in vivo requires B7 costimulation, but not the thymus. *J. Exp. Med.* **201,** 127–137.

Liang, B., Workman, C., Lee, J., Chew, C., Dale, B. M., Colonna, L., Flores, M., Li, N., Schweighoffer, E., Greenberg, S., Tybulewicz, V., Vignali, D., *et al.* (2008). Regulatory T cells inhibit dendritic cells by lymphocyte activation gene-3 engagement of MHC class II. *J. Immunol.* **180,** 5916–5926.

Lim, H. W., Hillsamer, P., Banham, A. H., and Kim, C. H. (2005). Cutting edge: direct suppression of B cells by CD4+ CD25+ regulatory T cells. *J. Immunol.* **175,** 4180–4183.

Ling, K. L., Pratap, S. E., Bates, G. J., Singh, B., Mortensen, N. J., George, B. D., Warren, B. F., Piris, J., Roncador, G., Fox, S. B., Banham, A. H., and Cerundolo, V. (2007). Increased frequency of regulatory T cells in peripheral blood and tumour infiltrating lymphocytes in colorectal cancer patients. *Cancer Immun.* **7,** 7.

Liston, A., Lesage, S., Wilson, J., Peltonen, L., and Goodnow, C. C. (2003). Aire regulates negative selection of organ-specific T cells. *Nat. Immunol.* **4,** 350–354.

Liu, H., Komai-Koma, M., Xu, D., and Liew, F. Y. (2006a). Toll-like receptor 2 signaling modulates the functions of CD4+ CD25+ regulatory T cells. *Proc. Natl. Acad. Sci. USA* **103,** 7048–7053.

Liu, W., Putnam, A. L., Xu-Yu, Z., Szot, G. L., Lee, M. R., Zhu, S., Gottlieb, P. A., Kapranov, P., Gingeras, T. R., de St, Fazekas, Groth, B., Clayberger, C., *et al.* (2006b). CD127 expression inversely correlates with FoxP3 and suppressive function of human CD4+ T reg cells. *J. Exp. Med.* **203,** 1701–1711.

Liu, V. C., Wong, L. Y., Jang, T., Shah, A. H., Park, I., Yang, X., Zhang, Q., Lonning, S., Teicher, B. A., and Lee, C. (2007). Tumor evasion of the immune system by converting CD4+CD25- T cells into CD4+CD25+ T regulatory cells: role of tumor-derived TGF-beta. *J. Immunol.* **178,** 2883–2892.

Liu, Y., Zhang, P., Li, J., Kulkarni, A. B., Perruche, S., and Chen, W. (2008). A critical function for TGF-beta signaling in the development of natural CD4+CD25+Foxp3+ regulatory T cells. *Nat. Immunol.* **9,** 632–640.

Liyanage, U. K., Moore, T. T., Joo, H. G., Tanaka, Y., Herrmann, V., Doherty, G., Drebin, J. A., Strasberg, S. M., Eberlein, T. J., Goedegebuure, P. S., and Linehan, D. C. (2002). Prevalence of regulatory T cells is increased in peripheral blood and tumor microenvironment of patients with pancreas or breast adenocarcinoma. *J. Immunol.* **169,** 2756–2761.

Lohr, J., Knoechel, B., and Abbas, A. K. (2006). Regulatory T cells in the periphery. *Immunol. Rev.* **212,** 149–162.

Lopes, J. E., Soper, D. M., and Ziegler, S. F. (2007). Foxp3 is required throughout the life of a regulatory T cell. *Sci. STKE 2007* pe36.

Lopez, M. N., Pereda, C., Segal, G., Munoz, L., Aguilera, R., Gonzalez, F. E., Escobar, A., Ginesta, A., Reyes, D., Gonzalez, R., Mendoza-Naranjo, A., Larrondo, M., *et al.* (2009). Prolonged survival of dendritic cell-vaccinated melanoma patients correlates with tumor-specific delayed type IV hypersensitivity response and reduction of tumor growth factor beta-expressing T cells. *J. Clin. Oncol.* **27,** 945–952.

Lundqvist, A., Palmborg, A., Pavlenko, M., Levitskaya, J., and Pisa, P. (2005). Mature dendritic cells induce tumor-specific type 1 regulatory T cells. *J. Immunother.* **28,** 229–235.

Lutsiak, M. E., Semnani, R. T., De Pascalis, R., Kashmiri, S. V., Schlom, J., and Sabzevari, H. (2005). Inhibition of CD4(+)25+ T regulatory cell function implicated in enhanced immune response by low-dose cyclophosphamide. *Blood* **105,** 2862–2868.

Mahic, M., Yaqub, S., Johansson, C. C., Tasken, K., and Aandahl, E. M. (2006). FOXP3+CD4+CD25+ adaptive regulatory T cells express cyclooxygenase-2 and suppress effector T cells by a prostaglandin E2-dependent mechanism. *J. Immunol.* **177,** 246–254.

Mahnke, K., Ring, S., Johnson, T. S., Schallenberg, S., Schonfeld, K., Storn, V., Bedke, T., and Enk, A. H. (2007a). Induction of immunosuppressive functions of dendritic cells in vivo by CD4+CD25+ regulatory T cells: role of B7-H3 expression and antigen presentation. *Eur. J. Immunol.* **37,** 2117–2126.

Mahnke, K., Schonfeld, K., Fondel, S., Ring, S., Karakhanova, S., Wiedemeyer, K., Bedke, T., Johnson, T. S., Storn, V., Schallenberg, S., and Enk, A. H. (2007b). Depletion of CD4+CD25+ human regulatory T cells in vivo: kinetics of Treg depletion and alterations in immune functions in vivo and in vitro. *Int. J. Cancer* **120,** 2723–2733.

Maker, A. V., Attia, P., and Rosenberg, S. A. (2005). Analysis of the cellular mechanism of antitumor responses and autoimmunity in patients treated with CTL A-4 blockade. *J. Immunol.* **175**, 7746–7754.

Malek, T. R., Yu, A., Vincek, V., Scibelli, P., and Kong, L. (2002). CD4 regulatory T cells prevent lethal autoimmunity in IL-2Rbeta-deficient mice. Implications for the nonredundant function of IL-2. *Immunity* **17**, 167–178.

Malmberg, K. J., Arulampalam, V., Ichihara, F., Petersson, M., Seki, K., Andersson, T., Lenkei, R., Masucci, G., Pettersson, S., and Kiessling, R. (2001). Inhibition of activated/memory (CD45RO(+)) T cells by oxidative stress associated with block of NF-kappaB activation. *J. Immunol.* **167**, 2595–2601.

Mantel, P. Y., Ouaked, N., Ruckert, B., Karagiannidis, C., Welz, R., Blaser, K., and Schmidt-Weber, C. B. (2006). Molecular mechanisms underlying FOXP3 induction in human T cells. *J. Immunol.* **176**, 3593–3602.

Marshall, N. A., Christie, L. E., Munro, L. R., Culligan, D. J., Johnston, P. W., Barker, R. N., and Vickers, M. A. (2004). Immunosuppressive regulatory T cells are abundant in the reactive lymphocytes of Hodgkin lymphoma. *Blood* **103**, 1755–1762.

McHugh, R. S., Whitters, M. J., Piccirillo, C. A., Young, D. A., Shevach, E. M., Collins, M., and Byrne, M. C. (2002). CD4(+)CD25(+) immunoregulatory T cells: gene expression analysis reveals a functional role for the glucocorticoid-induced TNF receptor. *Immunity* **16**, 311–323.

Mehrotra, S., Mougiakakos, D., Johansson, C. C., Voelkel-Johnson, C., and Kiessling, R. (2009). Oxidative stress and lymphocyte persistence: implications in immunotherapy. *Adv. Cancer Res.* **102**, 197–227.

Miller, J. F. (1961). Immunological function of the thymus. *Lancet* **2**, 748–749.

Miller, A. M., Lundberg, K., Ozenci, V., Banham, A. H., Hellstrom, M., Egevad, L., and Pisa, P. (2006). CD4+CD25high T cells are enriched in the tumor and peripheral blood of prostate cancer patients. *J. Immunol.* **177**, 7398–7405.

Mills, K. H. (2004). Regulatory T cells: friend or foe in immunity to infection? *Nat. Rev. Immunol.* **4**, 841–855.

Misra, N., Bayry, J., Lacroix-Desmazes, S., Kazatchkine, M. D., and Kaveri, S. V. (2004). Cutting edge: human CD4+CD25+ T cells restrain the maturation and antigen-presenting function of dendritic cells. *J. Immunol.* **172**, 4676–4680.

Mizukami, Y., Kono, K., Kawaguchi, Y., Akaike, H., Kamimura, K., Sugai, H., and Fujii, H. (2008a). CCL17 and CCL22 chemokines within tumor microenvironment are related to accumulation of Foxp3+ regulatory T cells in gastric cancer. *Int. J. Cancer* **122**, 2286–2293.

Mizukami, Y., Kono, K., Kawaguchi, Y., Akaike, H., Kamimura, K., Sugai, H., and Fujii, H. (2008b). Localisation pattern of Foxp3+ regulatory T cells is associated with clinical behaviour in gastric cancer. *Br. J. Cancer* **98**, 148–153.

Modigliani, Y., Bandeira, A., and Coutinho, A. (1996). A model for developmentally acquired thymus-dependent tolerance to central and peripheral antigens. *Immunol. Rev.* **149**, 155–174.

Moller, G. (1988). Do suppressor T cells exist? *Scand. J. Immunol.* **27**, 247–250.

Molling, J. W., de Gruijl, T. D., Glim, J., Moreno, M., Rozendaal, L., Meijer, C. J., van den Eertwegh, A. J., Scheper, R. J., von Blomberg, M. E., and Bontkes, H. J. (2007). CD4(+)CD25hi regulatory T-cell frequency correlates with persistence of human papillomavirus type 16 and T helper cell responses in patients with cervical intraepithelial neoplasia. *Int. J. Cancer* **121**, 1749–1755.

Morgan, D. A., Ruscetti, F. W., and Gallo, R. (1976). Selective in vitro growth of T lymphocytes from normal human bone marrows. *Science* **193**, 1007–1008.

Morse, M. A., Hobeika, A. C., Osada, T., Serra, D., Niedzwiecki, D., Lyerly, H. K., and Clay, T. M. (2008). Depletion of human regulatory T cells specifically enhances antigen-specific immune responses to cancer vaccines. *Blood* **112**, 610–618.

Mosier, D. E. (1967). A requirement for two cell types for antibody formation in vitro. *Science* **158**, 1573–1575.

Motta, M., Rassenti, L., Shelvin, B. J., Lerner, S., Kipps, T. J., Keating, M. J., and Wierda, W. G. (2005). Increased expression of CD152 (CTLA-4) by normal T lymphocytes in untreated patients with B-cell chronic lymphocytic leukemia. *Leukemia* **19**, 1788–1793.

Mougiakakos, D., Johansson, C. C., and Kiessling, R. (2009). Naturally occurring regulatory T cells show reduced sensitivity toward oxidative stress-induced cell death. *Blood* **113**, 3542–3545.

Mougiakakos, D., Johansson, C. C., Trocme, E., All-Ericsson, C., Economou, M. A., Larsson, O., Seregard, S., and Kiessling (2010). *Cancer* Epub, PMID: 20209608.

Mucida, D., Kutchukhidze, N., Erazo, A., Russo, M., Lafaille, J. J., and Curotto de Lafaille, M. A. (2005). Oral tolerance in the absence of naturally occurring Tregs. *J. Clin. Invest.* **115**, 1923–1933.

Munn, D. H., and Mellor, A. L. (2007). Indoleamine 2,3-dioxygenase and tumor-induced tolerance. *J. Clin. Invest.* **117**, 1147–1154.

Munn, D. H., Sharma, M. D., Hou, D., Baban, B., Lee, J. R., Antonia, S. J., Messina, J. L., Chandler, P., Koni, P. A., and Mellor, A. L. (2004). Expression of indoleamine 2,3-dioxygenase by plasmacytoid dendritic cells in tumor-draining lymph nodes. *J. Clin. Invest.* **114**, 280–290.

Nakamura, K., Kitani, A., Fuss, I., Pedersen, A., Harada, N., Nawata, H., and Strober, W. (2004). TGF-beta 1 plays an important role in the mechanism of CD4+CD25+ regulatory T cell activity in both humans and mice. *J. Immunol.* **172**, 834–842.

Nicholaou, T., Ebert, L. M., Davis, I. D., McArthur, G. A., Jackson, H., Dimopoulos, N., Tan, B., Maraskovsky, E., Miloradovic, L., Hopkins, W., Pan, L., Venhaus, R., *et al.* (2009). Regulatory T-cell-mediated attenuation of T-cell responses to the NY-ESO-1 ISCOMATRIX vaccine in patients with advanced malignant melanoma. *Clin. Cancer Res.* **15**, 2166–2173.

Nishikawa, H., Jager, E., Ritter, G., Old, L. J., and Gnjatic, S. (2005). CD4+ CD25+ regulatory T cells control the induction of antigen-specific CD4+ helper T cell responses in cancer patients. *Blood* **106**, 1008–1011.

Nocentini, G., and Riccardi, C. (2005). GITR: a multifaceted regulator of immunity belonging to the tumor necrosis factor receptor superfamily. *Eur. J. Immunol.* **35**, 1016–1022.

North, R. J. (1982). Cyclophosphamide-facilitated adoptive immunotherapy of an established tumor depends on elimination of tumor-induced suppressor T cells. *J. Exp. Med.* **155**, 1063–1074.

Olkhanud, P. B., Baatar, D., Bodogai, M., Hakim, F., Gress, R., Anderson, R. L., Deng, J., Xu, M., Briest, S., and Biragyn, A. (2009). Breast cancer lung metastasis requires expression of chemokine receptor CCR4 and regulatory T cells. *Cancer Res.* **69**, 5996–6004.

Olsen, E., Duvic, M., Frankel, A., Kim, Y., Martin, A., Vonderheid, E., Jegasothy, B., Wood, G., Gordon, M., Heald, P., Oseroff, A., Pinter-Brown, L., *et al.* (2001). Pivotal phase III trial of two dose levels of denileukin diftitox for the treatment of cutaneous T-cell lymphoma. *J. Clin. Oncol.* **19**, 376–388.

Onizuka, S., Tawara, I., Shimizu, J., Sakaguchi, S., Fujita, T., and Nakayama, E. (1999). Tumor rejection by in vivo administration of anti-CD25 (interleukin-2 receptor alpha) monoclonal antibody. *Cancer Res.* **59**, 3128–3133.

Ormandy, L. A., Hillemann, T., Wedemeyer, H., Manns, M. P., Greten, T. F., and Korangy, F. (2005). Increased populations of regulatory T cells in peripheral blood of patients with hepatocellular carcinoma. *Cancer Res.* **65**, 2457–2464.

Pandiyan, P., Zheng, L., Ishihara, S., Reed, J., and Lenardo, M. J. (2007). CD4+CD25+Foxp3+ regulatory T cells induce cytokine deprivation-mediated apoptosis of effector CD4+ T cells. *Nat. Immunol.* **8**, 1353–1362.

Peggs, K. S., Quezada, S. A., Chambers, C. A., Korman, A. J., and Allison, J. P. (2009). Blockade of CTLA-4 on both effector and regulatory T cell compartments contributes to the antitumor activity of anti-CTLA-4 antibodies. *J. Exp. Med.* **206**, 1717–1725.

Peng, G., Guo, Z., Kiniwa, Y., Voo, K. S., Peng, W., Fu, T., Wang, D. Y., Li, Y., Wang, H. Y., and Wang, R. F. (2005). Toll-like receptor 8-mediated reversal of CD4+ regulatory T cell function. *Science* **309,** 1380–1384.

Penhale, W. J., Farmer, A., McKenna, R. P., and Irvine, W. J. (1973). Spontaneous thyroiditis in thymectomized and irradiated Wistar rats. *Clin. Exp. Immunol.* **15,** 225–236.

Penna, G., Vulcano, M., Roncari, A., Facchetti, F., Sozzani, S., and Adorini, L. (2002). Cutting edge: differential chemokine production by myeloid and plasmacytoid dendritic cells. *J. Immunol.* **169,** 6673–6676.

Perrone, G., Ruffini, P. A., Catalano, V., Spino, C., Santini, D., Muretto, P., Spoto, C., Zingaretti, C., Sisti, V., Alessandroni, P., Giordani, P., Cicetti, A., *et al.* (2008). Intratumoural FOXP3-positive regulatory T cells are associated with adverse prognosis in radically resected gastric cancer. *Eur. J. Cancer* **44,** 1875–1882.

Petersen, R. P., Campa, M. J., Sperlazza, J., Conlon, D., Joshi, M. B., Harpole, D. H., Jr., and Patz, E. F., Jr. (2006). Tumor infiltrating Foxp3+ regulatory T-cells are associated with recurrence in pathologic stage I NSCLC patients. *Cancer* **107,** 2866–2872.

Phan, G. Q., Yang, J. C., Sherry, R. M., Hwu, P., Topalian, S. L., Schwartzentruber, D. J., Restifo, N. P., Haworth, L. R., Seipp, C. A., Freezer, L. J., Morton, K. E., Mavroukakis, S. A., *et al.* (2003). Cancer regression and autoimmunity induced by cytotoxic T lymphocyte-associated antigen 4 blockade in patients with metastatic melanoma. *Proc. Natl. Acad. Sci. USA* **100,** 8372–8377.

Piccirillo, C. A., and Shevach, E. M. (2001). Cutting edge: control of CD8+ T cell activation by CD4+CD25+ immunoregulatory cells. *J. Immunol.* **167,** 1137–1140.

Prabhala, R. H., Neri, P., Bae, J. E., Tassone, P., Shammas, M. A., Allam, C. K., Daley, J. F., Chauhan, D., Blanchard, E., Thatte, H. S., Anderson, K. C., and Munshi, N. C. (2006). Dysfunctional T regulatory cells in multiple myeloma. *Blood* **107,** 301–304.

Pulko, V., Liu, X., Krco, C. J., Harris, K. J., Frigola, X., Kwon, E. D., and Dong, H. (2009). TLR3-stimulated dendritic cells up-regulate B7-H1 expression and influence the magnitude of CD8 T cell responses to tumor vaccination. *J. Immunol.* **183,** 3634–3641.

Qin, X. J., Shi, H. Z., Deng, J. M., Liang, Q. L., Jiang, J., and Ye, Z. J. (2009). CCL22 recruits CD4-positive CD25-positive regulatory T cells into malignant pleural effusion. *Clin. Cancer Res.* **15,** 2231–2237.

Rech, A. J., and Vonderheide, R. H. (2009). Clinical use of anti-CD25 antibody daclizumab to enhance immune responses to tumor antigen vaccination by targeting regulatory T cells. *Ann. N. Y. Acad. Sci.* **1174,** 99–106.

Ren, X., Ye, F., Jiang, Z., Chu, Y., Xiong, S., and Wang, Y. (2007). Involvement of cellular death in TRAIL/DR5-dependent suppression induced by CD4(+)CD25(+) regulatory T cells. *Cell Death Differ.* **14,** 2076–2084.

Rifa'i, M., Kawamoto, Y., Nakashima, I., and Suzuki, H. (2004). Essential roles of CD8+CD122+ regulatory T cells in the maintenance of T cell homeostasis. *J. Exp. Med.* **200,** 1123–1134.

Roncador, G., Brown, P. J., Maestre, L., Hue, S., Martinez-Torrecuadrada, J. L., Ling, K. L., Pratap, S., Toms, C., Fox, B. C., Cerundolo, V., Powrie, F., and Banham, A. H. (2005). Analysis of FOXP3 protein expression in human CD4+CD25+ regulatory T cells at the single-cell level. *Eur. J. Immunol.* **35,** 1681–1691.

Roncarolo, M. G., and Gregori, S. (2008). Is FOXP3 a bona fide marker for human regulatory T cells? *Eur. J. Immunol.* **38,** 925–927.

Roncarolo, M. G., Gregori, S., Battaglia, M., Bacchetta, R., Fleischhauer, K., and Levings, M. K. (2006). Interleukin-10-secreting type 1 regulatory T cells in rodents and humans. *Immunol. Rev.* **212,** 28–50.

Sakaguchi, S. (2001). Policing the regulators. *Nat. Immunol.* **2,** 283–284.

Sakaguchi, S. (2004). Naturally arising CD4+ regulatory t cells for immunologic self-tolerance and negative control of immune responses. *Annu. Rev. Immunol.* **22,** 531–562.

Sakaguchi, S., Sakaguchi, N., Asano, M., Itoh, M., and Toda, M. (1995). Immunologic self-tolerance maintained by activated T cells expressing IL-2 receptor alpha-chains (CD25). Breakdown of a single mechanism of self-tolerance causes various autoimmune diseases. *J. Immunol.* **155,** 1151–1164.

Sakaguchi, S., Yamaguchi, T., Nomura, T., and Ono, M. (2008). Regulatory T cells and immune tolerance. *Cell* **133,** 775–787.

Salama, P., Phillips, M., Grieu, F., Morris, M., Zeps, N., Joseph, D., Platell, C., and Iacopetta, B. (2009). Tumor-infiltrating FOXP3+ T regulatory cells show strong prognostic significance in colorectal cancer. *J. Clin. Oncol.* **27,** 186–192.

Sasada, T., Kimura, M., Yoshida, Y., Kanai, M., and Takabayashi, A. (2003). CD4+CD25+ regulatory T cells in patients with gastrointestinal malignancies: possible involvement of regulatory T cells in disease progression. *Cancer* **98,** 1089–1099.

Sato, E., Olson, S. H., Ahn, J., Bundy, B., Nishikawa, H., Qian, F., Jungbluth, A. A., Frosina, D., Gnjatic, S., Ambrosone, C., Kepner, J., Odunsi, T., *et al.* (2005). Intraepithelial CD8+ tumor-infiltrating lymphocytes and a high CD8+/regulatory T cell ratio are associated with favorable prognosis in ovarian cancer. *Proc. Natl. Acad. Sci. USA* **102,** 18538–18543.

Schaefer, C., Kim, G. G., Albers, A., Hoermann, K., Myers, E. N., and Whiteside, T. L. (2005). Characteristics of CD4+CD25+ regulatory T cells in the peripheral circulation of patients with head and neck cancer. *Br. J. Cancer* **92,** 913–920.

Schioppa, T., Uranchimeg, B., Saccani, A., Biswas, S. K., Doni, A., Rapisarda, A., Bernasconi, S., Saccani, S., Nebuloni, M., Vago, L., Mantovani, A., Melillo, G., *et al.* (2003). Regulation of the chemokine receptor CXCR4 by hypoxia. *J. Exp. Med.* **198,** 1391–1402.

Schmielau, J., and Finn, O. J. (2001). Activated granulocytes and granulocyte-derived hydrogen peroxide are the underlying mechanism of suppression of t-cell function in advanced cancer patients. *Cancer Res.* **61,** 4756–4760.

Seddiki, N., Santner-Nanan, B., Martinson, J., Zaunders, J., Sasson, S., Landay, A., Solomon, M., Selby, W., Alexander, S. I., Nanan, R., Kelleher, A., and Fazekas de St Groth, B. (2006). Expression of interleukin (IL)-2 and IL-7 receptors discriminates between human regulatory and activated T cells. *J. Exp. Med.* **203,** 1693–1700.

Seo, N., Hayakawa, S., Takigawa, M., and Tokura, Y. (2001). Interleukin-10 expressed at early tumour sites induces subsequent generation of CD4(+) T-regulatory cells and systemic collapse of antitumour immunity. *Immunology* **103,** 449–457.

Setoguchi, R., Hori, S., Takahashi, T., and Sakaguchi, S. (2005). Homeostatic maintenance of natural Foxp3(+) CD25(+) CD4(+) regulatory T cells by interleukin (IL)-2 and induction of autoimmune disease by IL-2 neutralization. *J. Exp. Med.* **201,** 723–735.

Sharma, S., Yang, S. C., Zhu, L., Reckamp, K., Gardner, B., Baratelli, F., Huang, M., Batra, R. K., and Dubinett, S. M. (2005). Tumor cyclooxygenase-2/prostaglandin E2-dependent promotion of FOXP3 expression and CD4+ CD25+ T regulatory cell activities in lung cancer. *Cancer Res.* **65,** 5211–5220.

Shevach, E. M. (2004). Fatal attraction: tumors beckon regulatory T cells. *Nat. Med.* **10,** 900–901.

Shimizu, J., Yamazaki, S., and Sakaguchi, S. (1999). Induction of tumor immunity by removing CD25+CD4+ T cells: a common basis between tumor immunity and autoimmunity. *J. Immunol.* **163,** 5211–5218.

Shimizu, J., Yamazaki, S., Takahashi, T., Ishida, Y., and Sakaguchi, S. (2002). Stimulation of CD25(+)CD4(+) regulatory T cells through GITR breaks immunological self-tolerance. *Nat. Immunol.* **3,** 135–142.

Simpson, E. (2008). Special regulatory T-cell review: Regulation of immune responses–examining the role of T cells. *Immunology* **123,** 13–16.

Sojka, D. K., Bruniquel, D., Schwartz, R. H., and Singh, N. J. (2004). IL-2 secretion by CD4+ T cells in vivo is rapid, transient, and influenced by TCR-specific competition. *J. Immunol.* **172,** 6136–6143.

Somasundaram, R., Jacob, L., Swoboda, R., Caputo, L., Song, H., Basak, S., Monos, D., Peritt, D., Marincola, F., Cai, D., Birebent, B., Bloome, E., *et al.* (2002). Inhibition of cytolytic T lymphocyte proliferation by autologous CD4+/CD25+ regulatory T cells in a colorectal carcinoma patient is mediated by transforming growth factor-beta. *Cancer Res.* **62,** 5267–5272.

Stanzer, S., Dandachi, N., Balic, M., Resel, M., Samonigg, H., and Bauernhofer, T. (2008). Resistance to apoptosis and expansion of regulatory T cells in relation to the detection of circulating tumor cells in patients with metastatic epithelial cancer. *J. Clin. Immunol.* **28,** 107–114.

Stephens, G. L., McHugh, R. S., Whitters, M. J., Young, D. A., Luxenberg, D., Carreno, B. M., Collins, M., and Shevach, E. M. (2004). Engagement of glucocorticoid-induced TNFR family-related receptor on effector T cells by its ligand mediates resistance to suppression by CD4+CD25+ T cells. *J. Immunol.* **173,** 5008–5020.

Strauss, L., Bergmann, C., Szczepanski, M. J., Lang, S., Kirkwood, J. M., and Whiteside, T. L. (2008). Expression of ICOS on human melanoma-infiltrating CD4+CD25highFoxp3+ T regulatory cells: implications and impact on tumor-mediated immune suppression. *J. Immunol.* **180,** 2967–2980.

Suciu-Foca, N., and Cortesini, R. (2007). Central role of ILT3 in the T suppressor cell cascade. *Cell. Immunol.* **248,** 59–67.

Suciu-Foca, N., Manavalan, J. S., Scotto, L., Kim-Schulze, S., Galluzzo, S., Naiyer, A. J., Fan, J., Vlad, G., and Cortesini, R. (2005). Molecular characterization of allospecific T suppressor and tolerogenic dendritic cells: review. *Int. Immunopharmacol.* **5,** 7–11.

Sun, C. M., Hall, J. A., Blank, R. B., Bouladoux, N., Oukka, M., Mora, J. R., and Belkaid, Y. (2007). Small intestine lamina propria dendritic cells promote de novo generation of Foxp3 T reg cells via retinoic acid. *J. Exp. Med.* **204,** 1775–1785.

Sutmuller, R. P., den Brok, M. H., Kramer, M., Bennink, E. J., Toonen, L. W., Kullberg, B. J., Joosten, L. A., Akira, S., Netea, M. G., and Adema, G. J. (2006). Toll-like receptor 2 controls expansion and function of regulatory T cells. *J. Clin. Invest.* **116,** 485–494.

Sutmuller, R. P., van Duivenvoorde, L. M., van Elsas, A., Schumacher, T. N., Wildenberg, M. E., Allison, J. P., Toes, R. E., Offringa, R., and Melief, C. J. (2001). Synergism of cytotoxic T lymphocyte-associated antigen 4 blockade and depletion of CD25(+) regulatory T cells in antitumor therapy reveals alternative pathways for suppression of autoreactive cytotoxic T lymphocyte responses. *J. Exp. Med.* **194,** 823–832.

Suzuki, H., Kundig, T. M., Furlonger, C., Wakeham, A., Timms, E., Matsuyama, T., Schmits, R., Simard, J. J., Ohashi, P. S., Griesser, H., *et al.* (1995). Deregulated T cell activation and autoimmunity in mice lacking interleukin-2 receptor beta. *Science* **268,** 1472–1476.

Suzuki, H., Onishi, H., Wada, J., Yamasaki, A., Tanaka, H., Nakano, K., Morisaki, T., and Katano, M. (2009). Vascular endothelial growth factor receptor 2 (VEGFR2) is selectively expressed by FOXP3(high)CD4(+) regulatory T cells. *Eur. J. Immunol.* **40,** 197–203.

Szatrowski, T. P., and Nathan, C. F. (1991). Production of large amounts of hydrogen peroxide by human tumor cells. *Cancer Res.* **51,** 794–798.

Tai, X., Cowan, M., Feigenbaum, L., and Singer, A. (2005). CD28 costimulation of developing thymocytes induces Foxp3 expression and regulatory T cell differentiation independently of interleukin 2. *Nat. Immunol.* **6,** 152–162.

Taieb, J., Chaput, N., Schartz, N., Roux, S., Novault, S., Menard, C., Ghiringhelli, F., Terme, M., Carpentier, A. F., Darrasse-Jeze, G., Lemonnier, F., and Zitvogel, L. (2006). Chemoimmunotherapy of tumors: cyclophosphamide synergizes with exosome based vaccines. *J. Immunol.* **176,** 2722–2729.

Takahashi, T., Kuniyasu, Y., Toda, M., Sakaguchi, N., Itoh, M., Iwata, M., Shimizu, J., and Sakaguchi, S. (1998). Immunologic self-tolerance maintained by CD25+CD4+ naturally anergic and suppressive T cells: induction of autoimmune disease by breaking their anergic/suppressive state. *Int. Immunol.* **10,** 1969–1980.

Tan, M. C., Goedegebuure, P. S., Belt, B. A., Flaherty, B., Sankpal, N., Gillanders, W. E., Eberlein, T. J., Hsieh, C. S., and Linehan, D. C. (2009). Disruption of CCR5-dependent homing of regulatory T cells inhibits tumor growth in a murine model of pancreatic cancer. *J. Immunol.* **182,** 1746–1755.

Tanaka, H., Tanaka, J., Kjaergaard, J., and Shu, S. (2002). Depletion of CD4+ CD25+ regulatory cells augments the generation of specific immune T cells in tumor-draining lymph nodes. *J. Immunother.* **25,** 207–217.

Tang, H. L., and Cyster, J. G. (1999). Chemokine Up-regulation and activated T cell attraction by maturing dendritic cells. *Science* **284,** 819–822.

Tang, Q., Boden, E. K., Henriksen, K. J., Bour-Jordan, H., Bi, M., and Bluestone, J. A. (2004). Distinct roles of CTLA-4 and TGF-beta in CD4+CD25+ regulatory T cell function. *Eur. J. Immunol.* **34,** 2996–3005.

Thornton, A. M., and Shevach, E. M. (1998). CD4+CD25+ immunoregulatory T cells suppress polyclonal T cell activation in vitro by inhibiting interleukin 2 production. *J. Exp. Med.* **188,** 287–296.

Thornton, A. M., Piccirillo, C. A., and Shevach, E. M. (2004). Activation requirements for the induction of CD4+CD25+ T cell suppressor function. *Eur. J. Immunol.* **34,** 366–376.

Tiemessen, M. M., Mitchell, T. J., Hendry, L., Whittaker, S. J., Taams, L. S., and John, S. (2006). Lack of suppressive CD4+CD25+FOXP3+ T cells in advanced stages of primary cutaneous T-cell lymphoma. *J. Invest. Dermatol.* **126,** 2217–2223.

Tiemessen, M. M., Jagger, A. L., Evans, H. G., van Herwijnen, M. J., John, S., and Taams, L. S. (2007). CD4+CD25+Foxp3+ regulatory T cells induce alternative activation of human monocytes/macrophages. *Proc. Natl. Acad. Sci. USA* **104,** 19446–19451.

Toscano, M. A., Bianco, G. A., Ilarregui, J. M., Croci, D. O., Correale, J., Hernandez, J. D., Zwirner, N. W., Poirier, F., Riley, E. M., Baum, L. G., and Rabinovich, G. A. (2007). Differential glycosylation of TH1, TH2 and TH-17 effector cells selectively regulates susceptibility to cell death. *Nat. Immunol.* **8,** 825–834.

Turner, M. S., Kane, L. P., and Morel, P. A. (2009). Dominant role of antigen dose in CD4+Foxp3+ regulatory T cell induction and expansion. *J. Immunol.* **183,** 4895–4903.

Tzankov, A., Meier, C., Hirschmann, P., Went, P., Pileri, S. A., and Dirnhofer, S. (2008). Correlation of high numbers of intratumoral FOXP3+ regulatory T cells with improved survival in germinal center-like diffuse large B-cell lymphoma, follicular lymphoma and classical Hodgkin's lymphoma. *Haematologica* **93,** 193–200.

Vaclavkova, P., Cao, Y., Wu, L. K., Michalek, J., and Vitetta, E. S. (2006). A comparison of an anti-CD25 immunotoxin, Ontak and anti-CD25 microbeads for their ability to deplete alloreactive T cells in vitro. *Bone Marrow Transplant.* **37,** 559–567.

van der Burg, S. H., Piersma, S. J., de Jong, A., van der Hulst, J. M., Kwappenberg, K. M., van den Hende, M., Welters, M. J., Van Rood, J. J., Fleuren, G. J., Melief, C. J., Kenter, G. G., and Offringa, R. (2007). Association of cervical cancer with the presence of CD4+ regulatory T cells specific for human papillomavirus antigens. *Proc. Natl. Acad. Sci. USA* **104,** 12087–12092.

van der Vliet, H. J., Koon, H. B., Yue, S. C., Uzunparmak, B., Seery, V., Gavin, M. A., Rudensky, A. Y., Atkins, M. B., Balk, S. P., and Exley, M. A. (2007). Effects of the administration of high-dose interleukin-2 on immunoregulatory cell subsets in patients with advanced melanoma and renal cell cancer. *Clin. Cancer Res.* **13,** 2100–2108.

van Maren, W. W., Jacobs, J. F., de Vries, I. J., Nierkens, S., and Adema, G. J. (2008). Toll-like receptor signalling on Tregs: to suppress or not to suppress? *Immunology* **124,** 445–452.

van Santen, H. M., Benoist, C., and Mathis, D. (2004). Number of T reg cells that differentiate does not increase upon encounter of agonist ligand on thymic epithelial cells. *J. Exp. Med.* **200**, 1221–1230.

Vence, L., Palucka, A. K., Fay, J. W., Ito, T., Liu, Y. J., Banchereau, J., and Ueno, H. (2007). Circulating tumor antigen-specific regulatory T cells in patients with metastatic melanoma. *Proc. Natl. Acad. Sci. USA* **104**, 20884–20889.

Venet, F., Pachot, A., Debard, A. L., Bohe, J., Bienvenu, J., Lepape, A., Powell, W. S., and Monneret, G. (2006). Human CD4+CD25+ regulatory T lymphocytes inhibit lipopolysaccharide-induced monocyte survival through a Fas/Fas ligand-dependent mechanism. *J. Immunol.* **177**, 6540–6547.

Vignali, D. A., Collison, L. W., and Workman, C. J. (2008). How regulatory T cells work. *Nat. Rev. Immunol.* **8**, 523–532.

Viguier, M., Lemaitre, F., Verola, O., Cho, M. S., Gorochov, G., Dubertret, L., Bachelez, H., Kourilsky, P., and Ferradini, L. (2004). Foxp3 expressing CD4+CD25(high) regulatory T cells are overrepresented in human metastatic melanoma lymph nodes and inhibit the function of infiltrating T cells. *J. Immunol.* **173**, 1444–1453.

von Boehmer, H. (2005). Mechanisms of suppression by suppressor T cells. *Nat. Immunol.* **6**, 338–344.

Vukmanovic-Stejic, M., Zhang, Y., Cook, J. E., Fletcher, J. M., McQuaid, A., Masters, J. E., Rustin, M. H., Taams, L. S., Beverley, P. C., Macallan, D. C., and Akbar, A. N. (2006). Human CD4+ CD25hi Foxp3+ regulatory T cells are derived by rapid turnover of memory populations in vivo. *J. Clin. Invest.* **116**, 2423–2433.

Wada, J., Suzuki, H., Fuchino, R., Yamasaki, A., Nagai, S., Yanai, K., Koga, K., Nakamura, M., Tanaka, M., Morisaki, T., and Katano, M. (2009). The contribution of vascular endothelial growth factor to the induction of regulatory T-cells in malignant effusions. *Anticancer Res.* **29**, 881–888.

Walker, L. S. (2004). CD4+ CD25+ Treg: divide and rule? *Immunology* **111**, 129–137.

Walker, M. R., Kasprowicz, D. J., Gersuk, V. H., Benard, A., Van Landeghen, M., Buckner, J. H., and Ziegler, S. F. (2003). Induction of FoxP3 and acquisition of T regulatory activity by stimulated human CD4+CD25- T cells. *J. Clin. Invest.* **112**, 1437–1443.

Wang, H. Y., Lee, D. A., Peng, G., Guo, Z., Li, Y., Kiniwa, Y., Shevach, E. M., and Wang, R. F. (2004). Tumor-specific human CD4+ regulatory T cells and their ligands: implications for immunotherapy. *Immunity* **20**, 107–118.

Wang, H. Y., Peng, G., Guo, Z., Shevach, E. M., and Wang, R. F. (2005a). Recognition of a new ARTC1 peptide ligand uniquely expressed in tumor cells by antigen-specific CD4+ regulatory T cells. *J. Immunol.* **174**, 2661–2670.

Wang, X., Zheng, J., Liu, J., Yao, J., He, Y., Li, X., Yu, J., Yang, J., Liu, Z., and Huang, S. (2005b). Increased population of CD4(+)CD25(high), regulatory T cells with their higher apoptotic and proliferating status in peripheral blood of acute myeloid leukemia patients. *Eur. J. Haematol.* **75**, 468–476.

Wang, L., Pino-Lagos, K., de Vries, V. C., Guleria, I., Sayegh, M. H., and Noelle, R. J. (2008). Programmed death 1 ligand signaling regulates the generation of adaptive Foxp3+CD4+ regulatory T cells. *Proc. Natl. Acad. Sci. USA* **105**, 9331–9336.

Wang, W., Lau, R., Yu, D., Zhu, W., Korman, A., and Weber, J. (2009). PD1 blockade reverses the suppression of melanoma antigen-specific CTL by CD4+ CD25(Hi) regulatory T cells. *Int. Immunol.* **21**, 1065–1077.

Wei, S., Kryczek, I., Zou, L., Daniel, B., Cheng, P., Mottram, P., Curiel, T., Lange, A., and Zou, W. (2005). Plasmacytoid dendritic cells induce CD8+ regulatory T cells in human ovarian carcinoma. *Cancer Res.* **65**, 5020–5026.

Wei, S., Kryczek, I., and Zou, W. (2006). Regulatory T-cell compartmentalization and trafficking. *Blood* **108**, 426–431.

Wei, S., Kryczek, I., Edwards, R. P., Zou, L., Szeliga, W., Banerjee, M., Cost, M., Cheng, P., Chang, A., Redman, B., Herberman, R. B., and Zou, W. (2007). Interleukin-2 administration alters the CD4+FOXP3+ T-cell pool and tumor trafficking in patients with ovarian carcinoma. *Cancer Res.* **67,** 7487–7494.

Welters, M. J., Kenter, G. G., Piersma, S. J., Vloon, A. P., Lowik, M. J., Berends-van der Meer, D. M., Drijfhout, J. W., Valentijn, A. R., Wafelman, A. R., Oostendorp, J., Fleuren, G. J., Offringa, R., *et al.* (2008). Induction of tumor-specific CD4+ and CD8+ T-cell immunity in cervical cancer patients by a human papillomavirus type 16 E6 and E7 long peptides vaccine. *Clin. Cancer Res.* **14,** 178–187.

Wieckowski, E. U., Visus, C., Szajnik, M., Szczepanski, M. J., Storkus, W. J., and Whiteside, T. L. (2009). Tumor-derived microvesicles promote regulatory T cell expansion and induce apoptosis in tumor-reactive activated CD8+ T lymphocytes. *J. Immunol.* **183,** 3720–3730.

Wieczorek, G., Asemissen, A., Model, F., Turbachova, I., Floess, S., Liebenberg, V., Baron, U., Stauch, D., Kotsch, K., Pratschke, J., Hamann, A., Loddenkemper, C., *et al.* (2009). Quantitative DNA methylation analysis of FOXP3 as a new method for counting regulatory T cells in peripheral blood and solid tissue. *Cancer Res.* **69,** 599–608.

Wilcox, R. A., Feldman, A. L., Wada, D. A., Yang, Z. Z., Comfere, N. I., Dong, H., Kwon, E. D., Novak, A. J., Markovic, S. N., Pittelkow, M. R., Witzig, T. E., and Ansell, S. M. (2009). B7-H1 (PD-L1, CD274) suppresses host immunity in T-cell lymphoproliferative disorders. *Blood* **114,** 2149–2158.

Wolf, M., Schimpl, A., and Hunig, T. (2001). Control of T cell hyperactivation in IL-2-deficient mice by CD4(+)CD25(-) and CD4(+)CD25(+) T cells: evidence for two distinct regulatory mechanisms. *Eur. J. Immunol.* **31,** 1637–1645.

Wolf, A. M., Wolf, D., Steurer, M., Gastl, G., Gunsilius, E., and Grubeck-Loebenstein, B. (2003). Increase of regulatory T cells in the peripheral blood of cancer patients. *Clin. Cancer Res.* **9,** 606–612.

Wolf, D., Wolf, A. M., Rumpold, H., Fiegl, H., Zeimet, A. G., Muller-Holzner, E., Deibl, M., Gastl, G., Gunsilius, E., and Marth, C. (2005). The expression of the regulatory T cell-specific forkhead box transcription factor FoxP3 is associated with poor prognosis in ovarian cancer. *Clin. Cancer Res.* **11,** 8326–8331.

Wolf, D., Rumpold, H., Koppelstatter, C., Gastl, G. A., Steurer, M., Mayer, G., Gunsilius, E., Tilg, H., and Wolf, A. M. (2006). Telomere length of in vivo expanded CD4(+)CD25(+) regulatory T-cells is preserved in cancer patients. *Cancer Immunol. Immunother.* **55,** 1198–1208.

Woo, E. Y., Chu, C. S., Goletz, T. J., Schlienger, K., Yeh, H., Coukos, G., Rubin, S. C., Kaiser, L. R., and June, C. H. (2001). Regulatory CD4(+)CD25(+) T cells in tumors from patients with early-stage non-small cell lung cancer and late-stage ovarian cancer. *Cancer Res.* **61,** 4766–4772.

Woo, E. Y., Yeh, H., Chu, C. S., Schlienger, K., Carroll, R. G., Riley, J. L., Kaiser, L. R., and June, C. H. (2002). Cutting edge: Regulatory T cells from lung cancer patients directly inhibit autologous T cell proliferation. *J. Immunol.* **168,** 4272–4276.

Wu, Y., Borde, M., Heissmeyer, V., Feuerer, M., Lapan, A. D., Stroud, J. C., Bates, D. L., Guo, L., Han, A., Ziegler, S. F., Mathis, D., Benoist, C., *et al.* (2006). FOXP3 controls regulatory T cell function through cooperation with NFAT. *Cell* **126,** 375–387.

Xystrakis, E., Cavailles, P., Dejean, A. S., Cautain, B., Colacios, C., Lagrange, D., van de Gaar, M. J., Bernard, I., Gonzalez-Dunia, D., Damoiseaux, J., Fournie, G. J., and Saoudi, A. (2004a). Functional and genetic analysis of two CD8 T cell subsets defined by the level of CD45RC expression in the rat. *J. Immunol.* **173,** 3140–3147.

Xystrakis, E., Dejean, A. S., Bernard, I., Druet, P., Liblau, R., Gonzalez-Dunia, D., and Saoudi, A. (2004b). Identification of a novel natural regulatory CD8 T-cell subset and analysis of its mechanism of regulation. *Blood* **104,** 3294–3301.

Yamagiwa, S., Gray, J. D., Hashimoto, S., and Horwitz, D. A. (2001). A role for TGF-beta in the generation and expansion of CD4+CD25+ regulatory T cells from human peripheral blood. *J. Immunol.* **166,** 7282–7289.

Yan, Z., Garg, S. K., Kipnis, J., and Banerjee, R. (2009). Extracellular redox modulation by regulatory T cells. *Nat. Chem. Biol.* **5,** 721–723.

Yang, Y., Huang, C. T., Huang, X., and Pardoll, D. M. (2004). Persistent Toll-like receptor signals are required for reversal of regulatory T cell-mediated CD8 tolerance. *Nat. Immunol.* **5,** 508–515.

Yang, Z. Z., Novak, A. J., Stenson, M. J., Witzig, T. E., and Ansell, S. M. (2006a). Intratumoral CD4+CD25+ regulatory T-cell-mediated suppression of infiltrating CD4+ T cells in B-cell non-Hodgkin lymphoma. *Blood* **107,** 3639–3646.

Yang, Z. Z., Novak, A. J., Ziesmer, S. C., Witzig, T. E., and Ansell, S. M. (2006b). Attenuation of CD8(+) T-cell function by CD4(+)CD25(+) regulatory T cells in B-cell non-Hodgkin's lymphoma. *Cancer Res.* **66,** 10145–10152.

Yang, Z. Z., Novak, A. J., Ziesmer, S. C., Witzig, T. E., and Ansell, S. M. (2007). CD70+ non-Hodgkin lymphoma B cells induce Foxp3 expression and regulatory function in intratumoral CD4+CD25 T cells. *Blood* **110,** 2537–2544.

Yaqub, S., Henjum, K., Mahic, M., Jahnsen, F. L., Aandahl, E. M., Bjornbeth, B. A., and Tasken, K. (2008). Regulatory T cells in colorectal cancer patients suppress anti-tumor immune activity in a COX-2 dependent manner. *Cancer Immunol. Immunother.* **57,** 813–821.

Yates, J., Rovis, F., Mitchell, P., Afzali, B., Tsang, J. Y., Garin, M., Lechler, R. I., Lombardi, G., and Garden, O. A. (2007). The maintenance of human CD4+ CD25+ regulatory T cell function: IL-2, IL-4, IL-7 and IL-15 preserve optimal suppressive potency in vitro. *Int. Immunol.* **19,** 785–799.

Zenclussen, A. C. (2006). Regulatory T cells in pregnancy. *Springer Semin. Immunopathol.* **28,** 31–39.

Zhang, L., Bertucci, A. M., Ramsey-Goldman, R., Burt, R. K., and Datta, S. K. (2009). Regulatory T cell (Treg) subsets return in patients with refractory lupus following stem cell transplantation, and TGF-beta-producing CD8+ Treg cells are associated with immunological remission of lupus. *J. Immunol.* **183,** 6346–6358.

Zhao, D. M., Thornton, A. M., DiPaolo, R. J., and Shevach, E. M. (2006). Activated CD4+CD25+ T cells selectively kill B lymphocytes. *Blood* **107,** 3925–3932.

Zhou, G., Drake, C. G., and Levitsky, H. I. (2006). Amplification of tumor-specific regulatory T cells following therapeutic cancer vaccines. *Blood* **107,** 628–636.

Zhou, X., Bailey-Bucktrout, S., Jeker, L. T., and Bluestone, J. A. (2009). Plasticity of CD4(+) FoxP3(+) T cells. *Curr. Opin. Immunol.* **21,** 281–285.

Ziegler, S. F. (2007). FOXP3: not just for regulatory T cells anymore. *Eur. J. Immunol.* **37,** 21–23.

Zorn, E., Nelson, E. A., Mohseni, M., Porcheray, F., Kim, H., Litsa, D., Bellucci, R., Raderschall, E., Canning, C., Soiffer, R. J., Frank, D. A., and Ritz, J. (2006). IL-2 regulates FOXP3 expression in human CD4+CD25+ regulatory T cells through a STAT-dependent mechanism and induces the expansion of these cells in vivo. *Blood* **108,** 1571–1579.

Zou, W., Machelon, V., Coulomb-L'Hermin, A., Borvak, J., Nome, F., Isaeva, T., Wei, S., Krzysiek, R., Durand-Gasselin, I., Gordon, A., Pustilnik, T., Curiel, D. T., *et al.* (2001). Stromal-derived factor-1 in human tumors recruits and alters the function of plasmacytoid precursor dendritic cells. *Nat. Med.* **7,** 1339–1346.

Role of EBERs in the Pathogenesis of EBV Infection

Dai Iwakiri and Kenzo Takada

Department of Tumor Virology, Institute for Genetic Medicine, Hokkaido University, Sapporo, Japan

---

Epstein–Barr virus (EBV)-encoded small RNAs (EBERs) are noncoding RNAs that are expressed abundantly in latently EBV-infected cells. Previous studies demonstrated that EBERs (EBER1 and EBER2) play significant roles in various EBV-infected cancer cells. EBERs are responsible for malignant phenotypes of Burkitt's lymphoma (BL) cells including resistance to apoptosis. In addition, EBERs induce the expression of interleukin (IL)-10 in BL cells, insulin-like growth factor (IGF)-1 in gastric carcinoma and nasopharyngeal carcinoma cells, IL-9 in T cells that act as an autocrine growth factor. It was also reported that EBERs play critical roles in the B cell growth transformation including IL-6 induction by EBER2.

EBERs have been discovered to interact with cellular proteins that play a key role in antiviral innate immunity. They bind the protein kinase RNA-dependent (PKR) and inhibit its activation, leading to resistance to PKR-mediated apoptosis. Recently, it was demonstrated that EBERs bind RIG-I and activate its downstream signaling, which induces expression of type-I interferon (IFN)s. Furthermore, EBERs induce IL-10 through IRF3 but not NF-κB activation in BL cells, suggesting that modulation of innate immune signaling by EBERs contribute to EBV-mediated oncogenesis. Most recently, it was reported that EBERs are secreted from EBV-infected cells and are recognized by toll-like receptor (TLR)3, leading to induction of type-I IFNs and inflammatory cytokines, and subsequent immune activation. Furthermore, EBER1 could be detected in the sera of patients with active EBV infectious diseases, suggesting that activation of TLR3 signaling by EBER1 would be account for the pathogenesis of active EBV infectious diseases.

Advances in CANCER RESEARCH

0065-230X/10 $35.00
DOI: 10.1016/S0065-230X(10)07004-1

I. INTRODUCTION

Epstein–Barr virus (EBV) encodes small nonpolyadenylated, noncoding (nc) RNAs termed EBV-encoded small RNAs (EBERs). EBERs, EBER1 and EBER2 (Lerner *et al.*, 1981), are the most abundant viral transcripts in latently EBV-infected cells (Rymo, 1979), 167 and 172 nucleotides long, respectively, and transcribed by RNA polymerase III (pol III) (Rosa *et al.*, 1981). Because of their abundance, EBERs can be used as target molecules for detection of EBV-infected cells in tissues by *in situ* hybridization (ISH) (Chang *et al.*, 1992), and their existence is considered a reliable marker of the existence of EBV. Previous studies demonstrated the roles of EBERs in EBV-mediated oncogenesis. EBERs play a key role in the maintenance of malignant phenotypes of Burkitt's lymphoma (BL) cells (Komano *et al.*, 1999). In addition, they confer resistance to protein kinase RNA-dependent (PKR)-mediated apoptosis in BL and epithelial cells (Nanbo *et al.*, 2002, 2005). Furthermore, EBERs induce transcription of cytokines including interleukin (IL)-10 in BL cells, insulin-like growth factor (IGF)-1 in epithelial cells, and IL-9 in T cells that act as an autocrine growth factor of those EBV-infected cancer cells (Iwakiri *et al.*, 2003, 2005; Kitagawa *et al.*, 2000; Yang *et al.*, 2004). More recent studies reported that this ncRNA contributes to the pathogenesis of EBV infection through modulation of innate immune signals (Iwakiri *et al.*, 2009; Samanta *et al.*, 2006, 2008).

II. STRUCTURE OF EBERs

EBERs are encoded by the right-hand 1000 base pairs of the EcoRI J fragment of the EBV genome. EBER1 is 166 nucleotides long and EBER2 is 172 nucleotides long (Rosa *et al.*, 1981). The EBER genes are separated by 161 base pairs and are transcribed from left to right on the EBV map. Both EBER genes carry intragenic transcription control regions for RNA polymerase (pol) III, and can be transcribed by it. The primary sequence similarity between EBER1 and EBER2 is only 54%, but both EBER1 and EBER2 show striking similarity in their secondary structures with extensively base-paired structures containing a number of short stem loops (Fig. 1). Striking similarities in the secondary structures also exist between EBERs and adenovirus-associated RNAs (VAs) that are small nonpolyadenylated RNAs like EBERs (Rosa *et al.*, 1981).

The primary sequences of EBERs are strongly conserved among a number of EBV strains (Arrand *et al.*, 1989). Within 1 kilo base EBER region, 10 single base changes which group the strains into two families (1 and 2) have been

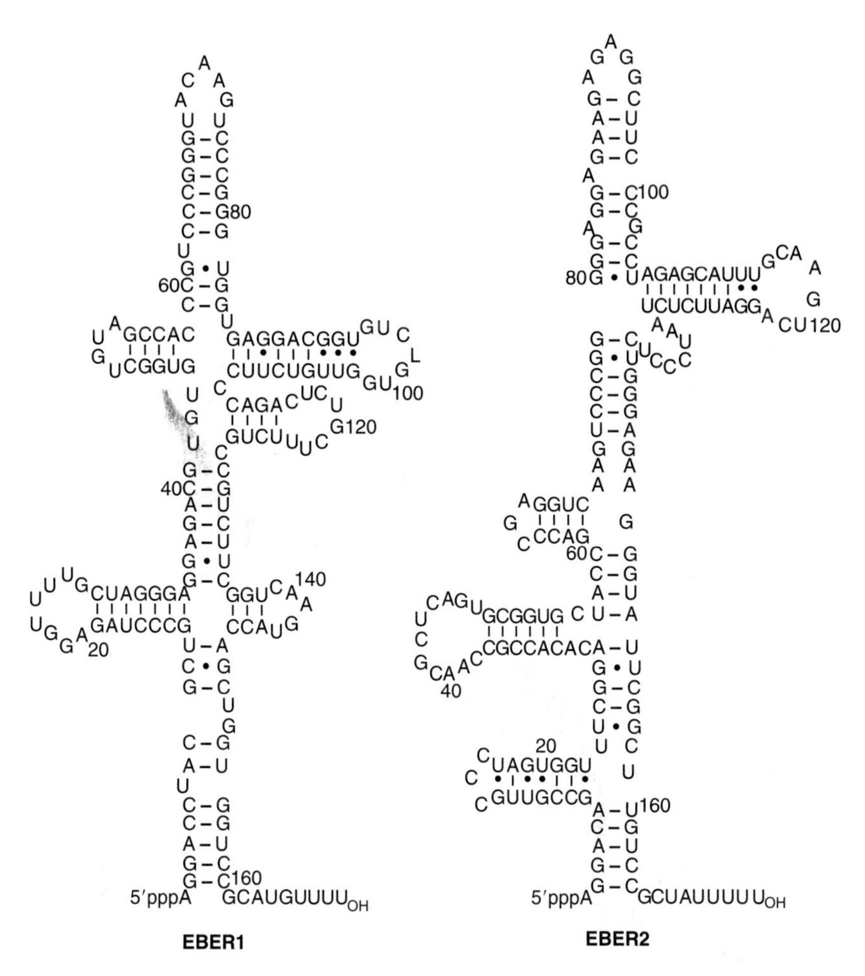

Fig. 1 Secondary structures of EBERs. Reproduced from Rosa *et al.* (1981).

identified. The EBER1 sequences are completely conserved, two base changes are within EBER2-coding sequence and eight are outside the coding regions. Family 1 and family 2 parallel type A and type B EBVs (also called type-I and type-II EBVs) as defined by variations in EBNA2 and EBNA3; however, some isolates appear to be intertypic recombinants, which are often observed in isolates from HIV patients (Yao *et al.*, 1996). The overall high conservation of the sequence suggests that EBERs are important for the virus life cycle.

III. TRANSCRIPTIONAL REGULATION AND EXPRESSION OF EBERs

EBERs are transcribed by RNA pol III (Rosa *et al.*, 1981). Class III promoters are characterized by their intragenic location and, in the case of EBERs, these sequences are nearly identical to the consensus sequences derived from boxes A and B. In addition, they contain three upstream elements that together stimulated *in vivo* expression 50-fold and contain a TATA box and ATF- and Sp1-like promoter elements, which resemble sites associated with typical class II promoters (Fig. 2, Howe and Shu, 1989). However, it is not known whether EBERs are transcribed by RNA polymerase II.

It appears that the copy number of EBERs is related to the copy number of EBV DNA molecules in each cell type (Lerner *et al.*, 1981). EBER plasmids that contain 10 tandem repeats of EBER genes give high number EBER expression in transected B lymphoma cells compared with EBER plasmid containing a single copy of the EBER gene (Komano *et al.*, 1999). Following EBV infection of primary B lymphocytes, EBV-determined nuclear antigen 2 (EBNA2) appears first at 6 h after infection, followed by other EBNAs, latent membrane proteins (LMPs) and EBERS. On the other hand, the nontransforming P3HR-1 strain, from which the EBNA2 gene is deleted, expresses only EBNA-leader protein (EBNA-LP) and trace amounts of EBER1 in primary B lymphocytes, while the same virus can express EBNA1, EBNA3, EBNA-LP, and EBERs in EBV-genome-negative BL cell lines (Rooney *et al.*, 1989). These findings suggest that EBER expression is dependent on the host cell, perhaps through products specific for the cell cycle or the state of B cell differentiation.

State of viral life cycle seems to influence EBER expression. Nuclear run-on assays showed downregulation of EBER transcription during the switch from latent infection to lytic replication of the virus (Greifenegger *et al.*, 1998). In contrast, the amounts of EBERs remain unaltered within 72 h after induction of lytic replication. Although both EBERs are transcribed at approximately equal rates, the steady-state level of EBER1 is 10-fold

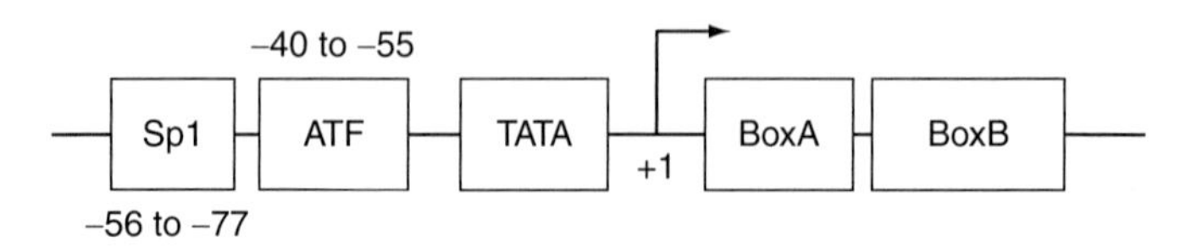

Fig. 2 Promoter structure of the Epstein–Barr virus-encoded small RNA (EBER)-2-gene, showing intragenic polymerase III control regions, box A and box B, and the upstream region containing TATA box and transcription factor ATF (−40 to −55), and Sp1 (−56 to −77) binding sites (Howe and Shu, 1989).

greater than that of EBER2. This is due to a much faster rate of turnover of EBER2. In the presence of actinomycin D, the half-lives of EBER1 and EBER2 are 8–9 h and 45 min, respectively (Clarke *et al.*, 1992).

EBERs are not expressed in the tissue of oral hairy leukoplakia, Sjogren's syndrome, salivary gland lymphoma, or oral papiloma, where EBV actively replicates (Gilligan *et al.*, 1990; Mizugaki *et al.*, 1998; Wen *et al.*, 1996, 1997). It is believed that EBERs are not expressed in permissively infected cells, but are consistently expressed in latently infected cells. Therefore, an ISH-targeted EBER1 has been extensively used as a reliable marker of EBV infection in tissue specimens (Chang *et al.*, 1992). On the other hand, previous studies have reported the evidences of EBV-positive but EBER-negative hepatocellular carcinoma and breast carcinoma (Bonnet *et al.*, 1999; Sugawara *et al.*, 1999). In addition, Yao *et al.* (2000) have reported heterogeneity of EBER expression in individual tumor cells of NPC; some cells express high levels of EBER, yet adjacent tumor cells express very little or none. However, in this case, it must be clarified whether EBER-negative cells are positive for the EBV genome. In conclusion, EBER-negative latent infection does exist, and by using the EBER ISH, a number of EBV-infected cells could remain undetected.

IV. LOCALIZATION OF EBERs AND THEIR INTERACTION WITH CELLULAR PROTEINS

EBERs are located in the nucleus as indicated by intense nuclear staining by EBER ISH (Chang *et al.*, 1992; Howe and Steiz, 1986). However, high-resolution ISH using confocal laser scanning microscopy has revealed that EBERs are found in cytoplasm as well as in the nuclei of interphase cells (Schwemmle *et al.*, 1992). The cytoplasmic staining is not homologous, with the perinuclear region being preferentially stained, which corresponds to the location of the rough ER and Golgi apparatus. Recent report demonstrated that EBERs are confined to the nucleus (Fok *et al.*, 2006a,b). On the other hand, Iwakiri *et al.* (2009) reported the new finding suggesting that EBERs are positively secreted in complex with La protein. In conclusion, EBERs should be localized not only to the nucleus but also to the cytoplasm, as EBERs are known to form complexes with a number of cytoplasmic proteins.

EBERs exist in nuclear ribonucleoprotein (RNP) complexes that are precipitated by anti-La antibodies associated with systemic lupus erythematosus (SLE) (Lerner *et al.*, 1981). In these complexes, La binds the oligouridylate stretch at the 3′-termini of all mammalian RNA pol III transcripts, transiently for most RNAs but stably in the case of the EBERs (Howe and Shu, 1988). Although the significance of their interaction is unknown, it is expected to affect the interaction between La and RNA pol III in EBV-infected cells. Most recent study suggests that EBER–La interaction is significant for secretion of EBER from EBV-infected cells since it was found that EBER is released mostly in complex with La (Iwakiri *et al.*, 2009).

VAs, VA1 and VA2, are small RNAs transcribed by RNA pol III (Akusjärvi *et al.*, 1980). Although there is no striking nucleotide sequence homology between EBERs and VAs, similarities exist in their size, degree of secondary structure, and genomic organization (Bhat and Thimmappaya, 1983). Like VA RNAs, EBERs bind PKR, the interferon (IFN)-inducible protein that is a key mediator of antiviral effect of IFN (Clarke *et al.*, 1991; Meurs *et al.*, 1990; Sharp *et al.*, 1993). It was reported that PKR binds to the stem-loop IV of EBER1 (Vuyisich *et al.*, 2002). Once activated by dsRNA, PKR phosphorylates the α-subunit of protein synthesis initiation factor eIF2, causing inhibition of translation at the level of initiation. *In vitro* assays have demonstrated that EBERs can inhibit PKR activation and block phosphorylation of eIF2a thus resulting in the blockage of inhibition of protein synthesis by eIF2a (Clarke *et al.*, 1990; Katze *et al.*, 1991; Sharp *et al.*, 1993). More recently, Nanbo *et al.* (2002) demonstrated that in BL cells, EBERs confer resistance to IFN-α-induced apoptosis by directly binding to PKR and inhibiting its phosphorylation. EBERs also block Fas-mediated apoptosis in epithelial cells through PKR inhibition (Nanbo *et al.*, 2005). Most recently, the interactions between PKR and EBERs/VAs have been analyzed. Mckenna *et al.* (2007) reported that EBERs/VAs bind preferentially to the latent dephosphorylated form of PKR with similar affinity as dsRNA activators. However, EBERs/VAs prevent the dimerization of PKR, which is required for efficient trans-autophosphorylation of PKR. This blocks the phosphorylation of PKR substrates, allowing protein synthesis to proceed.

A second highly abundant protein designated EAP (EBER-associated protein) was identified in La-containing RNP complexes (Toczyski and Steitz, 1991). It was reported that EBER1 mostly binds to EAP (Toczyski and Steitz, 1993) and EAP was subsequently shown to be the ribosomal protein L22 (Toczyski *et al.*, 1994). Although the functions of L22 are not well understood, L22 was identified as the target of chromosomal translocation in certain proteins with leukemia (Liu *et al.*, 1993; Nucifora *et al.*, 1993), suggesting that L22 levels may be a determinant in cell transformation. In uninfected human B lymphocytes, L22 is localized to the nucleoli and cytoplasm; however, following EBV infection, L22 binds to EBERs and relocalizes to the nucleoplasm (Toczyski *et al.*, 1994), suggesting a role of the EBER–L22 interaction is to sequester the cellular L22 molecules. Previous studies have demonstrated that EBER1 has multiple domains to bind to L22, including stem-loop III (Toczyski and Steitz, 1993), stem-loop IV (Dobbelstein and Shenk, 1995), and stem-loop I (Fok *et al.*, 2006a,b). These multiple binding domains for L22 provide the possibility that most of the EBERs form complexes with L22 *in vivo*, and thereby EBERs may modulate protein translation (Fok *et al.*, 2006a,b). Recently, Elia *et al.* (2004) reported that L22 and PKR compete for a common binding site on EBER-1. As a result of this competition, L22 interferes with the ability of EBERs to inhibit the activation of PKR by dsRNA. Although transient expression of EBER1 in murine embryonic fibroblasts stimulates reporter gene expression and partially reverse the inhibitory effect of PKR, EBER1 is also stimulatory when transfected into PKR-knockout cells, suggesting an additional, PKR-independent, mode of action of EBERs. Expression of L22 prevents both the PKR-dependent and -independent effects of EBER1 *in vivo*. These results suggest that the association of L22 with EBER1 in EBV-infected cells can attenuate the biological effect of the viral RNA. Such effects include both the inhibition of PKR and additional mechanism(s) by which EBER1 stimulates gene expression.

V. ROLE OF EBERs IN ONCOGENESIS

The lack of a suitable *in vitro* system that represents BL-type EBV infection, which is characterized by expression of a limited number of viral genes (termed type-I latency), including EBNA1, EBERs, and BARF0 (Rickinson and Kieff, 2002), has hampered study of the role of EBV in the genesis of BL. The Japanese BL-derived Akata cell line (Takada, 1984; Takada and Ono, 1989; Takada *et al.*, 1991) is unique in that it retains the *in vivo* phenotype of EBV expression even after long-term culture *in vitro* (Shimizu *et al.*, 1994). Isolation of EBV-negative cell clones from the parental Akata cell culture allowed for more systemic studies of the role of EBV in BL (Shimizu *et al.*, 1994). Comparison of EBV-positive and -negative cell clones revealed that the presence of EBV in Akata cells was required for the cells to be more malignant and apoptosis resistant (Chodosh *et al.*, 1998; Komano *et al.*, 1998; Ruf *et al.*, 1999; Shimizu *et al.*, 1994), which underlined the oncogenic role of EBV in the genesis of BL. Subsequent studies revealed that EBERs were responsible for these phenotypes (Komano *et al.*, 1999). Transfection of the EBER genes into EBV-negative Akata clones restored the capacity for growth in soft agar, tumorigenicity in SCID mice, resistance to apoptotic inducers, and upregulated expression of bcl-2 that was originally retained in parental EBV-positive Akata cells and lost in EBV-negative subclones. After their point, other studies presented essentially similar results (Maruo *et al.*, 2001; Ruf *et al.*, 2000; Yamamoto *et al.*, 2000).

More recently, it was demonstrated that EBERs induce human IL-10 expression in BL cells (Kitagawa *et al.*, 2000). It was found that EBV-positive Akata and Mutu cell clones expressed higher levels of IL-10 than their EBV-negative subclones at the transcriptional level. Transfection of an individual EBV latent gene into EBV-negative Akata cells revealed that EBERs were responsible for IL-10 induction. Recombinant IL-10 enabled EBV-negative Akata cells to grow in low (0.1%) serum conditions, while growth of EBV-positive Akata cells was blocked by treatment either with an anti-IL-10 antibody or antisense oligonucleotide against IL-10. EBV-positive BL biopsies consistently expressed IL-10, but EBV-negative BL biopsies did not. These results suggest that IL-10 induced by EBERs acts as an autocrine growth factor for BL. EBV associates with various T cell-proliferating diseases such as chronic active EBV infection (CAEBV) and nasal lymphoma. A human T cell line, MT-2, was susceptible to EBV infection, and EBV-infected cell clones showed type-II latency, which was identical with those seen in EBV-infected T cells *in vivo* (Yoshiyama *et al.*, 1995). It was found that EBV-positive MT-2 cells express higher levels of IL-9 than EBV-negative MT-2 cells at transcriptional level and EBERs were responsible for IL-9 expression (Yang *et al.*, 2004). The results of further study suggest that IL-9 induced by EBERs acts as an autocrine growth factor for EBV-infected T cells. Analysis of nasal lymphoma biopsies indicated that three of four specimens expressed IL-9. These findings suggest that EBERs directly affect the pathogenesis of EBV-associated T cell diseases.

About 5–10% of gastric carcinoma (GC) cases worldwide are associated with EBV (Takada, 2000). Iwakiri *et al.* (2003) reported that EBV infection induces expression of IGF-1 in the GC-derived EBV-negative cell line NU-GC-3, and that the secreted IGF-1 acts as an autocrine growth factor. Transfection of individual EBV latent gene into NU-GC-3 cells revealed that the EBERs were responsible for IGF-1 expression. These findings seem to be operative *in vivo*, as EBV-positive GC biopsies consistently express IGF-1, while EBV-negative GC biopsies do not. Therefore, EBERs would directly affect the pathogenesis of EBV-positive GC. Nasopharyngeal carcinoma (NPC) is strongly associated with EBV infection (Rickinson and Kieff, 2002). It was also reported that EBER induces IGF-1 expression in EBV-negative NPC-derived cell lines CNE1 and HONE1 (Iwakiri *et al.*, 2005). As observed in GC cells, IGF-1 acts as an autocrine growth factor. Moreover, it was demonstrated that the growth of EBV-positive NPC-derived line C666.1 is dependent on IGF-1 and NPC biopsies consistently express IGF-1, suggesting that EBERs contribute to the development of NPC *in vivo*. EBERs induce transcription of three different growth factors in different cell types and make key contribution to both lymphoid and epithelioid carcinogenesis. Studies on dominant-negative PKR and the PKR inhibitor suggested that PKR inhibition was not involved in transcriptional activation of these growth factors. Most recent study demonstrated that EBERs induce IL-10 expression through modulation of innate immune signals (Samanta *et al.*, 2008 see below).

Regarding the role of EBERs in the process of EBV-induced B cell transformation, Swaminathan *et al.* (1991) demonstrated that EBERs were not essential for the immortalization of B lymphocytes or for the replication of the virus. They restored the transformation-defect of the P3HR-1 strain EBV, having a deletion of the essential-transforming gene EBNA2, using homologous recombination between the P3HR-1 deleted genome and an EBER-deleted EBV DNA fragment spanning the EBNA2 locus. Their attempt resulted in obtaining lymphoblastoid cell lines (LCLs) harboring only EBER-deleted recombinant viruses, indicating that EBERs are dispensable for B cell transformation. However, they failed to produce a large quantity of pure EBER-deleted EBV. Instead, a cocultivation method was used to passage the EBER-deleted EBV from primary LCLs to secondary LCLs. Therefore, the transforming titer of EBER-deleted EBV has never been determined by using a pure recombinant virus. Recently, Yajima *et al.* (2005) revisited this issue by producing a large quantity of EBER-deleted EBV using an Akata cell system. Although the EBER-deleted virus efficiently infected B lymphocytes, its 50% transforming dose was approximately 100-fold less than that of the EBER-positive EBV. They then engineered the genome of EBER-deleted virus and generated a recombinant virus with the EBER genes reconstituted at their native locus. The resultant EBER-reconstituted EBV exhibited restored transforming ability. In addition, LCLs established with the EBER-deleted EBV grew significantly slower than those established with wild-type or EBER-reconstituted EBV, and the difference of growth rates was especially highlighted when the cells were plated at low cell densities. Thus, EBERs significantly contribute to efficient growth transformation of B lymphocytes by enhancing the growth potential of transformed lymphocytes. More recently, Wu *et al.* (2007) reported that EBER1 and EBER2 have distinct functions in latently infected LCLs. The transforming ability of recombinant EBVs expressing EBER2 was as high as that of EBVs expressing both EBER1 and EBER2. On the other hand, the transforming ability of recombinant EBVs carrying EBER1 was impaired and was similar to that of EBV lacking both EBER1 and EBER2. LCLs established with EBVs carrying EBER2 proliferated at low cell densities, while LCLs established with EBVs carrying EBER1 did not. IL-6 production in LCLs expressing EBER2 was more abundant than in those lacking EBER2. The growth of LCLs lacking EBER2 was enhanced by the addition of recombinant IL-6 to the cell culture, while the growth of EBER2-expressing LCLs was inhibited by a neutralizing anti-IL-6 antibody. These results demonstrate that EBER2, but not EBER1, contributes to efficient growth transformation of B lymphocytes.

VI. MODULATION OF INNATE IMMUNE SIGNALING BY EBERs AND ITS CONTRIBUTION TO EBV-MEDIATED PATHOGENESIS

The relationship between innate immunity and virus infection has been intensively studied in recent years. The host evokes innate immune responses to eliminate invading pathogens by detecting the presence of infection. Cells express a limited number of germ line-encoded pattern-recognition receptors (PRR) that specifically recognize pathogen-associated molecular patterns (PAMPs) within microbes. Retinoic acid-inducible gene (RIG)-I-like receptors (RLRs) including RIG-I (Yoneyama *et al.*, 2004), melanoma differentiation-associated gene (Mda)-5 (Kang *et al.*, 2002), and LGP2 (Yoneyama and Fujita, 2007), are cytoplasmic proteins that recognize viral RNA. RLRs are known as key molecules for IFN-inducible antiviral effects (Meylan and Tschopp, 2006). When RIG-I is activated by interaction with viral dsRNA, it initiates signaling pathways leading to induction of protective cellular genes, including type-I IFNs and inflammatory cytokines. RIG-I contains a C-terminal DExD/H-box RNA helicase domain and an N-terminal CARD. The helicase domain is responsible for dsRNA recognition, and the CARD domain activates downstream signaling cascades through a mitochondrial adaptor IFN-β promoter stimulator (IPS)-1, resulting in the activation of transcription factors, NF-κB and IRF3 (Kawai *et al.*, 2005; Yoneyama *et al.*, 2004). Samanta *et al.* (2006) reported that EBER, which is expected to form dsRNA structure, activates RIG-mediated signaling. In EBER-positive EBV-infected BL cells and EBER-transfected EBV-negative BL cells, overexpression of RIG-I induced type-I IFN and IFN-stimulating genes (ISGs), while RIG-I knockdown by siRNA resulted in reduction of IFN expression. In primary EBV infection, RIG-I recognized EBERs and activated signaling to express type-I IFN. Further study revealed that EBERs bind and coprecipitate with RIG-I. These results suggest that in BL cells, RIG-I is constitutively activated by EBERs, leading to activation of downstream signaling molecules NF-κB and IRF-3 to induce type-I IFN. Although IFN induction looks disadvantageous for virus, EBV can maintain latent infection. This would be because of the resistance to IFN, such as PKR inhibition by EBERs. Moreover, subsequent study demonstrated that EBER promotes the growth of BL cells through RIG-I signaling (Fig. 3) (Samanta *et al.*, 2008). Inhibition of NF-κB with IκB plasmid did not block IL-10 expression, whereas knockdown of IRF3 by siRNA dramatically reduced IL-10 expression in EBER-positive EBV-infected BL cells and EBER-expressing EBV-negative BL cells, but not in EBER-knockout EBV-infected or EBV-negative BL cell. These findings strongly suggest that in BL cells, EBERs induce the anti-inflammatory and growth-promoting cytokine IL-10 (Kitagawa *et al.*, 2000) through

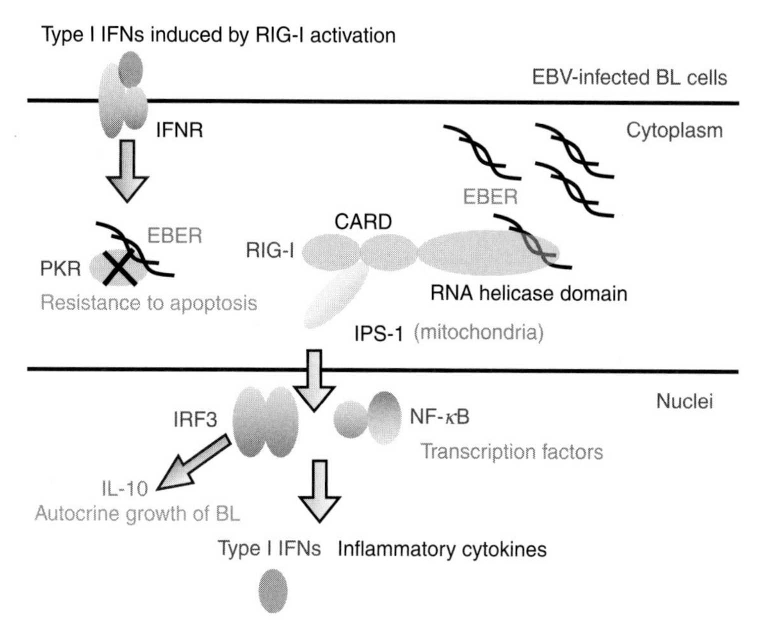

Fig. 3 Modulation of RIG-I signaling by EBERs contributes to EBV-mediated oncogenesis in BL cells. EBERs are recognized by RIG-I through the RNA helicase domain, and following recognition, RIG-I associates with the adaptor IPS-1 via CARD. IPS-1 is localized to mitochondria and initiates signaling leading to activation of IRF3 and NF-κB to induce type-I IFNs and inflammatory cytokines. Although type-I IFNs induce PKR expression through IFN receptor leading to induction of apoptosis, EBERs bind PKR and inhibit its phosphorylation, thus EBV could maintain latent infection. In addition to the induction of type-I IFNs, EBERs induce the growth-promoting cytokine IL-10 through RIG-I-mediated IRF3 but not NF-κB signaling, and may support the development of BL.(See Page 3 in Color Section at the back of the book.)

RIG-I-mediated IRF3 signaling independent of NF-κB (Samanta *et al.*, 2008).

Toll-like receptors (TLRs) constitute distinct families of PRRs that sense nucleic acids derived from viruses and trigger antiviral innate immune responses through activation of signaling cascades via Toll/IL-1 receptor (TIR) domain-containing adaptors (Akira and Takeda, 2004). The role of TLR3 in the recognition of dsRNA was demonstrated in a study of TLR3-deficient mice, which show reduced production of type-I IFN and inflammatory cytokines in response to genomic RNA purified from dsRNA viruses such as reovirus and a synthetic analog of dsRNA, poly IC that has been used to mimic viral infection (Alexopoulou *et al.*, 2001). Signal transduction from TLR3 induced by dsRNA leads to recruitment of TIR domain-containing adaptor inducing IFN-β (TRIF) and subsequent phosphorylation of downstream molecules such as IRF3 and NF-κB (Meylan and Tschopp, 2006).

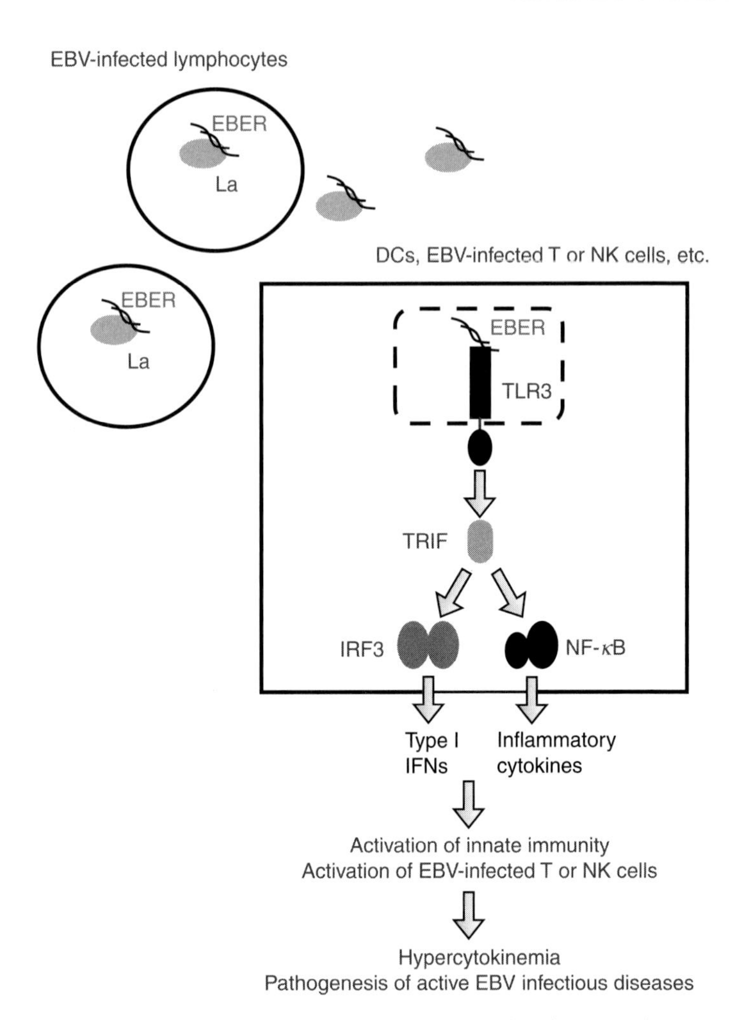

Fig. 4 Activation of innate immunity through TLR3 signaling by secreted EBER. During active EBV infection, EBER1 is released from EBV-infected lymphocytes mostly in complex with La. Circulating EBER would induce maturation of DCs via TLR3 signaling inducing type-I IFNs and inflammatory cytokines through activation of IRF3 and NF-κB. DC activation leads to T cell activation and systemic release of cytokines. Furthermore, TLR3-expressing T and NK cells including EBV-infected T or NK cells could be activated by EBER1 through TLR3 and produce inflammatory cytokines. Therefore, immunopathologic diseases that are caused by active EBV infections including activation of T or NK cells and hypercytokinemia could be attributed to TLR3-mediated T cell activation and cytokinemia by EBER1. (See Page 4 in Color Section at the back of the book.)

Iwakiri *et al.* (2009) reported that EBERs are released extracellulary and are recognized by TLR3, leading to induction of type-I IFN and inflammatory cytokines. A substantial amount of EBER, which was sufficient to induce

TLR3 signaling involving IRF3 and NF-κB activation, was released from EBV-infected cells. The majority of the released EBER existed as a complex with La, suggesting that EBER was released from the cells by active secretion of La. EBV has been known to cause active infectious diseases such as infectious mononucleosis (IM), CAEBV, and EBV-associated hemophagocytic lymphohistiocytosis (EBV-HLH). IM is characterized by the expansion of reactive T cells and is most likely to be an immunopathologic disease whose general symptoms are caused by inflammatory cytokines (Rickinson and Kieff, 2002). CAEBV and EBV-HLH are also active EBV infections with persistent or recurrent IM-like symptoms. EBV-HLH is characterized by an EBV infection in CD4-positive T cells or NK cells and the systemic release of inflammatory cytokines, which subsequently causes hemophagocytosis of blood cells through the activation of macrophages (Kasahara *et al.*, 2001; Kikuta *et al.*, 1993; Rickinson and Kieff, 2002). On the other hand, $CD8^+$-T cells are mainly infected with EBV in CAEBV (Kasahara *et al.*, 2001). Iwakiri *et al.* (2009) demonstrated that sera from patients with IM, CAEBV, and EBV-HLH contained EBER. Addition of RNA purified from the sera into culture medium activates TLR3 leading to induction of type-I IFN and inflammatory cytokines in peripheral blood mononuclear cells. Furthermore, dendritic cells (DCs) treated with EBER showed mature phenotype and antigen presentation capacity. These findings suggest that EBER, which is released from EBV-infected cells, is responsible for immune activation by EBV, inducing type-I IFN and inflammatory cytokines. Because $CD8^+$-T cells and NK cells express TLR3 and are activated by TLR3 signaling (Schmidt *et al.*, 2004; Tabiasco *et al.*, 2006), TLR3-expressing T and NK cells could be activated by EBER1 through TLR3 and produce inflammatory cytokines. Therefore, EBER-induced activation of innate immunity would account for immunopathologic diseases caused by active EBV infection (Fig. 4). In summary, EBERs contribute to the pathogenesis of EBV infection including cancer and active infectious diseases through interaction with RIG-I and TLR3. Indeed, released EBER in peripheral blood could be a novel therapeutic target for the treatment of CAEBV and EBV-HLH.

REFERENCES

Akira, S., and Takeda, K. (2004). Toll-like receptor signaling. *Nat. Rev. Immunol.* **4**, 499–511.

Akusjärvi, G., Mathews, M. B., Andersson, P., Vennström, B., and Pettersson, U. (1980). Structure of genes for virus-associated RNAI and RNAII of adenovirus type 2. *Proc. Natl. Acad. Sci. USA* 77, 2424–2428.

Alexopoulou, L., Holt, A. C., Medzhitov, R., and Flavell, R. A. (2001). Recognition of double-stranded RNA and activation of NF-kappaB by Toll-like receptor 3. *Nature* **413**, 732–738.

Arrand, J. R., Young, L. S., and Tugwood, J. D. (1989). Two families of sequences in the small RNA-encoding region of Epstein–Barr virus (EBV) correlate with EBV types A and B. *J. Virol.* **63,** 983–986.

Bhat, R. A., and Thimmappaya, B. (1983). Two small RNAs encoded by Epstein–Barr virus can functionally substitute for the virus-associated RNAs in the lytic growth of adenovirus 5. *Proc. Natl. Acad. Sci. USA* **80,** 4789–4793.

Bonnet, M., Guinebretiere, J. M., Kremmer, E., Grunewald, V., Benhamou, E., Contesso, G., and Joab, I. (1999). Detection of Epstein–Barr virus in invasive breast cancers. *J. Natl. Cancer Inst.* **91,** 1376–1381.

Chang, K. L., Chen, Y. Y., Shibata, D., and Weiss, L. M. (1992). Description of an *in situ* hybridization methodology for detection of Epstein–Barr virus RNA in paraffin-embedded tissues, with a survey of normal and neoplastic tissues. *Diagn. Mol./Pahol.* **1,** 246–255.

Chodosh, J., Holder, V. P., Gan, Y. J., Belgaumi, A., Sample, J., and Sixbey, J. W. (1998). Eradication of latent Epstein–Barr virus by hydroxyurea alters the growth-transformed cell phenotype. *J. Infect. Dis.* **177,** 1194–1201.

Clarke, P. A., Sharp, N. A., and Clemens, M. J. (1990). Translational control by the Epstein–Barr virus small RNA EBER-1. Reversal of the double-stranded RNA-induced inhibition of protein synthesis in reticulocyte lysates. *Eur. J. Biochem.* **193,** 635–641.

Clarke, P. A., Schwemmle, M., Schickinger, J., Hilse, K., and Clemens, M. J. (1991). Binding of Epstein–Barr virus small RNA EBER-1 to the double-stranded RNA-activated protein kinase DAI. *Nucleic Acids Res.* **25**(19), 243–248.

Clarke, P. A., Sharp, N. A., and Clemens, M. J. (1992). Expression of genes for the Epstein–Barr virus small RNAs EBER-1 and EBER-2 in Daudi Burkitt's lymphoma cells: Effects of interferon treatment. *J. Gen. Virol.* **73,** 3169–3175.

Dobbelstein, M., and Shenk, T. (1995). *In vitro* selection of RNA ligands for the ribosomal L22 protein associated with Epstein–Barr virus-expressed RNA by using randomized and cDNA-derived RNA libraries. *J. Virol.* **69,** 8027–8034.

Elia, A., Vyas, J., Laing, K. G., and Clemens, M. J. (2004). Ribosomal protein L22 inhibits regulation of cellular activities by the Epstein–Barr virus small RNA EBER-1. *Eur. J. Biochem.* **27,** 1895–1905.

Fok, V., Friend, K., and Steitz, J. A. (2006a). Epstein–Barr virus noncoding RNAs are confined to the nucleus, whereas their partner, the human La protein, undergoes nuclecytoplasmic shuttling. *J. Cell Biol.* **173,** 319–325.

Fok, V., Mitton-Fry, R. M., Grech, A., and Steitz, J. A. (2006b). Multiple domains of EBER 1, an Epstein–Barr virus noncoding RNA, recruit human ribosomal protein L22. *RNA* **12,** 872–882.

Gilligan, K., Rajadurai, P., Resnick, L., and Raab-Traub, N. (1990). Epstein–Barr virus small nuclear RNAs are not expressed in permissively infected cells in AIDS-associated leukoplakia. *Proc. Natl. Acad. Sci. USA* **87,** 8790–8794.

Greifenegger, N., Jäger, M., Kunz-Schughart, L. A., Wolf, H., and Schwarzmann, F. (1998). Epstein–Barr virus small RNA (EBER) genes: Differential regulation during lytic viral replication. *J. Virol.* **72,** 9323–9328.

Howe, J. G., and Steitz, J. A. (1986). Localization of Epstein-Barr virus-encoded small RNAs by *in situ* hybridization. *Proc. Natl. Acad. Sci. USA.* **83,** 9006–9010.

Howe, J. G., and Shu, M. D. (1988). Isolation and characterization of the genes for two small RNAs of herpesvirus papio and their comparison with Epstein–Barr virus-encoded EBER RNAs. *J. Virol.* **62,** 2790–2798.

Howe, J. G., and Shu, M. D. (1989). Epstein–Barr virus small RNA (EBER) genes: Unique transcription units that combine RNA polymerase II and III promoter elements. *Cell* **57,** 825–834.

Iwakiri, D., Eizuru, Y., Tokunaga, M., and Takada, K. (2003). Autocrine growth of Epstein–Barr virus-positive gastric carcinoma cells mediated by an Epstein–Barr virus-encoded small RNA. *Cancer Res.* **63,** 7062–7067.

Iwakiri, D., Sheen, T. S., Chen, J. Y., Huang, D. P., and Takada, K. (2005). Epstein–Barr virus-encoded small RNA induces insulin-like growth factor 1 and supports growth of nasopharyngeal carcinoma-derived cell lines. *Oncogene* **24,** 1767–1773.

Iwakiri, D., Zhou, L., Samanta, M., Matsumoto, M., Ebihara, T., Seya, T., Imai, S., Fujieda, M., Kawa, K., and Takada, K. (2009). Epstein–Barr virus (EBV)-encoded small RNA is released from EBV-infected cells and activates signaling from Toll-like receptor 3. *J. Exp. Med.* **206,** 2091–2099.

Kang, D.-C., Gopalkrishnan, R. V., Wu, Q., Jankowsky, E., Pyle, A. M., and Fisher, P. B. (2002). mda-5: An interferon-inducible putative RNA helicase with double-stranded RNA-dependent ATPase activity and melanoma growth-suppressive properties. *Proc. Natl. Acad. Sci. USA* **99,** 637–642.

Kasahara, Y., Yachie, A., Takei, K., Kanegane, C., Okada, K., Ohta, K., Seki, H., Igarashi, N., Maruhashi, K., Katayama, K., Katoh, E., Terao, G., *et al.* (2001). Differential cellular targets of Epstein–Barr virus (EBV) infection between acute EBV-associated hemophagocytic lymphohistiocytosis and chronic active EBV infection. *Blood* **98,** 1882–1888.

Katze, M. G., Wambach, M., Wong, M. L., Garfinkel, M., Meurs, E., Chong, K., Williams, B. R., Hovanessian, A. G., and Barber, G. N. (1991). Functional expression and RNA binding analysis of the interferon-induced, double-stranded RNA-activated, 68, 000-Mr protein kinase in a cell-free system. *Mol. Cell. Biol.* **11,** 5497–5505.

Kawai, T., Takahashi, K., Sato, S., Coban, C., Kumar, H., Kato, H., Ishii, K. J., Takeuchi, O., and Akira, S. (2005). IPS-1, an adaptor triggering RIG-I- and Mda5-mediated type I interferon induction. *Nat. Immunol.* **11,** 1074–1076.

Kikuta, H., Sakiyama, Y., Matsumoto, S., Oh-Ishi, T., Nakano, T., Nagashima, T., Oka, T., Hironaka, T., and Hirai, K. (1993). Fatal Epstein–Barr virus-associated hemophagocytic syndrome. *Blood* **82,** 3259–3264.

Kitagawa, N., Goto, M., Kurozumi, K., Maruo, S., Fukayama, N. T., Yasukaswa, M., Hino, K., Suzuki, T., Todo, S., and Takada, K. (2000). Epstein–Barr virus-encoded poly(A)(-) RNA supports Burkitt's lymphoma growth through interleukin-10 induction. *EMBO J.* **19,** 6742–6750.

Komano, J., Sugiura, M., and Takada, K. (1998). Epstein–Barr virus contributes to the malignant phenotype and to apoptosis resistance in Burkitt's lymphoma cell line Akata. *J. Virol.* **72,** 9150–9156.

Komano, J., Maruo, S., Kurozumi, K., Oda, T., and Takada, K. (1999). Oncogenic role of Epstein–Barr virus-encoded RNAs in Burkitt's lymphoma cell line Akata. *J. Virol.* **73,** 9827–9831.

Lerner, M. R., Andrews, N. C., Miller, G., and Steitz, J. A. (1981). Two small RNAs encoded by Epstein–Barr virus and complexed with protein are precipitated by antibodies from patients with systemic lupus erythematosus. *Proc. Natl. Acad. Sci. USA* **78,** 805–809.

Liu, P., Tarlé, S. A., Hajra, A., Claxton, D. F., Marlton, P., Freedman, M., Siciliano, M. J., and Collins, F. S. (1993). Fusion between transcription factor CBF beta/PEBP2 beta and a myosin heavy chain in acute myeloid leukemia. *Science* **261,** 1041–1044.

Maruo, S., Nanbo, A., and Takada, K. (2001). Replacement of the Epstein–Barr virus plasmid with the EBER plasmid in Burkitt's lymphoma cells. *J. Virol.* **75,** 9977–9982.

Mckenna, S. A., Lindhout, D. A., Shimoike, T., Aitken, C. E., and Puglisis, J. D. (2007). Viral dsRNA inhibitors prevent self-association and autophosphorylation of PKR. *J. Mol. Biol.* **372,** 103–113.

Meurs, E., Chong, K., Galabru, J., Thomas, N. S., Kerr, I. M., Williams, B. R., and Hovanessian, A. G. (1990). Molecular cloning and characterization of the human double-stranded RNA-activated protein kinase induced by interferon. *Cell* **62,** 379–390.

Meylan, E., and Tschopp, J. (2006). Toll-like receptors and RNA helicases: Two parallel ways to trigger antiviral responses. *Mol. Cell* **22,** 561–569.

Mizugaki, Y., Sugawara, Y., Shinozaki, F., and Takada, K. (1998). Detection of Epstein–Barr virus in oral papilloma. *Jpn J. Cancer Res.* **89,** 604–607.

Nanbo, A., Inoue, K., Adachi-Takasawa, K., and Takada, K. (2002). Epstein–Barr virus RNA confers resistance to interferon-alpha-induced apoptosis in Burkitt's lymphoma. *EMBO J.* **21,** 954–965.

Nanbo, A., Yoshiyama, H., and Takada, K. (2005). Epstein-Barr virus-encoded poly(A)- RNA confers resistance to apoptosis mediated through Fas by blocking the PKR pathway in human epithelial intestine 407 cells. *J. Virol.* **79,** 12280–12285.

Nucifora, G., Begy, C. R., Erickson, P., Drabkin, H. A., and Rowley, J. D. (1993). The 3;21 translocation in myelodysplasia results in a fusion transcript between the AML1 gene and the gene for EAP, a highly conserved protein associated with the Epstein–Barr virus small RNA EBER 1. *Proc. Natl. Acad. Sci. USA* **90,** 7784–7788.

Rickinson, A. B., and Kieff, E. (2002). Epstein–Barr virus. *In* Fields Virology, (B. N. Fields, D. M. Knipe, and P. M. Hoeley, Eds.), pp. 2575–2627. Lippincott-Raven Publishers, Philadelphia, PA.

Rooney, C., Howe, J. G., Speck, S. H., and Miller, G. (1989). Influence of Burkitt's lymphoma and primary B cells on latent gene expression by the nonimmortalizing P3J-HR-1 strain of Epstein–Barr virus. *J. Virol.* **63,** 1531–1539.

Rosa, M. D., Gottlieb, E., Lerner, M. R., and Steitz, J. A. (1981). Striking similarities are exhibited by two small Epstein–Barr virus-encoded ribonucleic acids and the adenovirus-associated ribonucleic acids VAI and VAII. *Mol. Cell. Biol.* **9,** 785–796.

Ruf, I. K., Rhyne, P. W., Yang, H., Borza, C. M., Hutt-Fletcher, L. M., Cleveland, J. L., and Sample, J. T. (1999). Epstein–Barr virus regulates c-MYC, apoptosis, and tumorigenicity in Burkitt's lymphoma. *Mol. Cell. Biol.* **3,** 1651–1660.

Ruf, I. K., Rhyne, P. W., Yang, C., Cleveland, J. L., and Sample, J. T. (2000). Epstein–Barr virus small RNAs potentiate tumorigenicity of Burkitt's lymphoma cells independently of an effect on apoptosis. *J. Virol.* **74,** 10223–10228.

Rymo, L. (1979). Identification of transcribed regions of Epstein–Barr virus DNA in Burkitt's lymphoma-derived cells. *J. Virol.* **32,** 8–18.

Samanta, M., Iwakiri, D., Kanda, T. Imaizumi, T., and Takada, K. (2006). EB virus-encoded RNAs are recognized by RIG-I and activate signaling to induce type I IFN. *EMBO J.* **25,** 4207–4214.

Samanta, M., Iwakiri, D., and Takada, K. (2008). Epstein-Barr virus-encoded small RNA induces IL-10 through RIG-I-mediated IRF-3 signaling. *Oncogene* **27,** 4150–4160.

Schmidt, K. N., Leung, B., Kwong, M., Zarember, K. A., Satyal, S., Navas, T. A., Wang, F., and Godowski, P. J. (2004). APC-independent activation of NK cells by the Toll-like receptor 3 agonist double-stranded RNA. *J. Immunol.* **172,** 138–143.

Schwemmle, M., Clemens, M. J., Hilse, K., Pfeifer, K., Tröster, H., Müller, W. E., and Bachmann, M. (1992). Localization of Epstein–Barr virus-encoded RNAs EBER-1 anEBER-2 in interphase and mitotic Burkitt's lymphoma cells. *Proc. Natl. Acad. Sci. USA* **89,** 10292–10296.

Sharp, T. V., Schwemmle, M., Jeffrey, I., Laing, K., Mellor, H., Proud, C. G., Hilse, K., and Clemens, M. J. (1993). Comparative analysis of the regulation of the interferon-inducible protein kinase PKR by Epstein–Barr virus RNAs EBER-1 and EBER-2 and adenovirus VAI RNA. *Nucleic Acids Res.* **21,** 4483–4490.

Shimizu, N., Tanabe-Tochikura, A., Kuroiwa, Y., and Takada, K. (1994). Isolation of Epstein–Barr virus (EBV)-negative cell clones from the EBV-positive Burkitt's lymphoma (BL) line Akata: Malignant phenotypes of BL cells are dependent on EBV. *J. Virol.* **68,** 6069–6073.

Sugawara, Y., Mizugaki, Y., Uchida, T., Torii, T., Imai, S., Makuuchi, M., and Takada, K. (1999). Detection of Epstein–Barr virus (EBV) in hepatocellular carcinoma tissue: A novel EBV latency characterized by the absence of EBV-encoded small RNA expression. *Virology* **256,** 196–202.

Swaminathan, S., Tomkinson, B., and Kieff, E. (1991). Recombinant Epstein–Barr virus with small RNA (EBER) genes deleted transforms lymphocytes and replicates *in vitro*. *J. Virol.* **66,** 5133–5136.

Tabiasco, J., Devêvre, E., Rufer, N., Salaun, B., Cerottini, J. C., Speiser, D., and Romero, D. (2006). Human effector CD8+ T lymphocytes express TLR3 as a functional coreceptor. *J. Immunol.* **177,** 8708–8713.

Takada, K. (1984). Cross-linking of cell surface immunoglobulins induces Epstein–Barr virus in Burkitt's lymphoma lines. *Int. J. Cancer* **33,** 27–32.

Takada, K. (2000). Epstein–Barr virus and gastric carcinoma. *Mol. Pathol.* **53,** 255–261.

Takada, K., and Ono, Y. (1989). Synchronous and sequential activation of latently infected Epstein–Barr virus genomes. *J. Virol.* **63,** 445–449.

Takada, K., Horinouchi, K., Ono, Y., Aya, T., Osato, T., Takahashi, M., and Hayasaka, S. (1991). An Epstein–Barr virus-producer line Akata: Establishment of the cell line and analysis of viral DNA. *Virus Genes* **2,** 147–156.

Toczyski, D. P., and Steitz, J. A. (1991). EAP, a highly conserved cellular protein associated with Epstein–Barr virus small RNAs (EBERs). *EMBO J.* **10,** 459–466.

Toczyski, D. P., and Steitz, J. A. (1993). The cellular RNA-binding protein EAP recognizes a conserved stem-loop in the Epstein–Barr virus small RNA EBER 1. *Mol. Cell. Biol.* **1,** 703–710.

Toczyski, D. P., Matera, A. G., Ward, D. C., and Steitz, J. A. (1994). The Epstein–Barr virus (EBV) small RNA EBER1 binds and relocalizes ribosomal protein L22 in EBV-infected human B lymphocytes. *Proc. Natl. Acad. Sci. USA* **91,** 3463–3467.

Vuyisich, M., Spanggord, R. J., and Beal, P. A. (2002). The binding site of the RNA-dependent protein kinase (PKR) on EBER1 RNA from Epstein–Barr virus. *EMBO Rep.* **3,** 622–627.

Wen, S., Shimizu, N., Yoshiyama, H., Mizugaki, Y., Shinozaki, F., and Takada, K. (1996). Association of Epstein–Barr virus (EBV) with Sjögren's syndrome: Differential EBV expression between epithelial cells and lymphocytes in salivary glands. *Am. J. Pathol.* **149,** 1511–1517.

Wen, S., Mizugaki, Y., Shinozaki, F., and Takada, K. (1997). Epstein–Barr virus (EBV) infection in salivary gland tumors: Lytic EBV infection in nonmalignant epithelial cells surrounded by EBV-positive T-lymphoma cells. *Virology* **227,** 484–487.

Wu, Y., Maruo, S., Yajima, M., Kanda, T., and Takada, K. (2007). Epstein–Barr virus (EBV)-encoded RNA 2 (EBER2) but not EBER1 plays a critical role in EBV-induced B-cell growth transformation. *J. Virol.* **80,** 11236–11245.

Yajima, M., Kanda, T., and Takada, K. (2005). Critical role of Epstein–Barr Virus (EBV)-encoded RNA in efficient EBV-induced B-lymphocyte growth transformation. *J. Virol.* **79,** 4298–4307.

Yamamoto, N., Takizawa, T., Iwanaga, Y., Shimizu, N., and Yamamoto, N. (2000). Malignant transformation of B lymphoma cell line BJAB by Epstein–Barr virus-encoded small RNAs. *FEBS Lett.* **484,** 153–158.

Yang, L., Aozasa, K., Oshimi, K., and Takada, K. (2004). Epstein–Barr virus (EBV)-encoded RNA promotes growth of EBV-infected T cells through interleukin-9 induction. *Cancer Res.* **64,** 5332–5337.

Yao, Q. Y., Tierney, R. J., Croom-Carter, D., Cooper, G. M., Ellis, C. J., Rowe, M., and Rickinson, A. B. (1996). Isolation of intertypic recombinants of Epstein–Barr virus from T-cell-immunocompromised individuals. *J. Virol.* **70,** 4895–4903.

Yao, Y., Minter, H. A., Chen, X., Reynolds, G. M., Bromley, M., and Arrand, J. R. (2000). Heterogeneity of HLA and EBER expression in Epstein–Barr virus-associated nasopharyngeal carcinoma. *Int. J. Cancer* **88,** 949–955.

Yoneyama, M., and Fujita, T. (2007). Function of RIG-I-like receptors in antiviral innate immunity. *J. Biol. Chem.* **282,** 15315–15318.

Yoneyama, M., Kikuchi, M., Natsukawa, T., Shinobu, N., Imaizumi, T., Miyagishi, M., Taira, K., Akira, S., and Fujita, T. (2004). The RNA helicase RIG-I has an essential function in double-stranded RNA-induced innate antiviral responses. *Nat. Immunol.* **5**, 730–737.
Yoshiyama, H., Shimizu, N., and Takada, K. (1995). Persistent Epstein–Barr virus infection in a human T-cell line: Unique program of latent virus expression. *EMBO J.* **14**, 3706–3711.

Androgen Regulation of Gene Expression

Kristin R. Lamont and Donald J. Tindall

Departments of Urology, Biochemistry and Molecular Biology
Mayo Clinic College of Medicine, Rochester, Minnesota, USA

The biological action of androgenic male sex steroid hormones in prostate tissue is mediated by the androgen receptor, a nuclear transcription factor. The transcriptional program of androgenic signaling in the prostate consists of thousands of gene targets whose products play a role in almost all cellular functions, including cellular proliferation, survival, lipid metabolism, and differentiation. This review will provide a summary of the most recent data regarding androgen-regulated target genes and modulation of androgen receptor activity, especially with regard to androgen-dependent and castration-recurrent prostate cancer. © 2010 Elsevier Inc.

I. INTRODUCTION

The development of the male phenotype is due to the action of androgens in target tissues. The key androgens responsible for eliciting responses that lead to the development of male genitalia as well as secondary sex characteristics during puberty are primarily testosterone and its more active metabolite dihydrotestosterone (DHT) (reviewed in Wilson *et al.*, 1983). Testosterone is produced by the testes, and to a much smaller extent from adrenal androgens. The effects of testosterone and DHT are mediated through the androgen receptor (AR), a member of the superfamily of nuclear receptor transcription factors. Binding of DHT ligand within the ligand binding domain of the AR activates the receptor; the AR is then translocated

Advances in CANCER RESEARCH

0065-230X/10 $35.00
DOI: 10.1016/S0065-230X(10)07005-3

to the nucleus. Once in the nucleus, ligand-bound AR homodimerizes and binds chromatin at specific androgen response elements (AREs) in the regulatory regions of target genes. The AR also interacts with coregulator proteins and the basic transcriptional machinery to induce or inhibit transcription of a particular target gene, thereby driving cellular signaling that results in the differentiated phenotype of male organs.

Androgens and the AR have been implicated in a number of human diseases and conditions, most notably prostate cancer. In the United States, prostate cancer remains the most frequently diagnosed and second deadliest cancer in men (Jemal *et al.*, 2008). In the late 1800s, before the discovery of testosterone, an inverse correlation between prostate size and castration in men was reported. This was among the first lines of evidence to suggest that endocrine signals generated by the testes have significant effects on prostate tissue (So *et al.*, 2003). Approximately 50 years later, Charles Huggins and Clarence V. Hodges demonstrated that bilateral orchiectomy is an effective treatment for prostate cancer, which earned them the Nobel Prize in 1966 (Huggins and Hodges, 1972). Based upon their findings, different therapeutic approaches, including castration and antiandrogen therapy, have become the gold standard for prostate cancer treatment. These antiandrogenic therapies are quite successful, leading to regression of tumors in a majority of cases; however, 12–18 months later, most prostate cancers recur in a more aggressive, castration-recurrent form that is incurable. Median survival for these patients after recurrence is 24–36 months (McLeod *et al.*, 1997).

A seminal feature of castration-recurrent prostate cancer, and a main focus of ongoing research, is that aggressive disease relies on the expression and activity of the AR for survival and proliferation, despite the absence of normal levels of androgens (Zegarra-Moro *et al.*, 2002). Thus, the AR can be transactivated through a variety of mechanisms including increased expression of the receptor, which sensitizes the receptor to lower levels of ligand, mutation of the AR allowing for promiscuous activation, noncanonical variants of the AR, alterations in expression or activity of coregulator proteins, and cross-talk with other survival or proliferation pathways (reviewed in Debes and Tindall, 2004). Transactivation of the AR in castration-recurrent tumors by any one or a combination of these mechanisms results in expression of AR target genes, which have many different roles including those that are characteristic of normal prostate cells, such as stimulating proliferation, inhibiting apoptosis, or signaling normal, differentiated functions such as the production of secreted proteases (like the archetype androgen-regulated serine protease, PSA).

Two key areas of intriguing complexity are emerging with regard to androgenic regulation of target genes: (1) the vast number of genetic targets identified, and (2) the many levels of regulation of the activity of the AR. In order to gain a complete understanding of the repertoire of androgenic action, both aspects of the pathway must be considered. The purpose of this review is to summarize the recent advances in both categories. First, an overview will be provided of the recent findings pertaining to the targets of androgen and AR-mediated activity, especially those that might play a role in prostate cancer. Also, recent discoveries of expression, modification, and activity that affect how changes to the AR itself lead to changes in target gene expression will be discussed.

II. NOVEL ANDROGEN-REGULATED GENES (ARGs)

Studies performed to examine the gene expression program stimulated by androgens in prostate cells using high throughput expression microarray and other related techniques is still growing, and the current challenge lies in drawing conclusions from the huge amount of data in these reports (reviewed in Dehm and Tindall, 2006). The AR transcriptome is estimated to be between 10,570 and 23,448 polyadenylated RNAs but with other kinds of transcribed targets, like microRNAs, the full repertoire of androgen-regulated targets is even larger (Dehm and Tindall, 2006). A pressing challenge is to understand the role that each target contributes to androgenic effects, and the signaling networks responsible for cellular and tissue homeostasis. A recent microarray study of ARGs in the LNCaP cell line found that of the 619 genes regulated by androgens only approximately 75% of those have a known or inferred function, suggesting that a large number of androgen targets have yet to be fully described in terms of their functionality (Ngan *et al.*, 2009).

Given the massive amount of data generated from expression profiling studies, individual gene validations, and pathway analysis studies, it is difficult to create a simple summary of androgen regulation of gene expression. This section will highlight the recent advances in this field by exploring the generalized pathways regulated by androgen modulation of target genes.

A. Cell Proliferation and Survival

The dependence of neoplastic prostate cells upon androgenic signaling is most strikingly demonstrated by the dependency of these cells on the AR (Zegarra-Moro *et al.*, 2002). Conditional knockout of the AR in mouse epithelial prostate cells results in prostate tissue that is less differentiated and hyperproliferative, indicating that AR signaling regulates a signaling program controlling normal proliferation rates (Wu *et al.*, 2007). Many genes that regulate cell division and apoptosis have been identified to be androgen-regulated targets, and transactivation of the AR in castration-recurrent prostate cancer leads to increased cell proliferation and survival through these targets (Dehm and Tindall, 2006). When the AR is expressed in AR-null PC3 prostate cancer cells, androgenic treatment inhibits cell proliferation (Yuan *et al.*, 1993). Expression of AR alone (without androgen treatment) in PC3 cells results in changes of expression of 3452 genes compared to mock transfected cells (2235 downregulated, 1217 upregulated). Treatment with 1 nM androgen adds 232 genes to the list (133 decreased, 101 increased), but treatment with 10 nM androgen adds 482 to the cohort of AR-regulated genes (324 decreased, 159 increased). These data comprise 4166 genes that can be regulated by various concentrations of androgen. A large number of these (239) are part of cell survival/apoptosis pathways, that is, approximately 5.7% of the whole cohort (Lin *et al.*, 2009a).

Androgenic signaling that results in proliferation and/or survival in prostate cancer cells very commonly arises through cross-talk with pathways that are traditionally associated with these effects. The number of proteins belonging to these pathways whose expression is found to be androgen-sensitive is continually growing. Signaling initiated by the insulin-like growth factor-1 (IGF-1) results in transcriptional activity via c-Jun and c-Fos, promoting cell growth and survival. In addition to being a commonly overexpressed protein in prostate tumors, IGF-1 is also androgen-regulated (Hellawell *et al.*, 2002; Ngan *et al.*, 2009). Androgenic control of the signaling ligand renders the entire IGF-1 signaling pathway androgen-sensitive. Another factor, TGF-β, is strongly linked to the development and progression of many types of cancer, including prostate (Tian and Schiemann, 2009). The cellular effects of TGF-β signaling are complicated, depending upon cell type and context. The large family of signal transducers in the TGF-β pathway, the SMAD proteins, shows various degrees of androgenic regulation. Androgens inhibit the expression of TGF-β receptor-regulated SMADs-1 and -3, and the inhibitory SMADs-6 and -7 in LNCaP cells, and stimulate the transcriptional inhibitor, ID3 (Ngan *et al.*, 2009). Another family of transcription factors (i.e., FOX) also plays a key role in promoting cell survival. Approximately, 38 FOX transcription factors are regulated by androgens at the mRNA level in LNCaP cells (24 increase and 4 decrease; Takayama *et al.*, 2008). Cross-talk is also evident in this pathway as some factors like FOXP1 and FOXH1 interact with the AR to negatively regulate AR transcription, while others like FOXA1 interact with the AR to positively regulate AR activity (Chen *et al.*, 2005; Gao *et al.*, 2003; Takayama *et al.*, 2008). Another protein, FLIP, is a primary mode of androgen-mediated protection from death receptor-mediated apoptosis. The ability of the AR to regulate FLIP expression in response to androgens is altered in castration-recurrent cells compared to androgen-dependent cells (Raclaw *et al.*, 2008).

AR-mediated pro- or antiproliferation signaling is determined by cell context, such as stromal versus epithelial origin. Modulation of AR expression can promote aggressive phenotypes or reduce proliferation and aggressiveness, but in other scenarios, it can function as a tumor suppressor (Niu *et al.*, 2008a). Knockdown of epithelial AR in the TRAMP mouse model results in larger primary tumors with higher rates of proliferation. Stromal AR persisting in this model suggests that active AR signaling in this compartment can strongly support epithelial tumor progression (Niu *et al.*, 2008a). These authors propose that AR is a tumor suppressor in epithelial cells and an oncogene in stromal cells. However, this classification is controversial as the TRAMP model uses the probasin promoter to specifically drive both SV40 large-T and small-t antigen expression in the mouse prostate, giving rise to prostate tumors with phenotypes similar to human disease, but which arise from a very different origin (Greenberg *et al.*, 1994).

Androgens regulate the expression of some factors involved in driving and regulating cell division, including Cyclin D and Cdc6, a G1/S regulator (Balk and Knudsen, 2008; Dehm and Tindall, 2006; Jin and Fondell, 2009; Perry *et al.*, 1998). The regulation of these factors can override or bypass cell cycle checkpoints that normally would inhibit oncogenic proliferation, but as has been asserted throughout this review, it is becoming increasingly important to consider the larger effect of multiple targets of androgenic action. A recent study compared gene expression profiles of AR-regulated cistromes in LNCaP and LNCaP-abl cells, two cell models of androgen-dependent and castration-recurrent prostate cancer (Wang *et al.*, 2009). The researchers found that during cell cycle progression, M-phase genes, including *UBEC2*, an important checkpoint gene, are induced through AR-required mechanisms solely in castration-recurrent cells.

Another key cell messenger, cyclic AMP (cAMP), which modulates expression of a number of cyclin proteins, is also androgen regulated. Protein kinase A is the primary target for cAMP signaling in the cell. Expression of the C_β subunit of Protein kinase A is increased in response to androgen treatment in prostate cancer cells (Kvissel *et al.*, 2007). Furthermore, expression of a variant $C_\beta 2$ is increased in prostate tumor cells compared to normal (Kvissel *et al.*, 2007), suggesting that C_β isoforms may play different roles in proliferation and differentiation. As androgen-sensitive targets, they represent one point of cross-talk between androgen signaling and that of other hormones. Additional research is required to determine if these proteins have potential for use as markers in prostate cancer progression.

B. Lipid and Steroid Metabolism

Lipid metabolism and steroid biosynthesis are significant for proliferation and differentiated functions of prostate cells. Lipids provide growing and dividing cells with energy, act as membrane constituents, and help in oxidative metabolism.

A key enzyme in the lipid metabolic pathway is acetyl-CoA acyltransferase, which catalyzes the final step of β-oxidation. The action of acetyl-CoA acyltransferase metabolizes fatty acids to acetyl-CoA, the starting molecule of the Krebs cycle. Acetyl-CoA acyltransferase, and another peroxisomal enzyme fatty acyl-CoA oxidase 3 are both upregulated by androgens in LNCaP cells, while the mitochondrial enzyme, acetyl-CoA acyltransferase 2 is downregulated by androgens (Ngan *et al.*, 2009). Thus, at least in prostate cancer cells, androgens appear to preferentially stimulate peroxisomal over mitochondrial branched fatty acid β-oxidation, suggesting an energy preference in the oncogenic state. Prostaglandins are another class of lipids that function as growth-stimulatory signaling molecules, especially in tumor cells. Indeed, the name prostaglandin was derived from the relative abundance of these molecules in the prostate. A key enzyme in prostaglandin synthesis, hydroxy-prostaglandin dehydrogenase, is highly androgen-responsive in LNCaP cells (Ngan *et al.*, 2009).

Cholesterol and acyl-CoA transport regulates membrane production, which has profound effects upon prostate cell proliferation, growth, signaling, and metabolism. Many genes involved in these pathways have been shown to be regulated by androgens (Swinnen *et al.*, 2004). Thus, abrogation of key lipogenic genes induces apoptosis in prostate cancer cell lines and tumor models. Steroid biosynthesis is a fundamental aspect of prostate cancer as it is present in both early PIN lesions and after the development of castration-recurrent disease. A large number of genes involved in steroid precursor synthesis, especially the enzyme of the rate-limiting step converting squalene to squalene-2,3-epoxide, squalene monooxygenase, are upregulated in the more aggressive tumors (Holzbeierlein *et al.*, 2004). Androgens also regulate the expression of AZGP1, a soluble protein that regulates lipolysis, the breakdown of triglycerides to free fatty acids (Bohm *et al.*, 2009).

Expression of Sterol Response Element Binding Proteins-1 and -2 (SREBP-1 and -2) is dysregulated in the progression to castration-recurrent prostate cancer, in both human and murine models (Ettinger *et al.*, 2004). Their downstream effectors, acyl-CoA-binding protein/diazepam-binding inhibitor, fatty acid synthase (FAS), ELOVL7 (a fatty acid elongase enzyme), and farnesyl diphosphate synthase are also regulated by androgens and are dysregulated in prostate cancer (Ettinger *et al.*, 2004; Tamura *et al.*, 2009). SCAP (Sterol Regulatory Element-binding Protein Cleavage-Activating Protein) is the sensor that regulates SREBP action. SCAP stimulates the activity of S1P and S2P protease activity in the absence of sterols, thus maintaining inactive SREBP transcription factor in the membrane of the Golgi. Following proteolysis, SREBP is released and enters the nucleus where it stimulates the transcription of factors involved in fatty acid and cholesterol synthesis, such as the LDL receptor and HMG-CoA synthase (Brown and Goldstein, 1999). SCAP expression is also increased during prostate cancer progression

(Ettinger *et al.*, 2004), and is regulated by androgens via an ARE in intron 8 (Heemers *et al.*, 2001, 2004). Also, expression of Kruppel-like factor 5 (KLF5), a transcription factor associated with EGFR regulation, which is a positive regulator of SREBP-1 transcriptional activity, is increased following androgen treatment of LNCaP cells (Lee *et al.*, 2009; Ngan *et al.*, 2009).

C. TMPRSS2:ERG Fusions

Functional protein products resulting from gene fusions have been implicated in other cancers, the most well-known of these is the BCR-Abl oncogene in chronic myeloid leukemia. Although fusions have been described in other types of mesenchymal tumors, it has been only recently that they have been shown to be prevalent in epithelial tumors. The TMPRSS2:ERG family of prostate tumor fusions was first described via a cancer outlier profile analysis (COPA) of the public access Oncomine database, which catalogues gene expression profiles from multiple cancer cell lines and patient tissues (Tomlins *et al.*, 2005). Currently, these kinds of fusions have been found to occur in approximately half of all prostate cancers (Clark and Cooper, 2009). The *TMPRSS2* and *ERG* genes are less than 3 Mb apart on chromosome 21. Fusions occur equally through interchromosomal insertion and deletion of the intervening region. Although more than 20 other isoforms between *TMPRSS2* and members of the ETS family of transcription factors have been described, and heterogeneity exists in the location of the fusions, the most common fusion is exon 1 of *TMPRSS2* fused to exon 4 of *ERG* (Clark and Cooper, 2009; Hermans *et al.*, 2009; Tomlins *et al.*, 2009). The resulting fusion generates a truncated ERG protein that is constitutively active, and expressed from the androgen-regulated *TMPRSS2* promoter, instead of a chimeric protein, as in BCR-Abl. While the resulting protein products resemble each other to a significant degree, the variability in the noncoding regulatory region of the other fusion gene can provide unique regulatory characteristics (Hermans *et al.*, 2009). Interestingly, androgenic signaling itself may drive the formation of fusions by facilitating interactions of disparate genomic regions and stimulating DNA breaks (Lin *et al.*, 2009b; Mani *et al.*, 2009).

Overexpression of common truncated fusion partners *ERG*, *ETV1*, or *ETV5* in primary or immortalized benign prostate epithelial cells increases cell migration and invasion in all reported models, but does not induce transformation (Tomlins *et al.*, 2009). Also, tellingly, overexpression of truncated fusion partners in mouse prostate epithelia results in PIN lesions, rather than cancer (Cai *et al.*, 2009; Klezovitch *et al.*, 2008; Tomlins *et al.*, 2007, 2008; Zong *et al.*, 2009). Thus, these animal models suggest that the generation of the fusion is not an initiating event, but mediates the transition to invasive cancer. Indeed, numerous studies of fusion expression in patient samples support the concept that the rearrangement itself is an early event, and that expression continues through all stages of disease. However, no consistent correlations have been observed between fusion expression and clinical outcome (Clark and Cooper, 2009). Due to the specificity of the fusions to prostate tumors, they may provide information for diagnosis and monitoring of malignancy, with minimal invasiveness for the patient (Rostad *et al.*, 2009). TMPRSS2:ERG fusions are frequent in treatment-naive lymph node metastases, but the presence of fusions does not further correlate with duration of endocrine therapy (Boormans *et al.*, 2009). Therefore, fusions are not helpful in selecting candidates for endocrine therapy.

Fusions appear to cooperate with other events to drive proliferation and metastasis in later stages of prostate cancer progression. *PTEN* deletions have been found to be correlated with *ERG* rearrangements during progression from benign tissue, to aggressive, castration-recurrent, to metastatic prostate cancer (Han *et al.*, 2009). Expression of high levels of ERG or ETV1 alone in mouse prostate cells results in hyperplasia and PIN lesions (Zong *et al.*, 2009). However, when combined with Pten knockdown or AKT upregulation, high levels of ERG or ETV1 induce adenocarcinoma. Moreover, upregulation of AR expression, combined with increased expression of ERG, results in a poorly differentiated, invasive carcinoma (Zong *et al.*, 2009). This may result from a collaboration with AR signaling to activate gene expression, since ETV1 (ETS variant 1) and ERG can bind with the AR on the PSA enhancer in prostate cancer cells (Shin *et al.*, 2009).

D. MicroRNAs

MicoRNAs (miRNAs) are short (21–23 nucleotide) fragments of single-stranded RNA that can regulate gene expression. They exert their effects by binding to mRNA sequences, effectively preventing target gene expression either by inhibiting translation or promoting RNA degradation, and are responsible for modulating as much as 30% of human gene expression (Xie *et al.*, 2005). Profiles of miRNA expression in cancer versus normal are highly informative as they can provide clues regarding developmental lineage and differentiation state of tumors (Lu *et al.*, 2005). An examination of previously identified miRNAs in LNCaP and derivative castration-recurrent cell lines found that the expression of 10 miRNAs is increased and 7 decreased in castration-recurrent compared to androgen-dependent lines (deVere White *et al.*, 2009). miR-125b exhibits increased expression in prostate cancer cell lines compared to normal cell lines; AR-positive cell lines express more than AR-negative cell lines, and its expression is sensitive to antiandrogen treatment (deVere White *et al.*, 2009). Transfection of miR-125b into LNCaP cells allows their growth in androgen-free conditions and leads to the downregulation of expression of Bak1, an apoptosis inhibitor that binds to Bcl-2 (deVere White *et al.*, 2009; Shi *et al.*, 2007). Other bioinformatics approaches have identified additional targets of miR-125b that are upregulated specifically in prostate cancer. EIF4EBP1 is a target of miR-125b, and is increased in prostate cancer tissues (Ozen *et al.*, 2007). The expression of other miRNAs, for example, miR-21, is androgen-sensitive, not androgen-dependent. Enhanced expression of miR-21 promotes prostate tumor growth *in vivo* and is sufficient for androgen-dependent tumors to overcome castration-mediated growth arrest (Ribas *et al.*, 2009).

E. Miscellaneous

As the panel of ARGs grows, the number of targets whose regulation leads to unknown effects within the cell also grows. The myosin light chain kinase (MLCK) is a calcium/calmodulin-dependent kinase involved in regulation of smooth muscle contraction (Driska *et al.*, 1981). MLCK was identified as an androgen-regulated target via an Affymetrix GeneChip Human Genome U95 Set analysis of LNCaP cells stimulated with androgen. MLCK was decreased by 80% at 18 h after treatment, and was accompanied by a similar reduction in protein levels (Léveillé *et al.*, 2009). Although the consequence of this regulation is speculative, this kinase may promote apoptosis. $\mu\mu$-Crystallin (CRYM) is a vision-based cytoplasmic protein that binds thyroid-hormone T3 after NADPH-activation and enhances hormone concentration. The expression of CRYM is increased in prostate cancer, but is reduced in castration-recurrent prostate tumors (Malinowska *et al.*, 2009). *Per1*, a circadian clock factor is another newly identified androgen target, which is present in prostate cancer cells (Cao *et al.*, 2009). Levels of *Per1* are lower in prostate cancer tissues compared to normal, and Per1 inhibits AR transactivation in 293 T and LNCaP cells. Ectopic overexpression of Per1 in prostate cancer cells induces apoptosis and inhibits proliferation (Cao *et al.*, 2009). Androgens can also affect genes that are foreign to prostate cancer, in particular the Human cytomegalovirus major immediate early (HCMV MIE) promoter via activation of PKA activity. It has been speculated that this modulation enhances malignancy of the virus (Michaelis *et al.*, 2009; Moon *et al.*, 2008).

Androgenic regulation of other transcription factors lends another level of complexity in prostate cells. Regulation of the Nerve growth factor IB (NR4A1) is one such transcription factor with still incompletely categorized function (Ngan *et al.*, 2009). Androgens also modulate the expression of regulatory proteins such as FKBP51, an immunophilin with peptidyl-prolyl isomerase activity that is part of chaperone complexes associated with steroid hormone transcription factors, which may also lead to feedback regulation of AR activity itself (Febbo *et al.*, 2005). The effect of androgens on Runx2 leads to changes in a number of pathways, due to the pleiotropic effects of Runx2. Runx2 is repressed by binding of the DNA-binding domain of the AR, thus inhibiting the recruitment of Runx2 to DNA, in both osteoblasts and prostate cancer cells (Baniwal *et al.*, 2009). In clinical prostate cancer samples, Runx2 target genes show inverse expression in treatment naïve and samples from patients treated with androgen-ablation therapy (Baniwal *et al.*, 2009). Another pleiotropic factor regulated by androgens is the chemokine receptor CXCR4 (Ngan *et al.*, 2009). The expression of CXCR4 is stimulated in response to androgen treatment of LNCaP cells. This upregulation is an indirect effect, due to the upregulation KLF5 by androgen (Ngan *et al.*, 2009). The result of this increase is increased migration of LNCaP cells in a gradient of the ligand of CXCR4, CXCL12. This is possibly a metastasis mechanism, since expression of CXCL12 is elevated at metastatic sites, for example, bone (Frigo *et al.*, 2009).

III. NOVEL DISCOVERIES PERTAINING TO ANDROGEN RECEPTOR

The other difficult challenge in gaining a thorough understanding of how androgens regulate gene expression is the many different ways in which the activity of the AR can be regulated. Specific regulation of the receptor allows for modulation of its transcriptional targets, and has been demonstrated to play a role in prostate cancer pathology.

A. Modulation of Androgen Receptor Expression

Many studies have demonstrated that increased expression of AR sensitizes prostate cancer cells to lower levels of androgens (Chen *et al.*, 2008). When the expression of the AR is increased four to six times higher than endogenous levels in LNCaP cells, about 70–85% more genes are activated 2–4 h after treatment with the same concentrations of androgen (Waltering *et al.*, 2009). At this basic level of regulation, large-scale alterations in transcriptional programs can occur with simple changes in the expression of one transcription factor. Two categories exist for mechanisms that affect the levels to which AR protein is expressed: genomic/mRNA expression and protein turnover.

1. GENOMIC/mRNA EXPRESSION

The amplification of the AR gene in castration-recurrent prostate cancer, via chromosome alteration and other mechanisms has been reviewed elsewhere (Chen *et al.*, 2008), and many studies support the concept that increased AR expression is a common feature of prostate cancer compared to benign tissue (Lévesque *et al.*, 2009). Amplification of the AR gene leads to increased basal transcription of mRNA. In one study, an increase in AR mRNA was the only consistent change associated with the development of resistance to androgen-ablation therapy, with the caveat that the resulting overexpressed protein must have a functional ligand binding domain (Chen *et al.*, 2004). Increased expression of AR in prostate cancer cells alters assembly of coregulator complexes in chromatin immunoprecipitation experiments, thus changing the transcriptional activity of the AR and its response to agonists and antagonists (Chen *et al.*, 2004).

Increased expression of AR mRNA can result from increased transcription from endogenous promoters or from stabilization of mRNA. Posttranscriptional regulation of AR mRNA is partly due to the Heterogeneous nuclear ribonucleoprotein K (HnRNP-K), a poly-C RNA binding protein, which binds to a site in the 5′-UTR of the AR mRNA (Mukhopadhyay *et al.*, 2009). HnRNP-K binding inhibits translation of AR mRNA, thus reducing AR gene expression and slowing proliferation of prostate cancer cells. Activation of ErbB1/EGFR/HER1 and ErbB2/HER2/neu signaling has also been demonstrated to lead to decreased expression of endogenous AR and PSA expression in prostate cancer cells. Activation of signaling through these receptors enhances AR mRNA degradation, and is not inhibited by PI3K or MEK inhibitors (Cai *et al.*, 2009). Prolonged exposure of LNCaP cells and tumors with a prostate stroma-derived ErbB1 ligand, heparin-binding epidermal growth factor-like factor (HB-EGF), results in reduced AR protein expression and reduced sensitivity of androgen-responsive promoters to androgen (Adam *et al.*, 2002). HB-EGF reduces AR protein levels through mTOR independent of ErbB2, Erk1/2, and p38 MAPK, and regulates cap-dependent mRNA translation (Cinar *et al.*, 2005).

2. PROTEIN TURNOVER

Posttranslational regulation of AR protein is an additional powerful mechanism by which the expression levels of AR can be modulated. The association of AR with heat shock protein promotes maturation of AR and facilitates conformations of AR that are amenable to ligand binding. One of the key factors in the heat shock protein complex is Hsp90. Association of Hsp90 with the AR is regulated by the deacetylase activity of HDAC6, a member of the family of histone deacetylases. Sulforaphane inhibition of HDAC6 in prostate cancer cells leads to destabilization of the AR (Gibbs *et al.*, 2009). Castration-recurrent C4-2 cells with reduced expression of HDAC6 exhibit slower tumor growth in mouse xenografts, and do not establish tumors at all in castrated mice. These cells also require higher levels of androgen to reach their maximal proliferation rate, due to decreased nuclear AR and reduced expression of PSA, under both ligand-dependent and ligand-free conditions (Ai *et al.*, 2009). Neuroendocrine cells within the prostate secrete growth factors, which facilitate survival of surrounding tumor cells (Bonkhoff *et al.*, 1991). Parathyroid hormone related protein (PTHrP) is one such factor that has been demonstrated to promote androgen-dependent prostate cancer cell proliferation (Dougherty *et al.*, 1999). PTHrP binds EGFR, which activates Src kinase activation, thereby stimulating downstream phosphorylation of the AR on Tyr^{534}. Phosphorylation at this residue reduces interaction of the AR with CHIP, an E3 ligase, which targets AR for degradation with a polyubiquitin degradation signal (DaSilva *et al.*, 2009). Abrogation of the AR/CHIP complex also occurs in prostate cancer cells expressing a mutant form of AR (E255K), which promotes stability and ligand-independent nuclear localization of AR by inhibiting CHIP E3 ligase activity (Steinkamp *et al.*, 2009).

3. ALTERATIONS IN ANDROGEN RECEPTOR STRUCTURE

a. Androgen Receptor Mutations

Mutations in the AR can have profound effects on androgen-mediated signaling. Some of these mutants result in altered coregulator interactions (Brooke *et al.*, 2007), or changes in ligand specificity, thus leading to positive growth advantages of prostate cancer cells independent of ligand binding (Brooke and Bevan, 2009). The well-described T877A mutation found in the LNCaP cell line creates a promiscuous AR due to changes in the conformation of the helix 12 domain (Zhou *et al.*, 2009). Treatment with antiandrogens can select for gain-of-function AR mutations with altered stability, promoter preference, or ligand specificity. For example, AR-V716M creates a promiscuous receptor, AR23, which is cytoplasmically restricted but enhances the ligand response of wtAR, presumably through interacting with chaperone proteins and allowing wtAR to escape degradation, and the W435L amino-terminal domain mutant influences promoter and cell-selective AR transactivation (Steinkamp *et al.*, 2009).

b. Androgen Receptor Splice Variants

AR isoforms that result from alternative splicing of its mRNA have recently been discovered. These AR splice variants have varied activities due to alterations in their ligand binding domains. Some of these AR variants may stimulate a subset of castration-recurrent tumors in the absence of androgen. A constitutively active, carboxy-terminal truncated AR variant was identified in the 22Rv1 cell line, which is derived from the castration-recurrent CWR22 xenograft (Sramkoski *et al.*, 1999). This alternative splicing product includes a different novel exon, 2b, which encodes a stop codon, creating two protein isoforms that lack carboxy-terminal domains, $AR^{1/2/2b}$ and $AR^{1/2/3/2b}$ (Dehm *et al.*, 2008). These isoforms promote expression of endogenous AR-regulated genes and ligand-free proliferation of 22Rv1 cells, and were also present in LuCaP23.1 and LuCaP35 castration-recurrent xenografts (Dehm *et al.*, 2008). Also, in the 22Rv1 cell line, an AR variant with a duplication of exon 3, resulting in an AR with three zinc fingers was described, and can be cleaved to result in a constitutively active form (Libertini *et al.*, 2007; Tepper *et al.*, 2002). This calpain-dependent cleavage of the AR results in a constitutively active fragment, which may contribute to castration-recurrence as inhibition of calpain activity prevented xenograft growth following castration (Libertini *et al.*, 2007).

A BLAST search of the all AR intron sequences within the NCBI human EST database identified novel cryptic exons for the AR (Hu *et al.*, 2009). Cloning of these AR variants into expression plasmids and expression in prostate cancer cells showed that the two most abundant variants are constitutively active. Moreover, these variants are expressed 20-fold higher in castration-recurrent tumors than hormone naïve samples, on average (Hu *et al.*, 2009). AR3 (AR-V7) is another constitutively active AR variant, whose transcriptional activity is not regulated by androgens or antiandrogens (Guo *et al.*, 2009; Hu *et al.*, 2009). Immunohistochemical analysis of AR3 expression in a prostate cancer tissue microarray revealed that AR3 is upregulated in prostate cancer progression and is associated with an increased risk of tumor recurrence after prostatectomy (Guo *et al.*, 2009). AR3 is also able to specifically regulate *AKT1* as evidenced by binding to its promoter region, but not the PSA enhancer region; suggesting that AR3, and possibly other AR isoforms, have unique and overlapping transcriptional programs compared to wild-type AR (Guo *et al.*, 2009).

B. Regulation of Androgen Receptor Activity

The AR is able to form myriad interactions with regulatory factors including the basal transcription machinery, other transcription factors, signaling molecules, and many others. This section will review some of the mechanisms that modulate AR activity.

1. INTERACTION WITH TRANSCRIPTIONAL MACHINERY

While the canonical ARE is still an important criteria used for searching and identifying genes that are regulated by androgens, recently new AR motifs have been identified through which binding can occur (Lin *et al.*, 2009a). This discovery may account for the expanded list of androgen-regulated targets found in many cell line models, including the AR-null

PC3 cell line after exogenous expression of the AR (Lin *et al.*, 2009a). The presence of multiple AREs in the regulatory regions of many androgen-regulated targets allow for coordination of transcriptional regulation based upon the recruitment of the AR, cofactors, and key transcriptional machinery. For example, on the PSA gene, full activation requires a complex formation including the AR, coactivators, and RNA polymerase II at both the enhancer and the promoter region, but repression of the gene can occur when factors are bound only at the promoter and not the enhancer. This evidence has led to the understanding of a mechanism of chromatin looping, which brings the enhancer region in close proximity to the promoter, thus creating a permissible AR transcription environment (Shang *et al.*, 2002). This mechanism may play an important role in the efficient regulation of other androgen-regulated targets. However, the individuality of each gene, that is, possessing different quantities of AR binding sequences in different locations, suggests a singular mechanism for each target. Enhancer elements have been described to be located as far as 90 kb away from transcription start sites, as is the case for *FKBP51* (Makkonen *et al.*, 2009).

Cell- and promoter-specific activity of the AR is also in part due to regulation of the intramolecular N/C interaction between FxxLF and WxxLF sequences in the amino terminus and the ligand binding domain (He *et al.*, 2000). Successful intramolecular interaction can promote aberrant AR transcriptional activity in addition to conferring ligand-independent activity (Dehm *et al.*, 2007). The ligand binding domain can also interact with similar LxxLL motifs of coactivator proteins, which can interfere with N/C formation (Savkur and Burris, 2004). Interaction of the ^{23}FQNLF^{27} region with the ligand binding domain affects the ability of the AR to activate transcription in the context of chromatin, but deletion or mutation in the region of amino acids 501–535 does not, while still impairing N/C interaction (Need *et al.*, 2009). These results suggest that mutations or alterations in this area may contribute to aberrant transcriptional activity of the AR in prostate cancer cells.

2. Effect of Coregulators

The identification and description of previously known AR coregulators as well as their roles in castration-recurrent prostate cancer have been reviewed previously (Heemers and Tindall, 2007; Heinlein and Chang, 2002). In an examination of the expression of 186 coregulator genes in LNCaP cells, by highly sensitive DASL array, approximately 30% were found to be androgen-regulated (Heemers *et al.*, 2009). Understanding the effect that androgens have upon the expression of AR coregulators is complicated by a number of factors: the extent of expression, dose dependency, kinetics, and androgen-dependency of cells. The androgen-sensitivity of some coregulators allows for complicated feedback and feed-forward regulation of coregulator expression as well as AR activity. Four and a half LIM domain protein (FHL-2) was identified as an AR coactivator about a decade ago (Muller *et al.*, 2000). Recently, it was found to be androgen-regulated by an indirect mechanism in prostate cancer cells via the transcriptional activation of serum response factor (SRF) (Heemers *et al.*, 2007). The resulting androgenic induction of FHL-2 leads to a feed-forward mechanism to promote the full expression of AR targets. The precise mechanism of the androgenic signaling is conveyed to SRF remains unclear, since SRF expression is not regulated by androgen. However, SRF was found to be constitutively recruited to CArG box in the FHL-2 promoter, and is thought to interact with the AR at this site to promote transcription (Heemers *et al.*, 2007). The action of numerous steroid receptor cofactors including ARA70, Tip60, and CBP have been known for a number of years, and their ability to alter AR activity and activation by agonists and antagonists have been well described (Edwards and Bartlett, 2005b). The p160 coactivators (SRC-1 and -3) and the steroid receptor RNA activator (SRA) exhibit different abilities to induce AR activation in response to androgen and to the partial agonist, cyproterone acetate (Agoulnik and Weigel, 2009). Furthermore, different promoters have individual requirements for coregulator combinations for effective androgenic expression (Agoulnik and Weigel, 2009). Thus it is important to examine transcription factor activity within the physiological context of chromatin, rather than with reporter genes. For example, CBP has been shown to act as a corepressor of AR when recruited to pericentric regions, but a coactivator of AR when recruited to euchromatin (Zhao *et al.*, 2009). Therefore action of coregulators depends upon the state of chromatin at target loci. Modification of histones is an integral part of altering chromatin organization, and an important cohort of AR regulators either possesses histone modification activity or is responsible for recruiting these enzymes to specific promoters. While treatment of prostate cancer cells with inhibitors of histone deacetylases, such as trichostatin A, sodium butyrate, SAHA, and LBH589, can lead to decreased AR mRNA expression (Ai *et al.*, 2009; Welsbie *et al.*, 2009), the effect of HDAC inhibition on AR transcriptional activity is profound. Inhibition of HDAC1 and HDAC3 widely suppresses the expression of many AR targets, including *TMPRSS2*. Inhibition of HDAC activity prevents the proper assembly of coactivator/ RNA Pol II complex after binding of AR to target genes, thus inhibiting AR activity (Welsbie *et al.*, 2009). HDACs retain their functionality in castration-recurrent prostate cancer, and so remain promising targets of potential therapy. Another AR coregulator Yin Yang 1 (YY1), which belongs to the GLI-Kruppel family of zinc-finger transcription factors, recruits HDAC1 and p300 to target promoters (Yao *et al.*, 2001). YY1 expression is increased in neoplastic and PIN lesions compared with matched benign tissues

(Seligson *et al.*, 2005). Reduced expression of YY1 leads to reduced expression of PSA; however, increased expression of YY1 does not affect PSA expression (Deng *et al.*, 2009). A corepressor of AR BAF60a, which is a subunit of the SWI/SNF chromatin remodeling complex, was identified by computational *in silico* screening. Knockdown of BAF60a in LNCaP cells reduces androgen-induced TMPRSS2 expression (van de Wijngaart *et al.*, 2009). GATA-2, a member of the GATA family of transcription factors, collaborates with the transcription factor Oct-1 to recruit AR to chromatin (Wang *et al.*, 2007). GATA-2 expression is associated with biochemical recurrence and metastatic progression. Modulation of its expression in prostate cancer cell lines affects AR gene expression (Bohm *et al.*, 2009). Pax6, a transcription factor with roles in development and glioblastoma, is expressed to a higher level in normal tissues compared with prostate cancer. The enforced expression of Pax6 in LNCaP suppresses cell proliferation and represses expression of PSA via direct interaction with the AR (Shyr *et al.*, 2009). The homeodomain protein, HOXB13 interacts with the AR and modulates the expression of AR target genes that also have a HOX element in proximity to an ARE (Norris *et al.*, 2009). Since HOXB13 has an important function in prostate differentiation and maturation, the collaboration of these two transcription factors can synergize to stimulate the expression of a subset of genes important in development, maturation, and differentiated function. Thus, the cross-talk of the AR with other transcription factors and gene expression machinery allows for cell and context-dependent activity.

Considered to be an archetype of an ARGs, PSA may also play a role in modulating AR function. PSA can cooperate with ARA70 to promote AR transactivation and growth of AR-positive cell lines (Niu *et al.*, 2008b). This action does not require the protease activity of PSA, and suggests that PSA may promote tumor growth via augmentation of AR activity.

The function of many coregulators is to modify the AR in some way, thus changing its conformation or regulating the interactions the receptor can form with other proteins. Some recently described AR coregulators that modify the AR structure include βArrestin2 and RNF6. βArrestin2 serves as an adapter molecule to bring the Ubiquitin-E3 ligase Mdm2 and the AR together, thus promoting polyubiquitination of the AR, and subsequent degradation by the proteasome. Expression of βArrestin2 inversely correlates with the expression of PSA in human prostate tissue, supporting the repressive function of βArrestin2 on androgen-mediating transcription (Lakshmikanthan *et al.*, 2009). Similarly, RNF6 has been identified as a coregulator of the AR (Xu *et al.*, 2009). RNF6 was found to interact with activated AR in both castration-recurrent (CWR-R1) and androgen-dependent (LNCaP) cells. RNF6 is also an ubiquitin E3 ligase, and its activity leads to polyubiquitination of the AR on K845. The resulting polyubiquitin chains are K6, K27 linked as opposed to K48, K63 linked, which signify proteasomal degradation. This unique polyubiquitin signal by RNF6 modulates AR activity in the presence and absence of androgen. A putative mechanism suggests that AR coregulator ARA54 binds this polyubiquitinated AR and promotes androgen-mediated transcription for a subset of targets. Indeed, RNF6 effects the expression of a number of AR target genes (i.e., *PPAP2A*, *TMEPAI*, *RLN1*, *KLK3*, *NKX3.1*, *BMF*). However, its activity is selective, since it does not affect expression of *PDIA5*, *SLC45A3*, *TMPRSS2*, or *SORD*. ReChIP analysis demonstrated that RNF6 binds to the AR at some target promoter sites but not others, supporting the concept that RNF6 mediates the regulation of a subset of AR targets. A role for RNF6-mediated AR regulation is supported by the observation of higher levels of RNF6 protein in castration-recurrent prostate cancer cell lines and tumors. Also, castration-recurrent prostate cancer mouse xenografts that do not express RNF6, or an AR form that is unable to be ubiquitinated at K845 show dramatically reduced growth rates (Xu *et al.*, 2009).

Many factors have been described to have coregulator activity in addition to their canonical functions, further complicating our understanding of androgenic signaling. The melanoma antigen gene protein-A11 (MAGE-11) interacts with AR at the amino-terminal FxxLF motif, and increased expression increases AR transcriptional activity in prostate cancer cells (Karpf *et al.*, 2009). The expression of MAGE-11 increases following androgen-ablation therapy and progression of the tumor, in both the CWR22 xenograft and in patient samples, due to CpG promoter hypomethylation and cAMP signaling (Karpf *et al.*, 2009). Another example is Myosin VI, the only myosin motor protein to translocate toward the pointed end of actin filaments, which forms a complex with the AR. When expression of Myosin VI is decreased in prostate cancer cells, AR expression decreases (Loikkanen *et al.*, 2009). Modulation of Myosin VI leads to changes in androgen-regulated reporter gene expression in prostate cancer cells (Loikkanen *et al.*, 2009). Another novel AR coregulator Peroxiredoxin 1 (Prx1) is a member of the mammalian peroxidase family of proteins whose major function is to combat reactive oxygen species (Fujii and Ikeda, 2002). As a transcriptional coregulator, it interacts with the AR and modulates expression of AR target genes in prostate cancer cell lines (Park *et al.*, 2007). The interaction of Prx1 with the AR enhances the amino-carboxy interaction of the AR dimer, promoting functionality of the receptor (Chhipa *et al.*, 2009). Low levels of Prx1 desensitize LNCaP cells to androgen, effectively making the cells acquire more androgen for full AR activity (Chhipa *et al.*, 2009).

3. POSTTRANSLATIONAL MODIFICATIONS

Modification of the AR, as with other signaling proteins, via covalent binding of molecular moieties affords another level of regulation. Phosphorylation of the AR at multiple sites both in the presence and absence of ligand promotes AR transactivation, regulates recruitment of cofactors, affects expression of target genes, and increases prostate cancer cell growth. The most prevalent phosphorylation signals are due to MAPK, Akt, and PKC signaling, and in many cases AR regulation by phosphorylation can promote castration-recurrent phenotypes (Edwards and Bartlett, 2005a). Prevalent Akt activity is a common feature of prostate cancer, especially the castration-recurrent stage, as deletion of PTEN leads to constitutive activity of Akt (Majumder and Sellers, 2005). Akt can phosphorylate the AR at Ser-210 and Ser-790, which results in suppression of AR target genes, such as p21, by inhibiting the binding of required cofactors (Lin *et al.*, 2001). Akt interaction with the E3 ubiquitin ligase Mdm2 leads to phosphorylation-dependent ubiquitylation, promoting AR degradation (Lin *et al.*, 2002). Another phosphorylation site of the AR (Ser-213) is found in normal prostate epithelial cells, but not stromal cells (Taneja *et al.*, 2005). This phosphorylation site is modulated through PI3K signaling and occurs in response to androgen, leading to inhibition of transcriptional activity, which is consistent with its presence in minimally proliferating tissue (Taneja *et al.*, 2005). The response of the AR to regulatory phosphorylation events is also mutable depending upon the context. The passage number of LNCaP cells in culture effects the response of AR activity following PI3K/Akt phosphorylation, perhaps due to growth factor signaling such as the IGF-1. At lower passages, the phosphorylation program results in suppression of the receptor, while at higher passages the same program enhances AR activity (Lin *et al.*, 2003). Removal of phosphate moieties can also affect activity of the AR. Protein phosphatase 1 (PP1) can remove Ser-650 phosphate groups from AR. Ser-650 resides in the hinge region of AR and is responsible for regulating nuclear export of the receptor. Inhibition of PP1 reduces accumulation of AR in the nucleus. Thus PP1 is able to enhance AR transcription by promoting cellular localization that is conducive to gene transcription (Chen *et al.*, 2009). Nuclear transport of AR is also achieved through the action of protein kinase D1, via a mechanism that includes Hsp27 (Hassan *et al.*, 2009).

IV. CONCLUSIONS

Enhanced sensitivity and capacity of microarray analyses have improved our understanding of androgen signaling in prostate cancer. Thus, systems biology and computational mathematics approaches are beginning to delineate the many variables of androgenic signaling, including how the AR is regulated and which target genes synergize to impart optimal cellular responses. These kinds of analyses have the potential to yield new insight into cellular responses to hormone signaling, and may provide improved methods for treating both androgen dependent and castration-recurrent prostate cancer.

REFERENCES

Adam, R. M., Kim, J., Lin, J., Orsola, A., Zhuang, L., Rice, D. C., and Freeman, M. R. (2002). Heparin-binding epidermal growth factor-like growth factor stimulates androgen-independent prostate tumor growth and antagonizes androgen receptor function. *Endocrinology* **143,** 4599–4608.

Agoulnik, I. U., and Weigel, N. L. (2009). Co-activator selective regulation of androgen receptor activity. *Steroids* **74,** 669–674.

Ai, J., Wang, Y., Dar, J. A., Liu, J., Liu, L., Nelson, J. B., and Wang, Z. (2009). HDAC6 regulates androgen receptor hypersensitivity and nuclear localization via modulating Hsp90 acetylation in castration-resistant prostate cancer. *Mol. Endocrinol,* me.2009-0188.

Balk, S. P., and Knudsen, K. (2008). AR, the cell cycle, and prostate cancer. *Nucl. Recept. Signal.* **6,** e001.

Baniwal, S. K., Khalid, O., Sir, D., Buchanan, G., Coetzee, G. A., and Frenkel, B. (2009). Repression of Runx2 by androgen receptor (AR) in osteoblasts and prostate cancer cells: AR binds Runx2 and abrogates its recruitment to DNA. *Mol. Endocrinol.* **23,** 1203–1214.

Bohm, M., Locke, W. J., Sutherland, R. L., Kench, J. G., and Henshall, S. M. (2009). A role for GATA-2 in transition to an aggressive phenotype in prostate cancer through modulation of key androgen-regulated genes. *Oncogene* **28**(43), 3847–3856.

Bonkhoff, H., Wernert, N., Dhom, G., and Remberger, K. (1991). Relation of endocrine-paracrine cells to cell proliferation in normal, hyperplastic, and neoplastic human prostate. *Prostate* **19,** 91–98.

Boormans, J. L., Hermans, K. G., Made, A. C. J. Z.-v.d., van Leenders, G. J. H. L., Wildhagen, M. F., Collette, L., Schröder, F. H., Trapman, J., and Verhagen, P. C. M. S. (2009). Expression of the androgen-regulated fusion gene TMPRSS2-ERG does not predict response to endocrine treatment in hormone-naïve, node-positive prostate cancer. *Eur. Urol.* (In Press).

Brooke, G., and Bevan, C. L. (2009). The role of androgen receptor mutations in prostate cancer progression. *Curr. Genomics* **9,** 18–25.

Brooke, G. N., Parker, M. G., and Bevan, C. L. (2007). Mechanisms of androgen receptor activation in advanced prostate cancer: Differential co-activator recruitment and gene expression. *Oncogene* **27,** 2941–2950.

Brown, M. S., and Goldstein, J. L. (1999). A proteolytic pathway that controls the cholesterol content of membranes, cells, and blood. *Proc. Natl. Acad. Sci. USA* **96,** 11041–11048.

Cai, C., Portnoy, D. C., Wang, H., Jiang, X., Chen, S., and Balk, S. P. (2009). Androgen receptor expression in prostate cancer cells is suppressed by activation of epidermal growth factor receptor and ErbB2. *Cancer Res.* **69,** 5202–5209.

Cao, Q., Gery, S., Dashti, A., Yin, D., Zhou, Y., Gu, J., and Koeffler, H. P. (2009). A role for the clock gene Per1 in prostate cancer. *Cancer Res.* **69,** 7619–7625.

Chen, C. D., Welsbie, D. S., Tran, C., Baek, S. H., Cheng, R., Vessella, R. L., Rosenfeld, M. G., and Sawyers, C. L. (2004). Molecular determinants of resistance to antiandrogen therapy. *Nat. Med.* **10,** 33–39.

Chen, G., Nomura, M., Morinaga, H., Matsubara, E., Okabe, T., Goto, K., Yanase, T., Zheng, H., Lu, J., and Nawata, H. (2005). Modulation of androgen receptor transactivation by FoxH1. *J. Biol. Chem.* **280,** 36355–36363.

Chen, Y., Sawyers, C. L., and Scher, H. I. (2008). Targeting the androgen receptor pathway in prostate cancer. *Curr. Opin. Pharmacol.* **8,** 440–448.

Chen, S., Kesler, C. T., Paschal, B. M., and Balk, S. P. (2009). Androgen receptor phosphorylation and activity are regulated by an association with protein phosphatase 1. *J. Biol. Chem.* **284,** 25576–25584.

Chhipa, R. R., Lee, K.-S., Onate, S., Wu, Y., and Ip, C. (2009). Prx1 enhances androgen receptor function in prostate cancer cells by increasing receptor affinity to dihydrotestosterone. *Mol. Cancer Res.* **7,** 1543–1552.

Cinar, B., De Benedetti, A., and Freeman, M. R. (2005). Post-transcriptional regulation of the androgen receptor by mammalian target of rapamycin. *Cancer Res.* **65,** 2547–2553.

Clark, J. P., and Cooper, C. S. (2009). ETS gene fusions in prostate cancer. *Nat. Rev. Urol.* **6,** 429–439.

DaSilva, J., Gioeli, D., Weber, M. J., and Parsons, S. J. (2009). The neuroendocrine-derived peptide parathyroid hormone-related protein promotes prostate cancer cell growth by stabilizing the androgen receptor. *Cancer Res.* 0008-5472.CAN-08-4687.

Debes, J. D., and Tindall, D. J. (2004). Mechanisms of androgen-refractory prostate cancer. *N. Engl. J. Med.* **351,** 1488–1490.

Dehm, S. M., and Tindall, D. J. (2006). Molecular regulation of androgen action in prostate cancer. *J. Cell. Biochem.* **99,** 333–344.

Dehm, S. M., Regan, K. M., Schmidt, L. J., and Tindall, D. J. (2007). Selective role of an NH2-terminal WxxLF motif for aberrant androgen receptor activation in androgen depletion independent prostate cancer cells. *Cancer Res.* **67,** 10067–10077.

Dehm, S. M., Schmidt, L. J., Heemers, H. V., Vessella, R. L., and Tindall, D. J. (2008). Splicing of a novel androgen receptor exon generates a constitutively active androgen receptor that mediates prostate cancer therapy resistance. *Cancer Res.* **68,** 5469–5477.

Deng, Z., Wan, M., Cao, P., Rao, A., Cramer, S. D., and Sui, G. (2009). Yin Yang 1 regulates the transcriptional activity of androgen receptor. *Oncogene* **28**(42), 3746–3757.

deVere White, R. W., Vinall, R. L., Tepper, C. G., and Shi, X.-B. (2009). MicroRNAs and their potential for translation in prostate cancer. *Urol. Oncol.: Semin. Orig. Investig.* **27,** 307–311.

Dougherty, K., Blomme, E., Koh, A., Henderson, J., Pienta, K., Rosol, T., and McCauley, L. (1999). Parathyroid hormone-related protein as a growth regulator of prostate carcinoma. *Cancer Res.* **59,** 6015–6022.

Driska, S. P., Aksoy, M. O., and Murphy, R. A. (1981). Myosin light chain phosphorylation associated with contraction in arterial smooth muscle. *Am. J. Physiol. Cell Physiol.* **240,** C222–C233.

Edwards, J., and Bartlett, J. M. S. (2005a). The androgen receptor and signal-transduction pathways in hormone-refractory prostate cancer. Part 1: Modifications to the androgen receptor. *BJU Int.* **95,** 1320–1326.

Edwards, J., and Bartlett, J. M. S. (2005b). The androgen receptor and signal-transduction pathways in hormone-refractory prostate cancer. Part 2: Androgen-receptor cofactors and bypass pathways. *BJU Int.* **95,** 1327–1335.

Ettinger, S. L., Sobel, R., Whitmore, T. G., Akbari, M., Bradley, D. R., Gleave, M. E., and Nelson, C. C. (2004). Dysregulation of sterol response element-binding proteins and downstream effectors in prostate cancer during progression to androgen independence. *Cancer Res.* **64,** 2212–2221.

Febbo, P. G., Lowenberg, M., Thorner, A. R., Brown, M., Loda, M., and Golub, T. R. (2005). Androgen mediated regulation and functional implications of FKBP51 expression in prostate cancer. *J. Urol.* **173,** 1772–1777.

Frigo, D. E., Sherk, A. B., Wittmann, B. M., Norris, J. D., Wang, Q., Joseph, J. D., Toner, A. P., Brown, M., and McDonnell, D. P. (2009). Induction of Kruppel-like factor 5 expression by androgens results in increased CXCR4-dependent migration of prostate cancer cells *in vitro*. *Mol. Endocrinol.* **23,** 1385–1396.

Fujii, J., and Ikeda, Y. (2002). Advances in our understanding of peroxiredoxin, a multifunctional, mammalian redox protein. *Redox Rep.* **7,** 123–130.

Gao, N., Zhang, J., Rao, M. A., Case, T. C., Mirosevich, J., Wang, Y., Jin, R., Gupta, A., Rennie, P. S., and Matusik, R. J. (2003). The role of hepatocyte nuclear factor-3{alpha} (Forkhead Box A1) and androgen receptor in transcriptional regulation of prostatic genes. *Mol. Endocrinol.* **17,** 1484–1507.

Gibbs, A., Schwartzman, J., Deng, V., and Alumkal, J. (2009). Sulforaphane destabilizes the androgen receptor in prostate cancer cells by inactivating histone deacetylase 6. *Proc. Natl. Acad. Sci.* **106,** 16663–16668.

Greenberg, N., DeMayo, F., Finegold, M., Medina, D., Tilley, W., Aspinalls, J., Cunha, G., Donjacour, A., Matusik, R., and Rosen, J. (1994). Prostate cancer in a transgenic mouse. *PNAS* **92,** 3439–3443.

Guo, Z., Yang, X., Sun, F., Jiang, R., Linn, D. E., Chen, H., Chen, H., Kong, X., Melamed, J., Tepper, C. G., Kung, H.-J., Brodie, A. M. H., *et al.* (2009). A novel androgen receptor splice variant is up-regulated during prostate cancer progression and promotes androgen depletion-resistant growth. *Cancer Res.* **69,** 2305–2313.

Han, B., Mehra, R., Lonigro, R. J., Wang, L., Suleman, K., Menon, A., Palanisamy, N., Tomlins, S. A., Chinnaiyan, A. M., and Shah, R. B. (2009). Fluorescence *in situ* hybridization study shows association of PTEN deletion with ERG rearrangement during prostate cancer progression. *Mod. Pathol.* **22,** 1083–1093.

Hassan, S., Biswas, M. H. U., Zhang, C., Du, C., and Balaji, K. C. (2009). Heat shock protein 27 mediates repression of androgen receptor function by protein kinase D1 in prostate cancer cells. *Oncogene* **28**(49), 4386–4396.

He, B., Kemppainen, J. A., and Wilson, E. M. (2000). FXXLF and WXXLF sequences mediate the NH2-terminal interaction with the ligand binding domain of the androgen receptor. *J. Biol. Chem.* **275,** 22986–22994.

Heemers, H., Maes, B., Foufelle, F., Heyns, W., Verhoeven, G., and Swinnen, J. V. (2001). Androgens stimulate lipogenic gene expression in prostate cancer cells by activation of the sterol regulatory element-binding protein cleavage activating protein/sterol regulatory element-binding protein pathway. *Mol. Endocrinol.* **15,** 1817–1828.

Heemers, H., Verrijdt, G., Organe, S., Claessens, F., Heyns, W., Verhoeven, G., and Swinnen, J. V. (2004). Identification of an androgen response element in intron 8 of the sterol regulatory element-binding protein cleavage-activating protein gene allowing direct regulation by the androgen receptor. *J. Biol. Chem.* **279,** 30880–30887.

Heemers, H. V., and Tindall, D. J. (2007). Androgen receptor (AR) co-regulators: A diversity of functions converging on and regulating the AR transcriptional complex. *Endocr. Rev.* **28,** 778–808.

Heemers, H. V., Regan, K. M., Dehm, S. M., and Tindall, D. J. (2007). Androgen induction of the androgen receptor co-activator four and a half LIM domain protein-2: Evidence for a role for serum response factor in prostate cancer. *Cancer Res.* **67,** 10592–10599.

Heemers, H. V., Regan, K. M., Schmidt, L. J., Anderson, S. K., Ballman, K. V., and Tindall, D. J. (2009). Androgen modulation of co-regulator expression in prostate cancer cells. *Mol. Endocrinol.* **23,** 572–583.

Heinlein, C. A., and Chang, C. (2002). Androgen receptor (AR) co-regulators: An overview. *Endocr. Rev.* **23,** 175–200.

Hellawell, G. O., Turner, G. D. H., Davies, D. R., Poulsom, R., Brewster, S. F., and Macaulay, V. M. (2002). Expression of the type 1 insulin-like growth factor receptor is up-regulated in primary prostate cancer and commonly persists in metastatic disease. *Cancer Res.* **62,** 2942–2950.

Hermans, K. G., Boormans, J. L., Gasi, D., van Leenders, G. J. H. L., Jenster, G., Verhagen, P. C. M. S., and Trapman, J. (2009). Overexpression of prostate-specific TMPRSS2(exon 0)-ERG fusion transcripts corresponds with favorable prognosis of prostate cancer. *Clin. Cancer Res.* **15**, 6398–6403.

Holzbeierlein, J., Lal, P., LaTulippe, E., Smith, A., Satagopan, J., Zhang, L., Ryan, C., Smith, S., Scher, H., Scardino, P., Reuter, V., and Gerald, W. L. (2004). Gene expression analysis of human prostate carcinoma during hormonal therapy identifies androgen-responsive genes and mechanisms of therapy resistance. *Am. J. Pathol.* **164**, 217–227.

Hu, R., Dunn, T. A., Wei, S., Isharwal, S., Veltri, R. W., Humphreys, E., Han, M., Partin, A. W., Vessella, R. L., Isaacs, W. B., Bova, G. S., and Luo, J. (2009). Ligand-independent androgen receptor variants derived from splicing of cryptic exons signify hormone-refractory prostate cancer. *Cancer Res.* **69**, 16–22.

Huggins, C., and Hodges, C. V. (1972). Studies on prostatic cancer: I. The effect of castration, of estrogen and of androgen injection on serum phosphatases in metastatic carcinoma of the prostate. *CA Cancer J. Clin.* **22**, 232–240.

Jemal, A., Siegel, R., Ward, E., Hao, Y., Xu, J., Murray, T., and Thun, M. J. (2008). Cancer statistics, 2008. *CA Cancer J. Clin.* **58**, 71–96.

Jin, F., and Fondell, J. D. (2009). A novel androgen receptor-binding element modulates Cdc6 transcription in prostate cancer cells during cell-cycle progression. *Nucleic Acids Res.* **37**, 4826–4838.

Karpf, A. R., Bai, S., James, S. R., Mohler, J. L., and Wilson, E. M. (2009). Increased expression of androgen receptor co-regulator MAGE-11 in prostate cancer by DNA hypomethylation and cyclic AMP. *Mol. Cancer Res.* **7**, 523–535.

Klezovitch, O., Risk, M., Coleman, I., Lucas, J. M., Null, M., True, L. D., Nelson, P. S., and Vasioukhin, V. (2008). A causal role for ERG in neoplastic transformation of prostate epithelium. *Proc. Natl. Acad. Sci.* **105**, 2105–2110.

Kvissel, A.-K., Ramberg, H., Eide, T., Svindland, A., Skålhegg, B. S., and Taskén, K. A. (2007). Androgen dependent regulation of protein kinase A subunits in prostate cancer cells. *Cell. Signal.* **19**, 401–409.

Lakshmikanthan, V., Zou, L., Kim, J. I., Michal, A., Nie, Z., Messias, N. C., Benovic, J. L., and Daaka, Y. (2009). Identification of Beta-Arrestin2 as a co-repressor of androgen receptor signaling in prostate cancer. *Proc. Natl. Acad. Sci.* **106**, 9379–9384.

Lee, M.-Y., Moon, J.-S., Park, S. W., Koh, Y.-k., Ahn, Y.-H., and Kim, K.-S. (2009). KLF5 enhances SREBP-1 action in androgen-dependent induction of fatty acid synthase in prostate cancer cells. *Biochem. J.* **417**, 313–322.

Léveillé, N., Fournier, A., and Labrie, C. (2009). Androgens down-regulate myosin light chain kinase in human prostate cancer cells. *J. Steroid Biochem. Mol. Biol.* **114**, 174–179.

Lévesque, M.-H., El-Alfy, M., Cusan, L., and Labrie, F. (2009). Androgen receptor as a potential sign of prostate cancer metastasis. *Prostate* **9999**, n/a.

Libertini, S. J., Tepper, C. G., Rodriguez, V., Asmuth, D. M., Kung, H.-J., and Mudryj, M. (2007). Evidence for calpain-mediated androgen receptor cleavage as a mechanism for androgen independence. *Cancer Res.* **67**, 9001–9005.

Lin, H.-K., Yeh, S., Kang, H.-Y., and Chang, C. (2001). Akt suppresses androgen-induced apoptosis by phosphorylating and inhibiting androgen receptor. *Proc. Natl. Acad. Sci. USA* **98**, 7200–7205.

Lin, H., Wang, L., Hu, Y., Altuwaijri, S., and Chang, C. (2002). Phosphorylation-dependent ubiquitylation and degradation of androgen receptor by Akt require Mdm2 E3 ligase. *EMBO J.* **21**, 4037–4048.

Lin, H.-K., Hu, Y.-C., Yang, L., Altuwaijri, S., Chen, Y.-T., Kang, H.-Y., and Chang, C. (2003). Suppression versus induction of androgen receptor functions by the phosphatidylinositol 3-kinase/Akt pathway in prostate cancer LNCaP cells with different passage numbers. *J. Biol. Chem.* **278**, 50902–50907.

Lin, B., Wang, J., Hong, X., Yan, X., Hwang, D., Cho, J.-H., Yi, D., Utleg, A. G., Fang, X., Schones, D. E., Zhao, K., Omenn, G. S., *et al.* (2009a). Integrated expression profiling and ChIP-seq analyses of the growth inhibition response program of the androgen receptor. *PLoS ONE* **4,** e6589.

Lin, C., Yang, L., Tanasa, B., Hutt, K., Ju, B.-g., Ohgi, K., Zhang, J., Rose, D. W., Fu, X.-D., Glass, C. K., and Rosenfeld, M. G. (2009b). Nuclear receptor-induced chromosomal proximity and DNA breaks underlie specific translocations in cancer. *Cell* **139,** 1069–1083.

Loikkanen, I., Toljamo, K., Hirvikoski, P., Vaisanen, T., Paavonen, T., and Vaarala, M. (2009). Myosin VI is a modulator on androgen-dependent gene expression. *Oncol. Rep.* **22,** 991–995.

Lu, J., Getz, G., Miska, E. A., Alvarez-Saavedra, E., Lamb, J., Peck, D., Sweet-Cordero, A., Ebert, B. L., Mak, R. H., Ferrando, A. A., Downing, J. R., Jacks, T., *et al.* (2005). MicroRNA expression profiles classify human cancers. *Nature* **435,** 834–838.

Majumder, P. K., and Sellers, W. R. (2005). Akt-regulated pathways in prostate cancer. *Oncogene* **24,** 7465–7474.

Makkonen, H., Kauhanen, M., Paakinaho, V., Jaaskelainen, T., and Palvimo, J. J. (2009). Long-range activation of FKBP51 transcription by the androgen receptor via distal intronic enhancers. *Nucleic Acids Res.* **37,** 4135–4148.

Malinowska, K., Cavarretta, I. T., Susani, M., Wrulich, O. A., Überall, F., Kenner, L., and Culig, Z. (2009). Identification of μ-crystallin as an androgen-regulated gene in human prostate cancer. *Prostate* **69,** 1109–1118.

Mani, R.-S., Tomlins, S. A., Callahan, K., Ghosh, A., Nyati, M. K., Varambally, S., Palanisamy, N., and Chinnaiyan, A. M. (2009). Induced chromosomal proximity and gene fusions in prostate cancer. *Science* **326,** 1230.

McLeod, D., Crawford, E., and DeAntoni, E. (1997). Combined androgen blockade: The gold standard for metastatic prostate cancer. *Eur. Urol.* **32,** 70–77.

Michaelis, M., Doerr, H. W., and Cinatl, J. J. (2009). Oncomodulation by human cytomegalovirus: Evidence becomes stronger. *Med. Microbiol. Immunol.* **198,** 79–81.

Moon, J.-S., Lee, M.-Y., Park, S. W., Han, W. K., Hong, S.-W., Ahn, J.-H., and Kim, K.-S. (2008). Androgen-dependent activation of human cytomegalovirus major immediate-early promoter in prostate cancer cells. *Prostate* **68,** 1450–1460.

Mukhopadhyay, N. K., Kim, J., Cinar, B., Ramachandran, A., Hager, M. H., Di Vizio, D., Adam, R. M., Rubin, M. A., Raychaudhuri, P., De Benedetti, A., and Freeman, M. R. (2009). Heterogeneous nuclear ribonucleoprotein K is a novel regulator of androgen receptor translation. *Cancer Res.* **69,** 2210–2218.

Muller, J. M., Isele, U., Metzger, E., Rempel, A., Moser, M., Pscherer, A., Breyer, T., Holubarsch, C., Buettner, R., and Schule, R. (2000). FHL2, a novel tissue-specific co-activator of the androgen receptor. *EMBO J.* **19,** 359–369.

Need, E. F., Scher, H. I., Peters, A. A., Moore, N. L., Cheong, A., Ryan, C. J., Wittert, G. A., Marshall, V. R., Tilley, W. D., and Buchanan, G. (2009). A novel androgen receptor amino terminal region reveals two classes of amino/carboxyl interaction-deficient variants with divergent capacity to activate responsive sites in chromatin. *Endocrinology* **150,** 2674–2682.

Ngan, S., Stronach, E. A., Photiou, A., Waxman, J., Ali, S., and Buluwela, L. (2009). Microarray coupled to quantitative RT-PCR analysis of androgen-regulated genes in human LNCaP prostate cancer cells. *Oncogene* **28,** 2051–2063.

Niu, Y., Altuwaijri, S., Lai, K.-P., Wu, C.-T., Ricke, W. A., Messing, E. M., Yao, J., Yeh, S., and Chang, C. (2008a). Androgen receptor is a tumor suppressor and proliferator in prostate cancer. *Proc. Natl. Acad. Sci.* **105,** 12182–12187.

Niu, Y., Yeh, S., Miyamoto, H., Li, G., Altuwaijri, S., Yuan, J., Han, R., Ma, T., Kuo, H.-C., and Chang, C. (2008b). Tissue prostate-specific antigen facilitates refractory prostate tumor progression via enhancing ARA70-regulated androgen receptor transactivation. *Cancer Res.* **68,** 7110–7119.

Norris, J. D., Chang, C.-Y., Wittmann, B. M., Kunder, R. S., Cui, H., Fan, D., Joseph, J. D., and McDonnell, D. P. (2009). The homeodomain protein HOXB13 regulates the cellular response to androgens. *Mol. Cell* **36**, 405–416.

Ozen, M., Creighton, C. J., Ozdemir, M., and Ittmann, M. (2007). Widespread deregulation of microRNA expression in human prostate cancer. *Oncogene* **27**, 1788–1793.

Park, S.-Y., Yu, X., Ip, C., Mohler, J. L., Bogner, P. N., and Park, Y.-M. (2007). Peroxiredoxin 1 interacts with androgen receptor and enhances its transactivation. *Cancer Res.* **67**, 9294–9303.

Perry, J. E., Grossmann, M. E., and Tindall, D. J. (1998). Epidermal growth factor induces cyclin D1 in a human prostate cancer cell line. *Prostate* **35**, 117–124.

Raclaw, K. A., Heemers, H. V., Kidd, E. M., Dehm, S. M., and Tindall, D. J. (2008). Induction of FLIP expression by androgens protects prostate cancer cells from TRAIL-mediated apoptosis. *Prostate* **68**, 1696–1706.

Ribas, J., Ni, X., Haffner, M., Wentzel, E. A., Salmasi, A. H., Chowdhury, W. H., Kudrolli, T. A., Yegnasubramanian, S., Luo, J., Rodriguez, R., Mendell, J. T., and Lupold, S. E. (2009). miR-21: An androgen receptor-regulated microRNA that promotes hormone-dependent and hormone-independent prostate cancer growth. *Cancer Res.* 0008-5472.CAN-09-1448.

Rostad, K., Hellwinkel, O. J. C., Haukaas, S. A., Halvorsen, O. J., ØYan, A. M., Haese, A., BudÄUs, L., Albrecht, H., Akslen, L. A., Schlomm, T., and Kalland, K.-H. (2009). TMPRSS2: ERG fusion transcripts in urine from prostate cancer patients correlate with a less favorable prognosis. *APMIS* **117**, 575–582.

Savkur, R. S., and Burris, T. P. (2004). The co-activator LXXLL nuclear receptor recognition motif. *J. Pept. Res.* **63**, 207–212.

Seligson, D., Horvath, S., Huerta-Yepez, S., Hanna, S., Garban, H., Robers, A., Shi, T., Liu, X., Chia, D., Goodglick, L., and Bonavida, B. (2005). Expression of transcription factor Yin Yang 1 in prostate cancer. *Int. J. Oncol.* **27**, 131–141.

Shang, Y., Myers, M., and Brown, M. (2002). Formation of the androgen receptor transcription complex. *Mol. Cell* **9**, 601–610.

Shi, X.-B., Xue, L., Yang, J., Ma, A.-H., Zhao, J., Xu, M., Tepper, C. G., Evans, C. P., Kung, H.-J., and deVere White, R. W. (2007). An androgen-regulated miRNA suppresses Bak1 expression and induces androgen-independent growth of prostate cancer cells. *Proc. Natl. Acad. Sci.* **104**, 19983–19988.

Shin, S., Kim, T.-D., Jin, F., van Deursen, J. M., Dehm, S. M., Tindall, D. J., Grande, J. P., Munz, J.-M., Vasmatzis, G., and Janknecht, R. (2009). Induction of prostatic intraepithelial neoplasia and modulation of androgen receptor by ETS variant 1/ETS-related protein 81. *Cancer Res.* **69**, 8102–8110.

Shyr, C.-R., Tsai, C.-R., Yeh, S., Kang, H.-Y., Chang, Y.-C., Wong, P.-L., Huang, C.-C., Huang, K.-E., and Chang, C. (2009). Tumor suppressor PAX6 functions as androgen receptor Co-repressor to inhibit prostate cancer growth. *Prostate* **9999**, n/a.

So, A., Hurtado-Coll, A., and Gleave, M. E. (2003). Androgens and prostate cancer. *World J. Urol.* **21**, 325–337.

Sramkoski, R. M., Pretlow, T. G., II, Giaconia, J. M., Pretlow, T. P., Schwartz, S., Sy, M.-S., Marengo, S. R., Rhim, S. R., Zhang, D., and Jacobberger, J. W. (1999). A new human prostate carcinoma cell line, 22R v1. *In Vitro Cell. Dev. Biol. Anim.* **35**, 403–409.

Steinkamp, M. P., O'Mahony, O. A., Brogley, M., Rehman, H., LaPensee, E. W., Dhanasekaran, S., Hofer, M. D., Kuefer, R., Chinnaiyan, A., Rubin, M. A., Pienta, K. J., and Robins, D. M. (2009). Treatment-dependent androgen receptor mutations in prostate cancer exploit multiple mechanisms to evade therapy. *Cancer Res.* **69**, 4434–4442.

Swinnen, J. V., Heemers, H., de Sande, T. V., Schrijver, E. D., Brusselmans, K., Heyns, W., and Verhoeven, G. (2004). Androgens, lipogenesis and prostate cancer. *J. Steroid Biochem. Mol. Biol.* **92**, 273–279.

Takayama, K., Horie-Inoue, K., Ikeda, K., Urano, T., Murakami, K., Hayashizaki, Y., Ouchi, Y., and Inoue, S. (2008). FOXP1 is an androgen-responsive transcription factor that negatively regulates androgen receptor signaling in prostate cancer cells. *Biochem. Biophys. Res. Commun.* **374,** 388–393.

Tamura, K., Makino, A., Hullin-Matsuda, F., Kobayashi, T., Furihata, M., Chung, S., Ashida, S., Miki, T., Fujioka, T., Shuin, T., Nakamura, Y., and Nakagawa, H. (2009). Novel lipogenic enzyme ELOVL7 is involved in prostate cancer growth through saturated long-chain fatty acid metabolism. *Cancer Res.* **69,** 8133–8140.

Taneja, S. S., Ha, S., Swenson, N. K., Huang, H. Y., Lee, P., Melamed, J., Shapiro, E., Garabedian, M. J., and Logan, S. K. (2005). Cell-specific regulation of androgen receptor phosphorylation *in vivo*. *J. Biol. Chem.* **280,** 40916–40924.

Tepper, C. G., Boucher, D. L., Ryan, P. E., Ma, A.-H., Xia, L., Lee, L.-F., Pretlow, T. G., and Kung, H.-J. (2002). Characterization of a novel androgen receptor mutation in a relapsed CWR22 prostate cancer xenograft and cell line. *Cancer Res.* **62,** 6606–6614.

Tian, M., and Schiemann, W. P. (2009). The TGF-Î paradox in human cancer: An update. *Future Oncol.* **5,** 259–271.

Tomlins, S. A., Rhodes, D. R., Perner, S., Dhanasekaran, S. M., Mehra, R., Sun, X.-W., Varambally, S., Cao, X., Tchinda, J., Kuefer, R., Lee, C., Montie, J. E., *et al.* (2005). Recurrent fusion of TMPRSS2 and ETS transcription factor genes in prostate cancer. *Science* **310,** 644–648.

Tomlins, S. A., Laxman, B., Dhanasekaran, S. M., Helgeson, B. E., Cao, X., Morris, D. S., Menon, A., Jing, X., Cao, Q., Han, B., Yu, J., Wang, L., *et al.* (2007). Distinct classes of chromosomal rearrangements create oncogenic ETS gene fusions in prostate cancer. *Nature* **448,** 595–599.

Tomlins, S. A., Laxman, B., Varambally, S., Cao, X., Yu, J., Helgeson, B. E., Cao, Q., Prensner, J. R., Rubin, M. A., Shah, R. B., Mehra, R., and Chinnaiyan, A. M. (2008). Role of the TMPRSS2-ERG gene fusion in prostate cancer. *Neoplasia* **10,** 177–188.

Tomlins, S. A., Bjartell, A., Chinnaiyan, A. M., Jenster, G., Nam, R. K., Rubin, M. A., and Schalken, J. A. (2009). ETS gene fusions in prostate cancer: From discovery to daily clinical practice. *Eur. Urol.* **56,** 275–286.

van de Wijngaart, D. J., Dubbink, H. J., Molier, M., de Vos, C., Trapman, J., and Jenster, G. (2009). Functional screening of FxxLF-like peptide motifs identifies SMARCD1/BAF60a as an androgen receptor cofactor that modulates TMPRSS2 expression. *Mol. Endocrinol,* me.2008-0280.

Waltering, K. K., Helenius, M. A., Sahu, B., Manni, V., Linja, M. J., Janne, O. A., and Visakorpi, T. (2009). Increased expression of androgen receptor sensitizes prostate cancer cells to low levels of androgens. *Cancer Res.* **69,** 8141–8149.

Wang, Q., Li, W., Liu, X. S., Carroll, J. S., Jänne, O. A., Keeton, E. K., Chinnaiyan, A. M., Pienta, K. J., and Brown, M. (2007). A hierarchical network of transcription factors governs androgen receptor-dependent prostate cancer growth. *Mol. Cell* **27,** 380–392.

Wang, Q., Li, W., Zhang, Y., Yuan, X., Xu, K., Yu, J., Chen, Z., Beroukhim, R., Wang, H., Lupien, M., Wu, T., Regan, M. M., *et al.* (2009). Androgen receptor regulates a distinct transcription program in androgen-independent prostate cancer. *Cell* **138,** 245–256.

Welsbie, D. S., Xu, J., Chen, Y., Borsu, L., Scher, H. I., Rosen, N., and Sawyers, C. L. (2009). Histone deacetylases are required for androgen receptor function in hormone-sensitive and castrate-resistant prostate cancer. *Cancer Res.* **69,** 958–966.

Wilson, J., Griffin, J., George, F., and Leshin, M. (1983). The endocrine control of male phenotypic development. *Aust. J. Biol. Sci.* **36,** 101–128.

Wu, C.-T., Altuwaijri, S., Ricke, W. A., Huang, S.-P., Yeh, S., Zhang, C., Niu, Y., Tsai, M.-Y., and Chang, C. (2007). Increased prostate cell proliferation and loss of cell differentiation in mice lacking prostate epithelial androgen receptor. *Proc. Natl. Acad. Sci. USA* **104,** 12679–12684.

Xie, X., Lu, J., Kulbokas, E. J., Golub, T. R., Mootha, V., Lindblad-Toh, K., Lander, E. S., and Kellis, M. (2005). Systematic discovery of regulatory motifs in human promoters and 3 [prime] UTRs by comparison of several mammals. *Nature* **434,** 338–345.

Xu, K., Shimelis, H., Linn, D. E., Jiang, R., Yang, X., Sun, F., Guo, Z., Chen, H., Li, W., Chen, H., Kong, X., Melamed, J., *et al.* (2009). Regulation of androgen receptor transcriptional activity and specificity by RNF6-induced ubiquitination. *Cancer Cell* **15,** 270–282.

Yao, Y.-L., Yang, W.-M., and Seto, E. (2001). Regulation of transcription factor YY1 by acetylation and deacetylation. *Mol. Cell. Biol.* **21,** 5979–5991.

Yuan, S., Trachtenberg, J., Mills, G. B., Brown, T. J., Xu, F., and Keating, A. (1993). Androgen-induced inhibition of cell proliferation in an androgen-insensitive prostate cancer cell line (PC-3) transfected with a human androgen receptor complementary DNA. *Cancer Res.* **53,** 1304–1311.

Zegarra-Moro, O. L., Schmidt, L. J., Huang, H., and Tindall, D. J. (2002). Disruption of androgen receptor function inhibits proliferation of androgen-refractory prostate cancer cells. *Cancer Res.* **62,** 1008–1013.

Zhao, Y., Takeyama, K.-i, Sawatsubashi, S., Ito, S., Suzuki, E., Yamagata, K., Tanabe, M., Kimura, S., Fujiyama, S., Ueda, T., Murata, T., Matsukawa, H., *et al.* (2009). Corepressive action of CBP on androgen receptor transactivation in pericentric heterochromatin in a *Drosophila* experimental model system. *Mol. Cell. Biol.* **29,** 1017–1034.

Zhou, J., Shirley, L., Geng, G., and Wu, J. H. (2009). Study of the impact of the T877A mutation on ligand-induced helix-12 positioning of the androgen receptor resulted in design and synthesis of novel antiandrogens. *Proteins: Struct. Funct. Bioinform.* **9999,** NA.

Zong, Y., Xin, L., Goldstein, A. S., Lawson, D. A., Teitell, M. A., and Witte, O. N. (2009). ETS family transcription factors collaborate with alternative signaling pathways to induce carcinoma from adult murine prostate cells. *Proc. Natl. Acad. Sci. USA* **106,** 12465–12470.

MYC in Oncogenesis and as a Target for Cancer Therapies

Ami Albihn,* John Inge Johnsen,† and Marie Arsenian Henriksson*

**Department of Microbiology, Tumor and Cell Biology (MTC) Karolinska Institutet, Stockholm, Sweden*

†Childhood Cancer Research Unit, Department of Woman & Children's Health Karolinska Institutet, Stockholm, Sweden

---

MYC proteins (c-MYC, MYCN, and MYCL) regulate processes involved in many if not all aspects of cell fate. Therefore, it is not surprising that the *MYC* genes are deregulated in several human neoplasias as a result from genetic and epigenetic alterations. The near "omnipotency" together with the many levels of regulation makes MYC

Advances in CANCER RESEARCH

0065-230X/10 $35.00
DOI: 10.1016/S0065-230X(10)07006-5

an attractive target for tumor intervention therapy. Here, we summarize some of the current understanding of MYC function and provide an overview of different cancer forms with *MYC* deregulation. We also describe available treatments and highlight novel approaches in the pursuit for MYC-targeting therapies. These efforts, at different stages of development, constitute a promising platform for novel, more specific treatments with fewer side effects. If successful a MYC-targeting therapy has the potential for tailored treatment of a large number of different tumors.

I. C-MYC, MYCN, AND MYCL: THREE VERSIONS OF A MULTIFUNCTIONAL PROTEIN

The *MYC* gene was originally identified in avian retroviruses as the oncogene responsible for inducing *my*elo*c*ytomatosis in birds (Sheiness and Bishop, 1979). The cellular homologue, *c-MYC*, was found to be evolutionarily conserved (Vennström *et al.*, 1982). Later, *MYCN* and *MYCL* were found amplified in neuroblastoma and in small cell lung cancer (SCLC), respectively (Henriksson and Luscher, 1996). These genes share the same general topography with the main open reading frame retained within the second and third exons. *c-MYC* is one of the most widely studied proto-oncogenes and it is localized to chromosome 8q24.21, a region that is translocated in Burkitt's lymphoma (BL) (Dalla-Favera *et al.*, 1982). The *MYC* genes encode short-lived nuclear phosphoproteins with a half-life of 20–30 min that are subsequently ubiquitinated for proteasomal degradation (Gregory and Hann, 2000). Human *c-MYC* encodes two major isoforms p67 (MYC-1) and p64 (MYC-2), with different expression patterns and biologically distinct functions (Hann *et al.*, 1994). Transcription of MYC-1 is initiated at a cryptic start codon at the end of exon 1, whereas the more abundant MYC-2 protein is transcribed from an ATG start codon in exon 2, yielding a 439-residue protein. MYC is a basic Helix–Loop–Helix Leucine Zipper (bHLHZip) protein that heterodimerizes with the small bHLHZip protein Max resulting in dimers with DNA-binding ability at CACGTG and similar E-box sequences. The basic region (b) promotes sequence-specific DNA binding; the HLHZip confers protein–protein interaction while the Zip domain functions in cooperation with the HLH to stabilize protein–protein interactions and to establish dimerization specificity. Studies of the c-MAX protein revealed that the bHLHZip region conferring Max heterodimerization and specific DNA binding to the E-box is critical for all known MYC functions (Conzen *et al.*, 2000; Luscher and Larsson, 1999). MYC is a multifunctional protein with the ability to regulate activities as distinct as cell cycle, growth and metabolism, differentiation, apoptosis, transformation, genomic instability, and angiogenesis (Fig. 1) (Meyer and Penn, 2008; Oster *et al.*, 2002). It is believed that the majority of these functions are

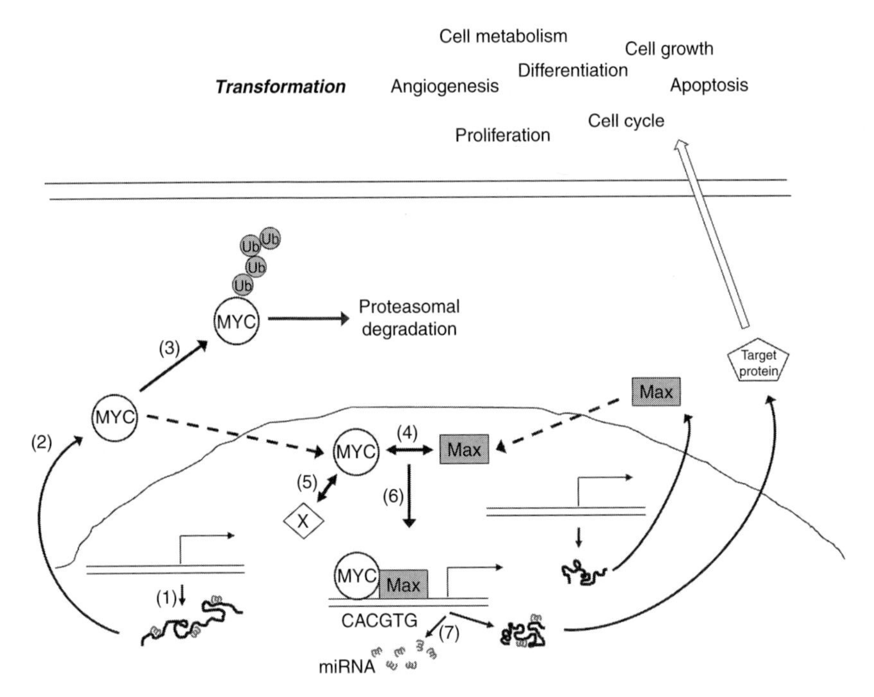

Fig. 1 *Different levels of MYC regulation and outcomes.* MYC activity can be regulated at the levels of [1] transcription, [2] translation, [3] ubiquitination and proteasomal degradation, [4] dimerization with Max, [5] dimerization with proteins other than Max (designated "X"), [6] DNA binding and target gene transcription, and [7] miRNA transcription. See text for details. Outcomes of MYC activation range from angiogenesis, proliferation, cell cycle, differentiation, cell growth, and metabolism, to apoptosis and if deregulated, transformation. MYC, c-MYC, MYCN, or MYCL protein; Ub, ubiquitin; miRNA, microRNA.

exerted through gene regulation, which is supported by the findings that MYC interacts with proteins essential for transcriptional regulation, for example, transformation/transcription domain-associated protein (TRRAP) and histone acetyltransferases (HATs). These interactions occur through an evolutionarily conserved region called MYC Box 2 (MB2) within the transcriptional activation domain (TAD) in the N-terminus of the protein (Fig. 2) (Henriksson and Luscher, 1996; Meyer and Penn, 2008; Oster *et al.*, 2002; Vita and Henriksson, 2006).

In addition to MB2 spanning residues 128–143 of the protein, the N-terminal TAD harbors another conserved region, MB1, encompassing residue 47–62 (Fig. 2). Several important functions have been ascribed to both MB1 and 2. As an example, MB1 encompasses the residues found to be required for MYC activity and breakdown (Henriksson *et al.*, 1993; Oster *et al.*, 2002) whereas MB2 is essential for cell transformation (Conzen *et al.*,

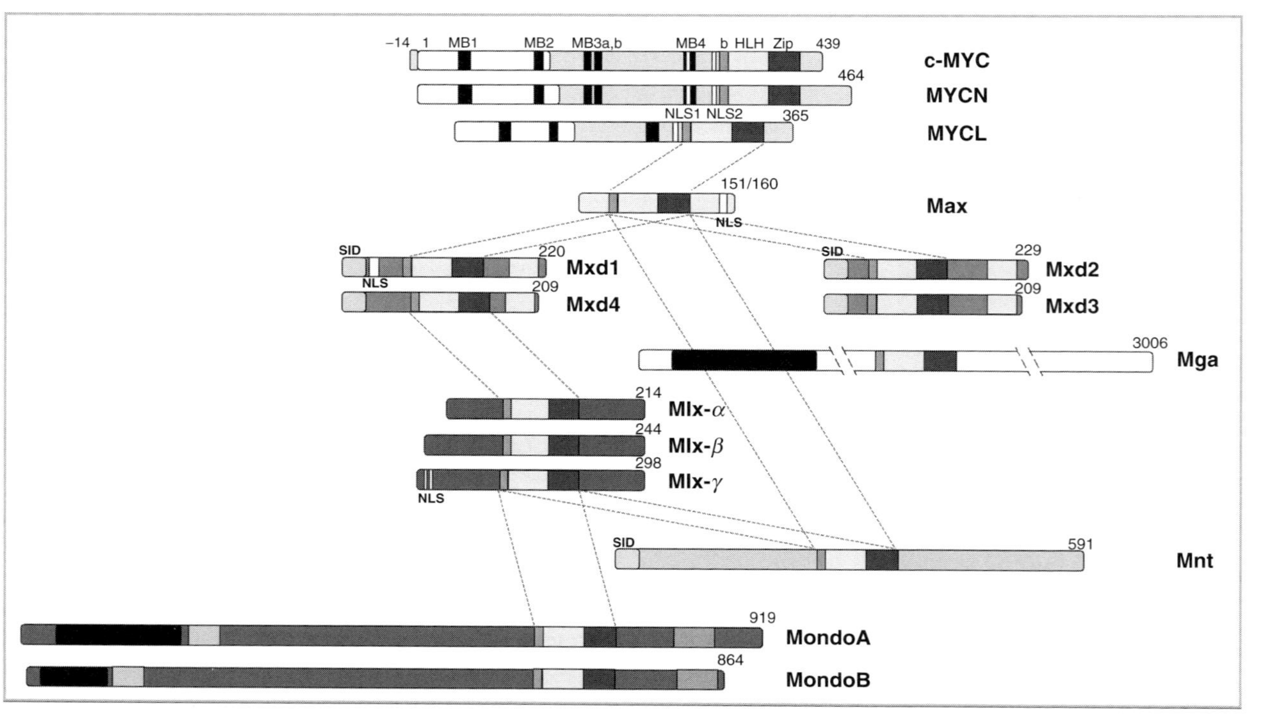

Fig. 2 *Members of the two parallel networks with the Mlx protein in the center.* The three MYC proteins are shown with MYC Boxes (MB) and bHLHZip domains. Dimerization partners are indicated by the dotted lines, connecting the bHLHZip regions of the respective proteins. Size and important structures of the proteins in the two networks are indicated. Mxd1 and -4 and the Mnt protein from the MYC/Max/Mxd network heterodimerize with both Max and Mlx, and the two Mondo proteins (A and B) belong to the Mlx network. NLS, nuclear localization signal; SID, Sin3-interacting domain.

2000; Gregory and Hann, 2000). More recently, two additional MYC boxes, MB3 (residues 188–199) and MB4 (304–324) were identified within the central region of the protein (reviewed in Meyer and Penn, 2008). MYC box 3 was found to play a role in cellular transformation (Herbst *et al.*, 2004, 2005). Finally, MB4 overlaps with the nuclear localization signal (NLS)-1 and is required for MYC-induced focus formation in immortalized Rat1a cells, but not for cotransformation of primary rat embryo fibroblasts by activated Ras and MYC (Cowling *et al.*, 2006). All four MYC boxes are conserved between species and are present in both c-MYC and MYCN whereas the more distantly related family member MYCL lacks MB3 (Ponzielli *et al.*, 2005). In addition to NLS1 (320–332) within MB4, there is also an NLS2 (364–374). However, only NLS1 confers complete nuclear localization while NLS2 provides only a partial nuclear targeting, probably because it overlaps with the basic DNA-binding region. c-MYC and MYCN contain both NLS domains while MYCL harbors only NLS2 (Henriksson and Luscher, 1996).

A. Expression Patterns of the *MYC* Family Genes

Mouse models have revealed that *c-MYC*, *MYCN* as well as *max* are essential for survival, thus placing the network in a central position in the regulation of cell growth and homeostasis (Henriksson and Luscher, 1996). During early embryogenesis, there is some redundancy between *c-MYC* and *MYCN* since *c-MYC*$^{-/-}$ and *MYCN* $^{-/-}$ embryos survive until day 9–10, and day 11, respectively (Davis *et al.*, 1993; Stanton *et al.*, 1992). Such compensatory mechanisms are proposed to be possible only until MYC expression becomes more tissue-restricted during organogenesis. Expression of *c-MYC* is generally high during early embryonic development where it is required for embryonic stem (ES) cell pluripotency and reprogramming in addition to proliferation (Cartwright *et al.*, 2005; Takahashi and Yamanaka, 2006). In differentiated adult tissues, however, the expression is low or undetectable consistent with the virtual absence of cell proliferation. In contrast to the almost ubiquitous expression of c-MYC, MYCN and MYCL expression levels are more restricted with respect to tissue and developmental stage (reviewed in Oster *et al.*, 2002; Ponzielli *et al.*, 2005). MYCN expression is very high early in embryogenesis in several tissues and declines dramatically during later development, generally coinciding with differentiation (Strieder and Lutz, 2002). The expression pattern of *MYCL* resembles that of *MYCN* but is even more restricted (Hatton *et al.*, 1996; Zimmerman *et al.*, 1986). After birth, MYCL is mainly expressed in the central nervous system, nasal epithelium, kidney, and lung. Neither MYCN nor MYCL expression correlates well with proliferation, further supporting the notion that their expression is characterizing the undifferentiated state rather than promoting cell growth and division (reviewed in Henriksson and Luscher, 1996; Oster *et al.*, 2002; Ponzielli *et al.*, 2005).

B. Several Levels of Regulation

In resting cells, *MYC* mRNA and protein are virtually undetectable but the MYC levels increase rapidly after serum stimulation followed by a relatively slow decline initiated before the onset of S phase (Henriksson and Luscher, 1996). Protein synthesis is not required for the rapid and transient MYC induction during the G0/G1 transition. However, in contrast to many of the early response genes, *MYC* levels are maintained at a constant intermediate level in continuously proliferating cells.

In normal cells, MYC expression and activity is regulated at multiple levels through transcriptional, posttranscriptional, translational, and posttranslational mechanisms (Fig. 1). On the transcriptional level, *MYC* is regulated by signal transduction pathways that are activated both during normal development and in cancer. The most important include Sonic hedgehog, Wnt, Notch, receptor tyrosine kinase signaling, and transforming growth factor (TGF)-β (see below). At the posttranslational level MYC protein expression can be controlled through sequential and reversible phosphorylation at two highly conserved sites, threonine 58 (T58) and serine 62 (S62), located in the amino-terminal TAD of the protein (Henriksson *et al.*, 1993 and reviewed in Hann, 2006). Phosphorylation of MYC at T58 and S62 regulates protein turnover through ubiquitination and 26 S proteasomal degradation (Bahram *et al.*, 2000; Salghetti *et al.*, 1999; Yeh *et al.*, 2004). Phosphorylation at S62 increases MYC stability whereas T58 phosphorylation stimulates ubiquitination and degradation by the SCF^{Fbw7} complex (Welcker *et al.*, 2004; Yada *et al.*, 2004; Yeh *et al.*, 2004). Point mutation of MYC at either T58 or S62 has been reported in BLs as well as other lymphomas resulting in increased MYC protein stability (Bahram *et al.*, 2000; Salghetti *et al.*, 1999). However, point mutations have not been reported in solid tumors, despite the fact that several of these tumors exhibit stabilized MYC protein (Schulein and Eilers, 2009). Hence, other mechanisms are believed to be responsible for the increased stabilization of MYC proteins in some solid tumors.

It has been hypothesized that the glycogen synthase kinase (GSK)-3β-mediated phosphorylation of c-MYC at T58 and the subsequent dephosphorylation of S62 allows for binding of the ubiquitin ligase Fbw7 and recruitment of the SCF^{FBW7} complex to direct MYC ubiquitination and proteasomal degradation (reviewed in Dai *et al.*, 2006). The Fbw7 tumor suppressor protein is lost in many carcinomas, most notably that of colon (Rajagopalan *et al.*, 2004). In contrast, the deubiquitinating enzyme USP28 that antagonizes the function of Fbw7, resulting in MYC protein stabilization, is overexpressed in breast and colon carcinoma (Popov *et al.*, 2007). USP28 was found to be required for MYC function by forming a ternary complex with MYC and Fbw7 in the nucleus thereby preventing proteasomal degradation. Axin1, a scaffold protein that facilitates the interaction of c-MYC with GSK-3β, protein phosphatase 2A (PP2A), and the prolyl-isomerase Pin1 resulting in increased c-MYC ubiquitination, is inactivated through mutation in several cancers with high MYC expression (Arnold *et al.*, 2009; Salahshor and Woodgett, 2005). In addition, a PP2A antagonist preventing S62 dephosphorylation of MYC was identified and designated cancerous inhibitor of PP2A (CIP2A) (Junttila *et al.*, 2007). CIP2A was established as an oncoprotein, providing MYC stabilization as part of its oncogenic repertoire. Furthermore, as CIP2A is overexpressed in head and neck squamous cell carcinoma (HNSCC) and in colon cancer, it has been suggested that targeting CIP2A may be a possible treatment opportunity (Junttila and Westermarck, 2008; Junttila *et al.*, 2007). Recently a number of reports have linked GSK-3β, Pin1, and PP2A as critical components of MYC protein degradation as GSK-3β phosphorylates MYC at T58 whereas Pin1 and PP2A cooperate to dephosphorylate S62 (Fig. 3) (Schulein and Eilers, 2009).

In addition to Fbw7, two other ubiquitin ligases are involved in regulating MYC protein turnover and/or activity, namely the F-box protein Skp2 and Hect H9, containing a Hect domain (Adhikary *et al.*, 2005; Kim *et al.*, 2003; von der Lehr *et al.*, 2003; and reviewed in Dai *et al.*, 2006). While Skp2 confers transcriptional activation as well as degradation of c-MYC (Kim *et al.*, 2003; von der Lehr *et al.*, 2003), HectH9 seems to only promote its transcriptional activity (Adhikary *et al.*, 2005).

High expression of *MYCN* is an important mediator of proliferation of neural precursor cells during the development of the central nervous system. In mice, these high MYCN levels were shown to be caused by the persistent activation of phosphatidylinositol-3-kinase (PI3K)/Akt signal transduction by insulin or insulin-like growth factor (IGF) signaling which result in the phosphorylation of GSK-3β (Fig. 3) (Knoepfler and Kenney, 2006). Other proteins regulating the PI3K/Akt pathway, such as Ras, are frequently altered in human cancer, resulting in inhibition of GSK-3β activity and subsequently stabilization of MYC proteins through loss of T58 phosphorylation (Cully *et al.*, 2006; Sears *et al.*, 2000). In addition, a number of recent reports have shown that inhibiting key proteins in the PI3K/Akt signal transduction pathway results in the activation of GSK-3β with subsequent destabilization of MYC proteins (Chesler *et al.*, 2006; Johnsen *et al.*, 2008; Meyer *et al.*, 2007; Mulholland *et al.*, 2006). GSK-3β is also a key protein in

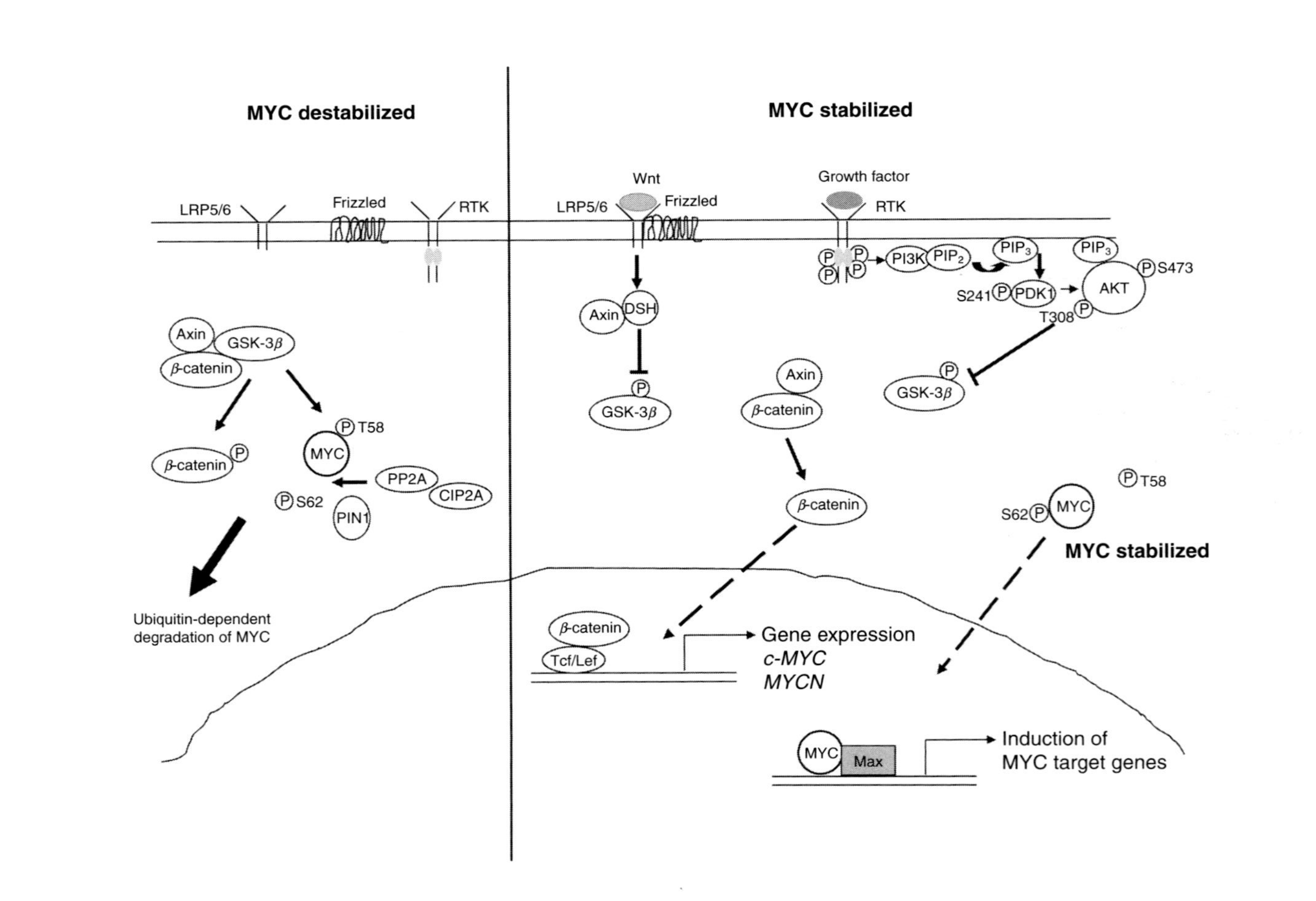

MYC destabilized
MYC stabilized
LRP5/6
Frizzled
RTK
Wnt
Growth factor
Axin
GSK-3β
β-catenin
MYC
T58
S62
PP2A
CIP2A
PIN1
Ubiquitin-dependent degradation of MYC
DSH
PI3K
PIP2
PIP3
PDK1
AKT
S241
S473
T308
Tcf/Lef
Gene expression
c-MYC
MYCN
Max
Induction of MYC target genes

the canonical Wnt/β-catenin signaling pathway by regulating the activity and nuclear translocation of β-catenin. In the absence of Wnt/Wingless ligand activation, β-catenin is sequestered in the cytoplasm by a multiprotein complex consisting of the adenomatous polyposis coli (APC) protein, Axin1, Axin2/Conductin, Casein kinase 1, and GSK-3β (Fig. 3). In this state, β-catenin is phosphorylated at amino-terminal serine and threonine residues by GSK-3β which targets it for ubiquitination and proteolytic degradation (Fodde and Brabletz, 2007). Activation of Wnt signaling by binding of Wnt ligands to a Frizzled receptor inhibits the formation of the multiprotein complex and GSK-3β-mediated phosphorylation of β-catenin resulting in an accumulation of hypophosphorylated β-catenin in the cytosol. Stabilized hypophosphorylated β-catenin eventually translocates to the nucleus where it interacts with members of the T cell factor/Lymphoid enhancer factor (Tcf/Lef) family of transcription factors, leading to increased transcription of a broad range of genes, including *MYC* (Fig. 3) (He *et al.*, 1998). Hence, agents that directly targets GSK-3β or key proteins in Wnt or PI3K/Akt signaling may have effects on MYC expression both through inhibition of *MYC* transcription and effects on MYC protein stability (Baryawno *et al.*, 2010; Johnsen *et al.*, 2008; Meyer *et al.*, 2007; Chesler *et al.*, 2006; Mulholland *et al.*, 2006). Despite the high MYC protein turnover rate, it has been reported that the highly unstable pool of the protein coexists with a metabolically stable pool within the cell (Tworkowski *et al.*, 2002). The difference in protein stability between these two pools is not due to cellular localization since they were both found within the nuclear compartment. In other cases, however, nuclear MYC protein has been detected predominantly in the cytoplasm. It has been suggested that hyperphosphorylation of MYC is one of the possible reasons for its redistribution to the cytoplasm (Oster *et al.*, 2002). Furthermore, it has been proposed that the transcription factor Miz-1, which is repressed by MYC, also in some circumstances regulates MYC activity by controlling its nuclear import (Peukert *et al.*, 1997).

Fig. 3 *Factors affecting MYC stability.* Phosphorylation of MYC is controlled by a complex interplay of key signal transduction molecules that have been shown to regulate the expression of MYC both at the posttranslational and transcriptional levels. Phosphorylation at two highly conserved sites, T58 and S62, regulates MYC protein turnover through ubiquitination and proteasomal degradation. MYC stability is increased by S62 phosphorylation whereas T58 phosphorylation stimulates ubiquitination and degradation. See text for details. LRP, low-density-related lipoprotein receptor; RTK, receptor tyrosine kinase; GSK-3β, glycogen synthase kinase-3β; PIN1, protein interacting with NIMA (never in mitosis A); PP2A, protein phosphatase 2A; CIP2A, cancerous inhibitor of PP2A; DSH, Dishevelled; PI3K, phosphatidylinositol 3-kinase; $PIP_{2,3}$, phosphatidylinositol (di-, tri-)phosphate; PDK1, phosphoinositide-dependent protein kinase-1; Tcf/Lef, T cell factor/Lymphoid enhancer factor.

II. NETWORKING IS KEY WITH MAX ACTING AS THE SPIDER IN THE WEB

As previously mentioned, the MYC dimerization partner Max, identified in 1991, was found to be an essential heterodimerization partner for all known c-MYC functions (Blackwood and Eisenman, 1991; Shen-Li *et al.*, 2000). It has an important role in embryonic development as mice lacking *max* die at day 5–6 of gestation (Gilladoga *et al.*, 1992; Shen-Li *et al.*, 2000). Max is highly conserved in vertebrate evolution and, with a half-life longer than 14 h, is constitutively expressed in a number of different cell types. The two major splice variants encode Max p21 and Max p22, which both form homodimers as well as heterodimers with other network members (Fig. 2). The homodimers possess lower affinity to DNA, seem less discriminating compared to Max heterodimer complexes and their DNA-binding properties may be negatively affected by phosphorylation (Banerjee *et al.*, 2006; Bousset *et al.*, 1993; Brownlie *et al.*, 1997). However, it is not clear whether the Max/Max homodimer has a function *in vivo*. In addition to the bHLHZip, Max contains an acidic region and a C-terminal NLS (Fig. 2) (reviewed in Henriksson and Luscher, 1996). Forced overexpression of *max* results in reduced growth (Zhang *et al.*, 1997) and induced differentiation (Canelles *et al.*, 1997).

In addition to the MYC proto-oncoproteins, the MYC/Max/Mxd network includes potential tumor suppressors (Mxd1–4 and Mnt) and Mga (Fig. 2). Whereas the Mxd proteins are mainly expressed in differentiated cells or tissues Mnt is expressed both in proliferating and differentiated cells and could function as a master regulator of MYC activity (reviewed in Wahlstrom and Henriksson, 2007). Some of the Mxd proteins interact with Mlx, another bHLHZip protein, suggested to be the center of a parallel network including the transcriptional activators Mondo A and B, proposed to regulate energy metabolism (reviewed in Billin and Ayer, 2006). However, the exact function of this parallel network is still poorly understood (Billin and Ayer, 2006). Thus the MYC network is in fact part of an intricate protein web. The crystal structure of the MYC/Max and the Mxd/Max complexes at the E-box revealed that there were marked structural differences in the dimerization patterns of the two heterodimers (Nair and Burley, 2003). While the MYC/Max heterodimers formed bivalent heterotetramers upon DNA binding, Mxd/Max complexes did not. The large size of the MYC/Max heterotetramers explains their ability to reach E-boxes spaced far apart and thereby to upregulate expression of such genes. Downstream effects are partially mediated through modifications of the chromatin structure to control DNA accessibility (Oster *et al.*, 2002; Ponzielli *et al.*, 2005).

A. Protein Interaction and Downstream Effects

Potential MYC-interacting proteins include the pRb-like p107 protein, the coactivator TRRAP, the multifunctional nucleolar protein, nucleophosmin (NPM), the tumor suppressor alternative reading frame (ARF), the transcriptional repressors TFII-I and Miz-1, as well as proteins responsible for MYC ubiquitination and proteasomal degradation discussed above (for review, see Dai *et al.*, 2006; Li and Hann, 2009; Oster *et al.*, 2002).

The interaction with p107 seems to involve a regulatory loop where the growth-inhibitory effects of p107 are counteracted by MYC while p107 significantly inhibits MYC-mediated transcriptional activation. However, p107 is unable to repress mutant c-MYC in BL (Gu *et al.*, 1994), possibly due to N-terminal alterations, preventing Cdk1-Cyclin A-mediated c-MYC phosphorylation. Such c-MYC mutations may be one way for the protein to escape regulation and contribute to oncogenesis (Hoang *et al.*, 1995). MYC also associates with HATs to acetylate histones and enable a transcription-permissive state of the chromatin at its target. The histone acetylation complexes are mainly recruited through the coactivator TRRAP, which was found to be essential for the transforming activity of MYC (McMahon *et al.*, 1998, 2000). Observed interactions with other HAT complexes and TRRAP-independent molecules with chromatin remodeling capacity raised the possibility that MYC could also recruit other complexes for controlling target gene transcription (Oster *et al.*, 2002). ARF has been shown to interact directly with c-MYC, leading to inhibition of its transforming activity while enhancing its apoptotic activity, independently of p53 (Li and Hann, 2009). NPM is another interactor that regulates c-MYC target gene expression by stimulating cell proliferation and transformation (reviewed in Li and Hann, 2009). c-MYC has also recently been suggested to have a function in chromatin dynamics since it was shown that its carboxy terminal region directly binds to the core subunit of the ATP-dependent chromatin remodeling complex: switching defective/sucrose nonfermenting (SWI/SNF), known as INI1/hSNF5. This hypothesis is further strengthened by the demonstrated interactions with other molecules with implications in chromatin remodeling, such as the ATPases/helicases TIP48/49 and the actin-related protein BAF53 (reviewed in Stojanova and Penn, 2009).

A number of MYC target genes have been described and documented in the MYC Target Gene Database: http://www.myc-cancer-gene.org/site/mycTargetDB.asp. Some examples are genes encoding Ornithine decarboxylase (Odc), p53, Carbamoylphosphate dihydroorotase (Cad), hTERT, and cell-cycle regulators such as Cdk4, Cyclin D, and E2Fs (for reviews, see Henriksson and Luscher, 1996; Meyer and Penn, 2008; Oster *et al.*, 2002; Ponzielli *et al.*, 2005). The Odc enzyme controls polyamine biosynthesis and is essential for progression into S phase. MYC-mediated Odc upregulation may contribute to the oncogenic phenotype since Odc overexpression in mouse fibroblasts results in transformation (Moshier *et al.*, 1993). The tumor suppressor protein p53, as previously described, is important in the cellular response to DNA damage with the ability to induce cell-cycle arrest or apoptosis (Sherr and Weber, 2000). MYC may activate p53 as a safeguard mechanism to prevent transformation by inducing apoptosis in MYC-overexpressing cells (Hermeking and Eick, 1994). Control of the G1/S transition is partially conferred by MYC through transcriptional induction of the Cad enzyme, required for *de novo* pyrimidine synthesis (Boyd and Farnham, 1997; Bush *et al.*, 1998; Miltenberger *et al.*, 1995). The catalytic subunit of telomerase (hTERT) also harbors E-box-elements to which MYC/Max as well as Mxd1/Max complexes have been shown to bind (Oh *et al.*, 2000; Wang *et al.*, 1998; Xu *et al.*, 2001). Indeed, activation or repression of MYC has been shown to alter hTERT activity both in normal and tumor cells (Grand *et al.*, 2002; Oh *et al.*, 1999; Wu *et al.*, 1999). Sustained MYC-induced telomerase activity has been reported in breast cancer, SCLC, and medulloblastoma (Geng *et al.*, 2003; Li *et al.*, 2002; Shalaby *et al.*, 2010). The gene encoding the cell cycle regulator Cdk4 was found to have four conserved MYC binding sites in its promoter and is a direct MYC target gene (Hermeking *et al.*, 2000). The incomplete *cdk4* induction observed in *MYC* deficient Rat1 cells presented a link between cell cycle regulation and the oncogenic effect of MYC. The MYC-driven cell-cycle regulation was shown to be deficient in MYC-overexpressing breast cancer cells where *CDK4* was no longer responsive to MYC (Pawar *et al.*, 2004). Transcriptional regulation of D-type Cyclins may be another way through which the cell cycle progression is delayed in *MYC* null Rat1 cells. This hypothesis is strengthened by the fact that MYC upregulated transcription of Cyclins *D1* and *D2* (Bouchard *et al.*, 1999; Perez-Roger *et al.*, 1999), and also to some degree *Cyclin D3* (Yu *et al.*, 2005). Thus, it was suggested that upregulation of individual D-type Cyclins was sufficient to mediate the oncogenic effect of MYC. *NBS1*, encoding the Nijmengen breakage syndrome (Nbs)-1 kinase, a component of the MRN complex (Mre11/Rad51/Nbs1) has also been identified as a MYC target (Chiang *et al.*, 2003). In a collaborative study, we recently showed that MYC-mediated control of transcriptional expression and nuclear translocation of Nbs1 is essential for regulating phosphorylation of the checkpoint response kinase ATM (Guerra *et al.*, 2010). In addition to targeting other proteins, it has been suggested that the *MYC* gene itself harbors MYC-responsive elements and can regulate its own expression (Facchini *et al.*, 1997). Taken together, these findings delineate a role for MYC in activating target gene transcription. MYC-mediated repression through interaction with TFII-I and Miz-1 is described below.

III. MYC-MEDIATED REPRESSION

Transcriptional repression by MYC is mainly mediated through protein–protein contacts, where MYC antagonizes the function of other transcriptional activators, without direct contact with the DNA (Kleine-Kohlbrecher *et al.*, 2006). For instance, c-MYC-mediated inhibition of transcription can be conferred through interaction with TFII-I in the transcription machinery, binding at initiator elements (Roy *et al.*, 1991). Together with observations that MYC-mediated repression by MYC-interacting zinc-finger protein-1 (Miz-1) also started from the initiator element (Seoane *et al.*, 2002; Staller *et al.*, 2001), it was originally believed that the initiator (Inr) element was a prerequisite for MYC-mediated transcriptional repression. In the case of Miz-1, a ternary complex with Max is required to mediate transcriptional repression of the Miz-1 target genes *p21* and *p15* (Herold *et al.*, 2002; Seoane *et al.*, 2002; Staller *et al.*, 2001; Wu *et al.*, 2003) as well as of *Mxd4* (Kime and Wright, 2003). Physical interaction between MYC and Miz-1 is conferred by binding of Miz-1 to the HLH domain of MYC. This interaction appears to be MYC-specific as both c-MYC and MYCN bind to Miz-1 while neither Max, nor the HLH protein USF can bind (Peukert *et al.*, 1997). It has been shown that MYC represses transcription of Miz-1 targets by displacing the Miz-1 coactivator p300 (Staller *et al.*, 2001). This in turn enables MYC to recruit the Dnmt3a DNA methyltransferase corepressor to Miz-1, thus mediating repression by DNA methylation (Brenner *et al.*, 2005).

MYC also controlled expression of *c/EBP-α*, one of the first MYC target genes shown to be repressed through an Inr element (Li *et al.*, 1994). However, MYC also represses genes without Inr elements, such as *p21*, through interaction with the Sp1 transcription factor (Gartel *et al.*, 2001). Furthermore, the MYC/Max complex was found to transrepress the *p27* gene by directly binding to an Inr-like element at the promoter (Yang *et al.*, 2001). Another mechanism for MYC-mediated repression of target genes occurs through recruitment of an mSin3/HDAC complex, when associated with MM-1 (Satou *et al.*, 2001). Smad2 and NF-Y are two other interaction partners involved in MYC-mediated repression (Feng *et al.*, 2002; Izumi *et al.*, 2001). Interestingly, it appears that the conserved N-terminal MB2, important for MYC-mediated transactivation, is also essential for its repressive function (Conzen *et al.*, 2000; Lee *et al.*, 1997).

IV. INDUCTION OF APOPTOSIS

As mentioned above, MYC is a multifunctional protein and one of its important functions is the potentiation of apoptosis in response to cellular stress (reviewed in Nilsson and Cleveland, 2003). Cyclin A and Odc are two potential mediators of MYC-induced apoptosis since Odc-blockage inhibits apoptosis in MYC-overexpressing cells and forced expression of Cyclin A is sufficient to induce apoptosis under low serum conditions (Hoang *et al.*, 1994; Packham and Cleveland, 1994). Ectopic expression of Cyclin A could also restore apoptosis in *c-MYC* null cells treated with etoposide (Adachi *et al.*, 2001). Induction of apoptosis by c-MYC has also been correlated with regulation of the Fas receptor and its ligand as well as proapoptotic Bax (Albihn *et al.*, 2006; Fulda *et al.*, 1998; Juin *et al.*, 1999; Mitchell *et al.*, 2000; Soucie *et al.*, 2001). Bax appears to be essential for signaling c-MYC-induced apoptosis although there are few reports describing changes in Bax levels in response to c-MYC overexpression (Brunelle *et al.*, 2004; Eischen *et al.*, 2001; Juin *et al.*, 2002; Mitchell *et al.*, 2000). Instead the effect on Bax may be indirect by regulating upstream molecules such as Caspase 8, which is frequently inactivated in childhood neuroblastomas with amplified *MYCN* (Teitz *et al.*, 2000). There is also the possibility of direct protein–protein interaction, since MYC under some circumstances can localize to the cytoplasm (Oster *et al.*, 2002). The relation between MYC and the tumor suppressor protein p53 is complex. Even though p53 has been found to be important but not required for c-MYC-induced apoptosis, there are numerous tumor cell lines with deregulated *c-MYC* that carry *p53* mutations or deletions (Gaidano *et al.*, 1991; Wagner *et al.*, 1994). In response to c-MYC activation and MYC/Ras-induced transformation, p53 is upregulated and stabilized to induce cell cycle arrest or, if the cell cycle blockade is overcome by c-MYC, apoptosis (Wagner *et al.*, 1994). However, MYC-induced apoptosis may also be indirect by accumulation of reactive oxygen species (ROS) as a consequence of NF-κB inhibition (Pelengaris and Khan, 2003).

The many and diverse effects of c-MYC in promoting pathways as distinct as proliferation and apoptosis has brought forth the proposal of a model where activated MYC promotes apoptosis as the preferred physiological response. In case of excessive amounts of survival factors or mutations in the apoptotic pathway, the cellular MYC response would instead be uncontrolled proliferation. This model has been coined "the dual signal model" (Harrington *et al.*, 1994; Hueber and Evan, 1998), and is supported by the observation that different regions of the c-MYC N-terminal domain can control distinct biological functions, including apoptosis (Chang *et al.*, 2000; Conzen *et al.*, 2000).

V. REGULATION OF STEMNESS

Analysis of transgenic mice with conditional expression of *c-MYC* or *MYCN* has shown that they are essential for normal developmental control of hematopoietic and neural stem cells, respectively (Knoepfler *et al.*, 2002;

Wilson *et al.*, 2004). *MYCN* has been shown to be required for normal neural stem cell function whereas *c-MYC* deficiency results in accumulation of defective hematopoietic stem cells (HSCs) due to niche-dependent differentiation defects (Baena *et al.*, 2007; Wilson *et al.*, 2004). It was recently shown that immature HSCs coexpress *c-MYC* and *MYCN* mRNA at similar levels and double knockout of *c-MYC* and *MYCN* results in pancytopenia and rapid lethality (Laurenti *et al.*, 2008). Moreover, *c-MYC* is crucial for self-renewal and maintenance of pluripotency in murine ES cells. Murine ES cells can be maintained as a pluripotent, self-renewing population by leukemia inhibitory factor (LIF)/STAT3-dependent signaling which directly regulates the expression of *c-MYC*. Following LIF withdrawal, *MYC* mRNA levels collapse and MYC protein becomes phosphorylated on threonine 58 (T58), triggering its GSK-3β-dependent degradation. However, forced expression of stable MYC (T58A) renders self-renewal and maintenance of pluripotency independently of LIF (Cartwright *et al.*, 2005).

Direct reprogramming of somatic cells provides an opportunity to generate patient- or disease-specific pluripotent stem cells. Murine and human somatic cells can be reprogammed to pluripotency through generation of induced pluripotent stem cells (iPS) by retrovirus-mediated introduction of *Oct3/4* (also known as *Pou5f1*), *Sox2*, *Klf4*, and *c-MYC* (Okita *et al.*, 2007; Wernig *et al.*, 2007). However, reactivation of the *c-MYC* retrovirus increases tumorigenicity in the chimeras and progeny mice, impeding clinical applications (Okita *et al.*, 2007). Although a recent report shows that *MYC* may be redundant in the conversion of somatic cells to iPS cells (Nakagawa *et al.*, 2008), there is still an implication for a novel stemness function of MYC that may be of importance for controlling tumor initiating cells.

Evidence suggests that MYC's role in pluripotency is connected to its ability to regulate the cell-cycle machinery (Singh and Dalton, 2009). Elevated c-MYC levels accelerate the progression of cells through G1 phase by positively regulating cyclin-cdk activity (Amati *et al.*, 1998). Conditional loss of *MYCN* in neural stem cells has been correlated with increased levels of the cdk inhibitors p18INK4c and p27KIP1 and decreased expression of cyclin D2 resulting in deregulation of the cell-cycle program (Knoepfler *et al.*, 2002; Singh and Dalton, 2009). In addition, cell-cycle changes that occur during differentiation of pluripotent cells coincide with downregulation of *MYC* levels and the regulatory subunit of telomerase (TERT) in murine ES cells (Cartwright *et al.*, 2005; Kim *et al.*, 2008; White and Dalton, 2005). Finally, microRNAs (miRNAs) have been suggested to contribute to MYC's role in pluripotency, either by coordinating the expression of cell-cycle molecules such as p21 and cyclin D2 in ES cells (Judson *et al.*, 2009), or by repressing mRNAs involved in differentiation (reviewed in Dang, 2009).

VI. ONCOGENIC PROPERTIES

A major fraction of all human cancers display deregulated MYC activity (Nilsson and Cleveland, 2003; Ponzielli *et al.*, 2005). Alterations include chromosomal translocations exemplified by the *c-MYC-Immunoglobulin* (*Ig*) fusion gene in BL (Hecht and Aster, 2000) and increased *c-MYC* expression due to gene amplification (Hogarty, 2003) as well as protein stabilization (Sears *et al.*, 2000). Other oncogenic features are induction of genomic destabilization (Felsher and Bishop, 1999b; Mai *et al.*, 1999), increased vascularization, and angiogenesis (Oster *et al.*, 2002).

Oncogenes that frequently synergize with *c-MYC* in transformation include *BCL-2*, *RAS*, *RAF*, and *c-ABL* (reviewed in Oster *et al.*, 2002; Pelengaris *et al.*, 2002). c-Abl and Bcl-2 have been proposed to negatively regulate MYC-induced apoptosis. Consequently, a large proportion of tumors with deregulated *c-MYC* expression overexpress Bcl-2 (Cory and Adams, 2002). There are also reports on MYC regulation of Cyclin D1 and D2, both of which seem essential for MYC-driven proliferation, the gene encoding the latter being a direct MYC target, activated in response to growth factor stimulation (Bouchard *et al.*, 2001). The nuclear zinc-finger protein encoded by *BMI-1* synergizes with c-MYC in lymphomagenesis, possibly by negative regulation of p19ARF in the ARF–p53–MDM2 pathway (Jacobs *et al.*, 1999). Prosurvival molecules that protect cells from c-MYC-induced apoptosis, such as the IGFs and platelet-derived growth factor (PDGF), may also facilitate transformation (Harrington *et al.*, 1994). In addition, one mediator of the IGF-1-antiapoptotic effect, the bHLH family member Twist, promotes oncogenesis by inhibiting the apoptotic function of p19ARF (Dupont *et al.*, 2001).

VII. NO TRANSFORMATION WITHOUT MYC?

Even though the generally accepted view is that cooperation of two oncoproteins such as MYC and ras are sufficient for cellular transformation (Land *et al.*, 1983), this only holds true for murine cells, whereas additional events are required in human cells (Boehm *et al.*, 2005; Hahn *et al.*, 1999). The initially identified four events were expanded to six when the viral element (SV40 large T oncoprotein) was substituted for c-MYC. It was found that inactivation of tumor suppressor genes (p53, pRb, and PTEN) and limitless ability to replicate (hTERT activation), together with oncogene activation, were prerequisites for human cell transformation (Boehm *et al.*, 2005). In fact, recent evidence has shown that the involvement of MYC may be necessary for oncogene-induced transformation (Soucek *et al.*, 2008;

Zhuang *et al.*, 2008). c-MYC overexpression was shown to partially overcome senescence induced by the oncoproteins B-Raf or N-Ras in a p53-independent manner, thereby contributing to the malignant phenotype in melanoma cells (Zhuang *et al.*, 2008). In a different setting, inhibition of endogenous MYC was sufficient to trigger regression of Ras-induced lung adenocarcinomas *in vivo*, suggesting that MYC, even at its endogenous state is of vital importance for maintaining Ras-dependent tumors (Soucek *et al.*, 2008). This newly identified importance of MYC, together with the finding that effects on normal regenerating tissues were well tolerated and completely reversible in the mouse, reinforces the prospect of targeting this nearly omnipotent oncogene as a feasible antitumor therapy. It was recently shown that the cooperation between MYC and ras in transformation of rodent cells required cdk2-induced phosphorylation of the MYC residue S62 (Hydbring *et al.*, 2009). This in turn indicates that pharmacological inhibition of Cdk2 may be considered as an important cancer therapy for *MYC*-driven tumors.

VIII. MYC-ASSOCIATED CANCERS AND THEIR TREATMENT

The *MYC* family genes are deregulated by different mechanisms in several human neoplasias of different origin, including diffuse large B cell lymphoma, multiple myeloma, colon cancer, glioblastoma, melanoma, ovarian cancer, and prostate cancer (Nesbit *et al.*, 1999; Vita and Henriksson, 2006). However, the extent of *MYC* involvement in these malignancies varies depending on the staging and the cancer form (Caccia *et al.*, 1984; Nesbit *et al.*, 1999; Pelengaris and Khan, 2003; Vita and Henriksson, 2006). Here, we focus on breast cancer, BL, lung cancer, medulloblastoma, neuroblastoma, and rhabdomyosarcoma (RMS); malignancies for which *MYC* has been shown to be important for development, progression, and/or patient risk-stratification. *MYC* status as well as current and future treatment approaches, some of which are already in clinical trials and others undergoing preclinical assessment, will be discussed (summarized in Table I).

A. Breast Cancer

Human breast carcinomas are heterogeneous, both in their pathology and molecular profiles and are the second most common cause of cancer deaths in the world. Molecular characterization of breast tumors have not revealed any common dominant pathway for the development or histological

Table I *MYC* Status, Current and Future Treatments for the Different Cancer Forms

Cancer	*MYC* status	Current treatment	Treatments in clinical trials
Breast cancer	*c-MYC* overexpression (45%) *c-MYC* amplification (9–48%) *MYCN* overexpression (25%)	Surgical resection, radiation, combinational chemotherapy, hormones, herceptin, lapatinib, antiangiogenic therapy	***Anti-HER2 therapy (pertuzumab, trastuzumab-DM1, KOS-953), aromatase inhibitors, tyrosine kinase inhibitors***, mTOR inhibitors, ***endocrine therapy, stem cell transplantation, antisense therapy***, imatinib+ vinorelbine, triptorelin, misteltoe
Burkitt's lymphoma	*c-MYC* translocation (100%) *c-MYC* overexpression (91%)	Short intensive multiagent chemotherapy, CNS prophylaxis, rituximab	***Different multiagent chemotherapy combinations+rituximab, stem cell transplantation***
Lung cancer	*c-MYC* amplification (20%) *MYCL* amplification (13%) *MYCN* amplification (10%)	Surgical resection, radiation, combinational chemotherapy, antiangiogenic therapy, EGFR-targeted therapies	***Tyrosine kinase inhibitors and antibodies combined with chemotherapy, antiangiogenic therapy***, HDAC inhibitors, ***mTOR inhibitors, EGFR inhibition (cetuximab), proteasome inhibitors***, NSAID, zileuton immunotherapy, cancer vaccines, talactoferrin, retinoids, mistletoe, green tea extract, nanoparticles
Medulloblastoma	*c-MYC* amplification (6%) *MYCN* amplification (4%)	Surgical resection, radiation, combinational chemotherapy	***Temozolamide***, irinotecan, immunotherapy, ***GDC-0449, erlotinib***, **dasatinib**, mTOR inhibitors,[a] Met inhibitor,[a] antiangiogenic therapy,[a] HDAC inhibitors,[a] Notch inhibitors,[a] PARP inhibitor,[a] nifurtimox[a]

Neuroblastoma	*MYCN* amplification (25–30%)	High-dose chemotherapy, surgical resection, myeloablative consolidation chemotherapy with autologous stem cell rescue, radiotherapy, 13-*cis* retinoic acid (RA), meta-iodobenzyl guanidine (MIBG)	***Ch14.18 delivery+IL-2 and GM-CSF,*** EBV-specific CTL,[a] ***temozolamide, fenretinide, vorinostat+cis-RA,***[a] ***TNP-470,***[a] ***PF-02341066,***[a] ABT-751, ***MLN8237,***[a] ***CEP-701,***[a] ***SF1126,***[a] ZD6474,[a] Proteasome inhibitors,[a] nifurtimox,[a] cixutumumab,[a] R(+) XK469,[a] OPT821+β-glucan[a]
Rhabdomyosarcoma	*MYCN* amplification (43–67%)	Radiation, multiagent chemotherapy, surgical resection	O-TIE (Etoposide, Idarubicin, Trofosfamide), temozolamide, ***anti-IGF-R1 therapy,***[a] ixabepilone, erlotinib, ***immunotherapy,*** donor peripheral stem cell transplantation, irinotecan and carboplatin as upfront therapy, antiangiogenic therapy[a]

Treatments in bold italics are mentioned in the text.
[a]Studies not concluded. Source; www.cancer.gov/clinicaltrials/search/.

presentation with the result that breast tumors historically have been categorized into at least 18 different subtypes. More recently, gene expression profiling offered a way of classifying breast carcinomas into five different subtypes based on their mRNA profiles; luminal A, luminal B, Erbb2, basal, and normal-like (Sorlie *et al.*, 2001). Both c-MYC (45%) and MYCN (25%) overexpression as well as *c-MYC* gene amplification (9–48%) have been described in breast carcinoma (Vita and Henriksson, 2006). Amplification of *c-MYC* is often correlated with a poor prognosis, in particular when it coincides with inactivation or mutation of the tumor suppressor gene *BRCA1* (Chen and Olopade, 2008). Besides this, the correlation between *MYC* status and clinical outcome is less consistent (reviewed in Chen and Olopade, 2008). Activation of *MYC* in breast carcinomas has been correlated with the activation of Wnt, Notch, and TGF-β signaling (Chen *et al.*, 2001; Klinakis *et al.*, 2006; Ozaki *et al.*, 2005; Stylianou *et al.*, 2006). For example, nuclear localization of β-catenin was shown to be both strongly correlated to MYC expression and significantly correlated with reduced APC levels in primary breast cancer samples (Ozaki *et al.*, 2005). MYC has also been shown to suppress DKK1 and SFRP1, two inhibitors of Wnt signaling thereby activating the Wnt pathway and promoting anchorage-independent growth of human epithelial mammary cells (Cowling *et al.*, 2007). Recently, prolyl-isomerase (PIN) 1-dependent activation of Notch signaling was shown to activate MYC and transform normal epithelial breast cells (Klinakis *et al.*, 2006; Rustighi *et al.*, 2009). Finally Smads, a family of transcription factors that are regulated by TGF-β signaling and inhibit MYC expression, were shown to be decreased in breast cancer (Chen *et al.*, 2001). The low expression of Smads significantly correlated with high tumor grade, larger tumor size, and hormone receptor negativity (Jeruss *et al.*, 2003).

At the posttranslational level, MYC has been shown to be stabilized through inhibition of T58 phosphorylation by the activation of the estrogen pathway in ER-α-responsive breast carcinoma cells or elevated phospholipase D activity in ER-negative breast cancer cells (Rodrik *et al.*, 2006). Detailed analyses of *MYC* amplification and elevated MYC expression in breast cancer samples have implicated *MYC* as a potential prognostic factor. However, more details on the expression levels and posttranslational modifications of MYC will be required in order to fully evaluate the importance of MYC expression in the initiation and progression of breast carcinomas.

B. Treatment of Breast Cancer

Different chemotherapy regimens, hormone therapy, and targeted therapy comprise the current standard (Engel and Kaklamani, 2007). Increasingly, treatment is becoming individualized and different prognostic and diagnostic markers are tested for reliability in choosing the most suitable therapy (Bouchalova *et al.*, 2009). Chemotherapy regimens include combinations of anthracyclines and taxanes, sometimes together with an antimetabolite such as 5-fluorouracil. The anthracyclines daunorubicin, doxorubicin, and epirubicin, all induce similar damage to cellular DNA and RNA (Rabbani *et al.*, 2005), while the taxanes paclitaxel and docetaxel are antimicrotubule agents interfering with cell division (Abal *et al.*, 2003). These treatments are sometimes combined with the CMF regimen which is a combination of the alkylating agent cyclophosphamide (C; or ifosfamide) with the antimetabolites methotrexate (M) and 5-fluorouracil (F) (Levine and Whelan, 2006). It has been hypothesized that c-MYC-amplified breast tumors are those that respond favorably to the CMF therapy (Bouchalova *et al.*, 2009), supported by the fact that *c-MYC* transcription is suppressed in response to 5-fluorouracil treatment. In contrast, tumors where c-MYC was coactivated with E2F responded poorly to TFAC (paclitaxel, 5-fluorouracil, doxorubicin, and cyclophosphamide) therapy, but were predicted to be more sensitive to docetaxel containing regimens (Salter *et al.*, 2008). Possibly, the sensitivity to docetaxel could be due to a decreased E2F activation.

Hormone therapy is used for treatment of hormone-receptor positive breast cancer patients. The hormonal therapeutics includes aromatase inhibitors, selective estrogen receptor modulators (SERMs), and estrogen receptor downregulators (ERDs) (Bush, 2007; Prat and Baselga, 2008). Aromatase inhibitors such as anastrozole, exemestane, and letrozole prevent the production of estrogen in postmenopausal women by blocking the activity of the enzyme aromatase that normally converts androgen into estrogen (Bush, 2007). It has been found that aromatase inhibitors are sometimes better than SERMs, such as tamoxifen, raloxifene, and toremifene in breast cancer treatment and prevention (Santen *et al.*, 2003). However, the benefit with SERM treatment is that it can be used in women both before and after menopause. Similarly to SERMs, the ERD fulvestrant blocks the estrogen receptor (ER) and prevents estradiol from binding and eliciting downstream effects (Bush, 2007). It also reduces the number of ERs and interferes with the function of existing ERs. c-MYC appears to be a target of estrogen action, specifically mediating its effects in cell growth (Musgrove *et al.*, 2008). Deregulated c-*MYC* was also shown to confer resistance to antiestrogen therapy in MCF7 cells by downregulating the expression of p21 (Mukherjee and Conrad, 2005). Consequently, breast tumors with deregulated c-MYC may not respond well to hormone therapy.

Targeted therapies include use of the monoclonal antibody herceptin (trastuzumab), blocking the effect of the growth factor protein Her-2. This treatment, combined with chemotherapy is useful in approximately 25% of women suffering from Her-2-positive breast cancer (Prat and Baselga, 2008). Likewise, the tyrosine kinase inhibitor lapatinib blocks the effect of the Her-2 protein and other tumor-specific proteins. Lapatinib is used in combination with capecitabine for treatment of Her-2-positive breast cancer patients that no longer respond to treatment with herceptin. Another type of targeted therapy is the prevention of tumor growth by blocking angiogenesis (Sirohi and Smith, 2008). The monoclonal antibody avastin (bevacizumab), used in combination with paclitaxel, exerts its antiangiogenic effect by blocking vascular endothelial growth factor (VEGF) required to stimulate angiogenesis. Recent results suggest that c-MYC is important in estrogen-induced VEGF transcription (Dadiani *et al.*, 2009), which would indicate that deregulated c-*MYC* would diminish the effect of avastin.

Other Her-2 targeting approaches are currently in clinical trials. The recombinant antibody pertuzumab prevents Her-2 dimerization, trastuzumab-DM1 combines the activity of trastuzumab with inhibition of tubulin polymerization, while KOS-953 is an Hsp90 inhibitor promoting ubiquitination and degradation of Her-2. Additional clinical trials exploit sentinel lymph node biopsy followed by surgery, high-dose chemotherapy with stem cell transplantation, and antisense therapy (reviewed in Prat and Baselga, 2008).

C. Burkitt's Lymphoma (BL)

BL is a non-Hodgkin's B cell lymphoma originally detected as an endemic form carrying a latent Epstein–Barr virus (EBV) infection. Later, sporadic and AIDS-associated BL, where the majority of tumors are negative for EBV, were described (Magrath, 1990). HIV infection and malaria are the most important risk factors for the development of BL through the induction of polyclonal B cell activation and hypergammaglobulinemia (Bornkamm, 2009). The common trait for BL is the development of tumors in extranodal sites in adolescents or young adults. The disease is classified as a distinct category of peripheral B cell lymphomas, and comprises a heterogenous group of highly aggressive B cell malignancies (reviewed in Hecht and Aster, 2000). It is invariably associated with chromosomal translocations, preferentially the t(8:14)(q24:q32) translocation, bringing the *c-MYC* proto-oncogene in proximity with the immunoglobulin heavy chain promoter (Dalla-Favera *et al.*, 1982; Taub *et al.*, 1982). Even though the chromosomal breakpoints are widely dispersed along the genes, the end result is a fusion gene where *c-MYC* is constitutively active. Because of the c-MYC overexpression, BL cells have the highest cell division rate observed in any human tumor. Although *c-MYC* rearrangements are observed in the majority of BL cases, this is not a BL-specific phenomenon as they are also observed in other types of lymphoma (reviewed in Hecht and Aster, 2000). However, translocated and activated *MYC* is the consistent feature of these tumors and not the presence of EBV. The exact mechanism of the *MYC* translocations seen in BL and other lymphomas has for a long time been a puzzle. Recently, AID (activation-induced cytidine deaminase) that is highly expressed in the lymph node germinal centre and can be induced by EBV late proteins was shown to be essential for the *MYC* translocations. AID activation induces deamination of cytidine residues, resulting in U:G mismatches and double-stranded DNA breaks (Roughan and Thorley-Lawson, 2009). Aberrant AID expression in IL-6 transgenic mice causes *Ig*-MYC translocations that mimics those detected in EBV-positive BL (Dorsett *et al.*, 2007).

D. Treatment of BL

Even though BL is a highly aggressive malignancy, it is highly treatable particularly in children and has a low level of relapse. The main treatment is chemotherapy, often consisting of various combinations of cyclophosphamide, vincristine, doxorubicin, methotrexate, cytarabine, ifosfamide, and etoposide (Aldoss *et al.*, 2008; Lacasce *et al.*, 2004; Mead *et al.*, 2002). Cyclophosphamide and ifosfamide are both nitrogen mustard-derived alkylating agents used in treatment of lymphomas, leukemias, and some solid tumors (Shanafelt *et al.*, 2007; Young *et al.*, 2006). The drugs cause cell death by inducing DNA cross-links in cells with low levels of aldehyde dehydrogenase. Cyclophosphamide-induced activation of p53 along with a decrease in c-MYC and Odc expression cause cell-cycle arrest, at least in cells from the gastrointestinal tract (Hui *et al.*, 2006). Doxorubicin and etoposide are both topoisomerase-II inhibitors (Hande, 1998; Rich *et al.*, 2000). While doxorubicin is an anthracycline antibiotic, analogous to daunorobucin, with broad-spectrum antitumor effects, the podophyllotoxin etoposide was the first anticancer drug demonstrated to inhibit topoisomerase-II. Etoposide causes p53 phosphorylation, possibly mediated by the DNA damage sensor DNA-PK, with subsequent upregulation of proapoptotic Bax and promotion of apoptosis through cytochrome *c* release (Karpinich *et al.*, 2002). Although poorly soluble in water and thus difficult to administer in effective doses, etoposide together with other topoisomerase-II-inhibiting agents are among the most effective chemotherapeutic drugs available for cancer therapy (Baldwin and Osheroff, 2005). The dihydrofolate reductase (DHFR) inhibitor methotrexate prevents DNA and RNA synthesis during the S phase of the cell cycle (Longo-Sorbello and Bertino, 2001). This antimetabolite is widely used in treatment of lymphomas, leukemias, some solid cancers, and also for autoimmune disorders. However, a high c-MYC expression was observed to cause methotrexate resistance in osteosarcoma and non-small cell lung cancer (NSCLC) cells (Scionti *et al.*, 2008; Serra *et al.*, 2008). The other antimetabolite, cytarabine (ara-C), causes DNA damage and inhibits DNA and

RNA synthesis in rapidly dividing cells (Grant, 1998). It is mainly used as an anticancer agent in treatment of lymphoma and leukemias, but also possesses antiviral activity (Gray *et al.*, 1972). In leukemia cells, the effect of cytarabine was enhanced by inhibition of MEK signaling. The resulting DNA damage caused enhanced expression of p21 and downregulation of c-MYC and Bcl-X_L, followed by growth arrest and apoptosis (Nishioka *et al.*, 2009). Finally, vincristine is a vinca alkaloid antimicrotubule agent that prevents polymerization of tubulin (Jordan *et al.*, 1991). Applications include treatment of lymphomas, childhood leukemia and it also functions as an immunosuppressant. Interestingly, the level of c-MYC does not appear to affect the cellular sensitivity to vincristine (Hirose and Kuroda, 1998; Ma *et al.*, 1992).

Adult BL patients are often treated with monoclonal antibodies such as rituximab, combined with chemotherapy (Thomas *et al.*, 2006). Rituximab targets the B cell surface antigen CD20 and is used for treatment of lymphomas, leukemias, and some autoimmune disorders. It is most efficient if the *MYC* gene is translocated to a non-Ig site, and if Bcl-2 protein expression is absent (Johnson *et al.*, 2009). In cases where very high doses of chemotherapy are used, sometimes combined with radiotherapy, stem cell replacement may be required (reviewed in Aldoss *et al.*, 2008). Stem cell transplantation may also be considered for patients with relapsed disease (Sweetenham *et al.*, 1996). The use of selective serotonin-reuptake inhibitors (SSRI) has also been considered for their reported proapoptotic effect on B cell-derived tumors. However, this effect has not been proven to be specific (Schuster *et al.*, 2007).

E. Lung Cancer

Lung cancer is one of the leading causes of cancer deaths in the industrialized world and has been correlated to cigarette smoking in over 80% of cases. Genetic factors and environmental exposures such as asbestos and radon contribute to the remaining 20% (Rom and Tchou-Wong, 2003). The disease has been divided into two histological subtypes in which the majority is NSCLC (80%) and the rest is SCLC. Both NSCLCs and SCLCs normally contain several numerical and structural chromosome alterations and epigenetic changes that result in aberrant expression of oncogenes and silencing of tumor suppressor genes. Lung cancer appears unique among epithelial tumors in that gene amplification and/or overexpression of each member of the *MYC* family, that is, c-*MYC*, *MYCN*, and *MYCL* can be detected in these tumors (Nau *et al.*, 1985, 1986; Wong *et al.*, 1986). *MYC* amplification occurs in 18–31% of SCLCs and in 8–20% of NSCLCs (Richardson and Johnson, 1993). Amplification of *MYC* genes has been shown to affect survival adversely in SCLC patients (Brennan *et al.*, 1991). The importance of MYC expression in the tumorigenesis of lung cancer was illustrated in transgenic mice expressing murine *c-MYC* under the control of the lung-specific surfactant protein C promoter. These mice invariably developed multifocal bronchiolo-alveolar adenocarcinomas from the alveolar epithelium (Ehrhardt *et al.*, 2001).

Moreover, endogenous c-*MYC* is involved in nonmetastatic *K-Ras*-induced NSCLC as was shown by use of a dominant-negative c-MYC mutant (Soucek *et al.*, 2008) and *c-MYC* in cooperation with c-Raf was recently shown to be a metastasis gene in NSCLC (Rapp *et al.*, 2009). *MYCN* that is frequently amplified and expressed in SCLC (Nau *et al.*, 1986) is crucial for normal lung organogenesis by maintaining a population of undifferentiated proliferating progenitor cells in the developing lung tissue. This is reflected by the findings that MYCN is highly expressed in embryonic lungs whereas adult lungs exhibit very low expression (Okubo *et al.*, 2005). This in turn suggests that MYCN is involved in the maintenance of a population of continuously proliferating cells in the tumor, possessing similar properties as lung stem cells. Hence, targeting MYC in lung cancer may be an adjuvant therapy for the eradication of potential lung cancer stem cells.

F. Treatment of Lung Cancer

To date, there is no efficient treatment for lung cancer and much effort is being invested in improving patient survival and reducing the adverse effects of standard treatment. Systemic chemotherapy remains the most important treatment option for these patients (reviewed in Higgins and Ettinger, 2009). Treatment of NSCLC includes a combination of surgical resection and adjuvant chemotherapy consisting of platinum-based compounds such as cisplatin and carboplatin in combination with the nucleoside analogue gemcitabine or the antimitotic drugs paclitaxel, vinorelbine, or docetaxel; or the newer antifolate, pemetrexate (Blackhall *et al.*, 2005; Higgins and Ettinger, 2009). In addition, the combination of chemotherapy and the anti-VEGF antibody bevacizumab is being investigated in relation to the effect of chemotherapy alone. Another approach that has received much attention is the use of small-molecule inhibitors of the epidermal growth factor receptor (EGFR) tyrosine kinase, overexpressed in more than 50% of all NSCLC cases (reviewed in Sharma *et al.*, 2007). As EGFR overexpression is correlated with a bad prognosis, several drugs are in clinical trials and two molecules, Gefitinib and erlotinib, have been approved for NSCLC treatment. There is also the monoclonal EGFR antibody cetuximab that has shown promise in clinical trials (Higgins and Ettinger, 2009). As second and third line treatment options for NSCLC, proteasome inhibitors and inhibitors of mTOR are being discussed along with sorafenib and sunitinib, kinase inhibitors with more general effects (Higgins and Ettinger, 2009). The traditional treatment of metastatic SCLC includes four different combinations: cyclophosphamide, doxorubicin, and vincristine; cisplatin and etoposide; ifosfamide and etoposide; and carboplatin and etoposide (reviewed in Wolf *et al.*, 2004). Similarly to NSCLC, pemetrexate is also successfully applied in SCLC, as is the novel anthracycline drug amrubicin (reviewed in Higgins and Ettinger, 2009). The advancement in identifying markers for molecular targeted therapeutics will bring about a new era of personalized medicine where it is no longer appropriate to differentiate between NSCLC and SCLC.

In addition to K-Ras and EGFR, MYC appears to be a suitable marker that can be targeted by gene therapy in treatment of SCLC where MYC is overexpressed. Cells transduced with the herpes simplex virus thymidine kinase (HSV-TK) expressed from the E-box sequence (CACGTG) had an increased sensitivity to ganciclovir in cells overexpressing MYC (Nishino *et al.*, 2001). Moreover, when these motifs were placed in a replication-deficient adenoviral vector (adMYCTK) and injected in MYC-overexpressing tumors in mice, followed by ganciclovir-administration, the tumor size was markedly reduced. This ganciclovir-induced shrinkage was not observed in adMYCTK-infected tumors not overexpressing MYC (Nishino *et al.*, 2001). As this treatment rendered no apparent side effects, it may be useful for clinical purposes in patients with MYC-overexpressing SCLC refractory to standard treatment.

G. Medulloblastoma

Medulloblastoma, a primitive neuroectodermal tumor, is the most common malignant pediatric brain tumor. It arises in the cerebellum and can originate from cerebellar granule neural precursor (GNP) cells located in the external granular layer (EGL) of the cerebellum (Schuller *et al.*, 2008; Yang *et al.*, 2008). The EGL contains actively proliferating progenitor cells derived from the rhombic lip during embryogenesis. While GNP cell proliferation requires Hedgehog signaling (Ho and Scott, 2002), their expansion and survival is also promoted by IGF signaling. Medulloblastoma cells retain many features resembling precursor cells of the embryonic brain (Schuller *et al.*, 2008) and more than half of these tumors contain abnormal activation of the Hedgehog or Wnt signaling pathways (Hambardzumyan *et al.*, 2008b). Moreover, activation of the PI3K/Akt signaling pathway has been shown to be important for proliferation of human medulloblastoma cells and cancer stem cells residing in the perivascular niche following radiation of medulloblastoma (Hambardzumyan *et al.*, 2008a; Hartmann *et al.*, 2006; Rao *et al.*, 2004).

Elevated MYCN expression is present in a significant proportion of human medulloblastoma (Eberhart *et al.*, 2004; Pomeroy *et al.*, 2002), and is required for Sonic hedgehog-driven medulloblastoma tumorigenesis (Hatton *et al.*, 2006). Activation of the *c*-MYC oncogene is frequently observed in medulloblastoma and has been shown to be one of the most reliable prognostic factors (Eberhart *et al.*, 2004; Herms *et al.*, 2000). Moreover, activation of both the Wnt/β-catenin pathway as well as of PI3K/Akt signaling have been shown to affect the expression of both *MYCN* and *c-MYC* in medulloblastoma cells (Baryawno *et al.*, 2010; Browd *et al.*, 2006; Momota *et al.*, 2008). Hence, MYC appears to play a central role in mediating the effects of aberrant Hedgehog, Wnt, and PI3K/Akt signaling in medulloblastoma.

H. Treatment of Medulloblastoma

The standard treatment of medulloblastoma still consists of surgery followed by high-grade craniospinal radiotherapy (reviewed in Mueller and Chang, 2009). However, as this treatment causes severe morbidity to the relatively few (<60%) surviving children, there are several ongoing trials in search of milder, but more efficient treatments. Single agent chemotherapy as well as treatment with a combination of drugs have been tested with varying success. Despite favorable initial responses to drugs such as methotrexate, cyclophosphamide, platinum drugs, vincristine, ifosfamide, etoposide, and temozolamide in most cases, the long-term disease control rate did not increase, suggesting that chemotherapy alone would not provide a cure for the disease. One obvious problem in using chemotherapy in treating medulloblastoma is the limited ability for many of the drugs to cross the blood–brain barrier. Therefore, chemotherapy is combined with radiation to give the best response and to reduce morbidity and mortality.

There are several preclinical and clinical trials in search of targeted therapies for medulloblastoma (reviewed in Rossi *et al.*, 2008). The two treatment approaches that have reached clinical trials target either the sonic hedgehog pathway or the EGF tyrosine kinases. The hedgehog targeting therapy is represented by the teratogen cyclopamine that binds to and inactivates the smoothened protein. The displayed *in vitro* outcomes of this treatment were cell-cycle arrest, initiation of neuronal cell differentiation, and consequently loss of the stem cell-like characteristics (Berman *et al.*, 2002; Romer and Curran, 2005). It appears that this effect occurs specifically in medulloblastoma-derived cells and not in cells resected from other brain tumors. Also, GDC-0449, a compound that inhibits Hedgehog signaling by deactivation of Smoothened has been used in the treatment of medulloblastoma with very good responses initially. However, the patient developed resistance against GDC-0449 caused by a point mutation in *Smoothened* preventing the compound to lock into the Smoothened protein (Rudin *et al.*,

2009; Yauch *et al.*, 2009). Targeting of epidermal growth factor tyrosine kinases (Erbb1 and -2) prevents the invasiveness of medulloblastomas. Therefore, small-molecule tyrosine kinase inhibitors have been designed to target Erbb1 and Erbb2. As such, the Erbb2 targeting drug erlotinib selectively blocks upregulation of prometastatic genes such as S100A4 (Hernan *et al.*, 2003), and is being tested for treatment of refractory solid brain tumors in combination with either radiation (clinicaltrials.gov: NCT00360854) or in combination with the alkylating agent temozolomide (Jakacki *et al.*, 2008). The latter combination is also a potential new therapy for neuroblastoma.

Another recent approach using the telomestatin derivative S2T1-6OTD, targeting G-quadruplex forming DNA sequences, proved to potently inhibit *c-MYC* transcription and its target gene *hTERT* (Shalaby *et al.*, 2010). The effective dose of the small molecule induced telomere shortening and cell-cycle arrest by downregulating Cdk2 protein expression in childhood brain cancer cell lines, including medulloblastoma. If this compound proves effective in animal models, it represents a promising new therapeutic approach. Effects of targeting c-MYC by RNA interference techniques have also been investigated (von Bueren *et al.*, 2009). In a panel of human medulloblastoma cell lines, it was found that in addition to inhibiting cellular proliferation and clonogenic growth, c-MYC downregulation also reduced the apoptotic response to chemotherapy and radiation. This approach would therefore require a timely introduction in combination with other therapies.

I. Neuroblastoma

Neuroblastoma is a malignant embryonal tumor of the peripheral nervous system that accounts for more than 10% of pediatric cancer deaths despite intensive treatment modalities (Johnsen *et al.*, 2009). Most cases are diagnosed during the first year of life at the peak of disease incidence (reviewed in Hogarty, 2003). The clinical outcome is heterogenous, ranging from spontaneously regressing tumors, to differentiating tumors and those that can be cured with chemotherapy, but also includes cases of aggressive metastatic tumors, often associated with a lethal outcome (Weinstein *et al.*, 2003). Neuroblastoma originates from neural crest cells and is linked to dysfunctional pathways, which are operative during normal development (Scotting *et al.*, 2005). The neural crest is a transient embryonal structure that arises from ectoderm during closure of the neural tube. A complex interplay between Hedgehog and Wnt signaling is important for proper neural crest formation (Fodde and Brabletz, 2007). Both Hedgehog and Wnt signaling have been shown to induce expression of MYCN. High MYCN expression stimulates proliferation and migration of neuroblasts, while reduced levels of this protein is associated with terminal differentiation.

Amplification of the *MYCN* gene, which occurs in 40–50% of high-risk neuroblastoma cases, remains the major key predictor of poor outcome and is associated with advanced-stage disease, rapid tumor progression, and a low survival rate (Maris *et al.*, 2007). This suggests an important function of MYCN in neuroblastoma. In fact, transgenic mice with targeted expression of MYCN to neural crest cells using the tyrosine hydroxylase promoter (*pTH-MYCN*) develop neuroblastoma that is histologically and genetically very similar to aggressive undifferentiated human neuroblastoma (Weiss *et al.*, 1997). Interestingly, in *MYCN* nonamplified neuroblastomas the level of MYCN transcripts and proteins do not correlate with outcome (Cohn and Tweddle, 2004; Cohn *et al.*, 2000; Tang *et al.*, 2006). Instead it was recently shown that neuroblastoma cells with low MYCN levels frequently overexpress c-MYC (Westermann *et al.*, 2008). Moreover, constitutive activation of PI3K/Akt as well as activation of Wnt signaling has recently been shown in primary neuroblastomas (Johnsen *et al.*, 2008; Liu *et al.*, 2008; Opel *et al.*, 2007; and reviewed in Gustafson and Weiss, 2010). Activation of both these signaling pathways is associated with increased MYCN expression in neuroblastoma (Johnsen *et al.*, 2008; Liu *et al.*, 2008). This suggests that a common MYC-dependent transcriptional profile may contribute to the pathogenesis and that therapies targeting MYC expression may have importance in clinical outcome. In fact, pathway-specific gene expression profiling using two large neuroblastoma datasets showed that patients with poor prognosis, as well as all *MYCN*-amplified cases, had elevated signaling through the MYC transcriptional network (*MYC*, *MYCN*, and *MYCL* target genes). This in turn suggests that overexpression of MYC target genes contributes to neuroblastoma aggressiveness (Fredlund *et al.*, 2008).

J. Treatment of Neuroblastoma

Depending on the disease stage, the treatment approaches for neuroblastoma consist of different combinations of surgery, radiation therapy, chemotherapy, and simply watchful waiting. Standard chemotherapy regimens used in treatment of neuroblastoma include different combinations of cisplatin, vincristine, carboplatin, etoposide, and cyclophosphamide (Pearson *et al.*, 2008). A recent study demonstrated a significantly better 5-year event-free survival in patients receiving myeloablative therapy with autologous bone marrow transplantation (Matthay *et al.*, 2009). Subsequent treatment with 13-*cis*-retinoic acid (RA), causing cell growth arrest and differentiation of tumor cells, further improves the overall survival in children suffering from neuroblastoma. Previous results showed that one mechanism of 13-*cis*-RA was to reduce *MYCN* mRNA expression (Reynolds and Lemons, 2001).

Treatments in clinical trials comprise a combination of the monoclonal antibody CH14.18, targeting the tumor antigen GD2, and cytokines IL-2 and GM-CSF together with 13-*cis*-RA. Systemic distribution of IL-2 cytokines is used to activate natural killer cells and a certain subpopulation of T cells into lysing the antibody-coated neuroblastoma cells (Verneris and Wagner, 2007). In the humanized antibody hu14.18-IL-2, a humanized version of CH14.18 has been coupled to recombinant IL-2. Other immune-based therapies in clinical trials include *ex vivo* activated and expanded T cells, tumor cell vaccines, tumor pulsed dendritic cells, and allogeneic HSC transplants; all requiring quite extensive further investigations before they can be approved for neuroblastoma treatment (reviewed in Verneris and Wagner, 2007). A synthetic vitamin A derivative, Fenretidine, reduces angiogenesis by inhibiting migration of endothelial cells and reduces the growth of neuroblastoma *in vitro* (Friedman and Castleberry, 2007). Another angiogenesis inhibitor undergoing clinical trials is TNP-470, a synthetic peptide that seems to work best in patients with a small tumor burden (Shusterman *et al.*, 2001). As such, TNP-470 has been suggested for use in treatment of minimal residual disease after chemotherapy. The recent discovery of anaplastic lymphoma kinase (Alk) mutations and/or amplification in high-risk neuroblastoma and the findings that small-molecule inhibitors of Alk suppress neuroblastoma growth *in vitro* and *in vivo* have resulted in a clinical trial using PF-02341066, a c-Met inhibitor that also has significant activity against Alk (Mosse *et al.*, 2009).

Among many other features causing drug resistance, MYCN may be responsible for neuroblastoma cell resistance to vincristine and cisplatin (Blaheta *et al.*, 2007). Therefore, *MYCN* together with oncogenes such as *MDM2* and *ALK* comprise potential new targets for molecular intervention in future neuroblastoma treatment (reviewed in Van Roy *et al.*, 2009). In addition to targeting *ALK*, trials are also ongoing where inhibitors of PI3K (SF1126), the Trk neurotropin receptor (CEP-701), and the Aurora A kinase (MLN8237) are being evaluated (reviewed in Gustafson and Weiss, 2010). One *MYCN* targeting approach still awaiting clinical trials is the employment of peptide nucleic acids (PNAs), DNA analogs modified for a higher stability and longer duration of activity (reviewed in Morgenstern and Anderson, 2006). An antisense *MYCN* PNA conjugated to a somatostatin analog was demonstrated to be rapidly internalized and significantly inhibited cell growth of neuroblastoma cells (Sun *et al.*, 2002). An even better outcome was observed by the use of antigene PNAs, designed to be complementary to the coding DNA strand (Tonelli *et al.*, 2005). These molecules block gene expression at the transcriptional level and were shown to inhibit cellular proliferation in a panel of neuroblastoma cell lines at a similar rate as the observed reduction in MYCN expression. Additional preclinical investigations involve the use of small-molecule inhibitors, antisense oligonucleotides, and miRNA (described below).

K. Rhabdomyosarcoma (RMS)

RMS represents the most common pediatric soft tissue sarcoma. The sarcomas resemble developing skeletal muscle and are, based on histology, divided into the two main subtypes alveolar and embryonic RMS (Anderson *et al.*, 1999). The tumors can be distributed to nearly any tissue in the body, except bone, but the head and neck area and the genitourinary tract are the most common locations in children (reviewed in Hayes-Jordan and Andrassy, 2009). Markers for RMS include transcription factors in skeletal muscle; differentiation and structural proteins normally seen in mature skeletal muscle. While embryonic RMS represents the majority of cases ($\sim$75%), the worst prognosis is observed in patients with alveolar RMS (reviewed in Morgenstern and Anderson, 2006). A complicating factor in diagnosing the disease is the lack of serum markers. Therefore, open biopsies are often required in order to confirm the RMS (Hayes-Jordan and Andrassy, 2009). One potential prognostic factor is *MYCN* that has been detected in increased copy numbers in both the embryonic and the alveolar subtype. In the alveolar subtype, overexpression or gain of genomic copies of *MYCN* has been significantly associated with adverse outcome (Williamson *et al.*, 2005). In contrast, high genomic copy number of *MYCN* did not necessarily lead to high protein expression and *MYCN* amplification did not correlate with clinical outcome in the embryonic variant.

L. Treatment of RMS

There are basically three different international therapeutic protocols, depending on the risk of recurrence: Low risk (estimated 3-year failure-free survival (FFS) rate of 88%), intermediate risk (estimated 3-year FFS rate: 55–76%), and high risk (estimated FFS rate $< 30\%$) (Hayes-Jordan and Andrassy, 2009). A multimodality approach is required, but while low-risk patients are treated with relatively low doses of radiation and chemotherapy those in the intermediate risk-group receive a combination of chemotherapeutic drugs combined with radiation when possible (reviewed in Hayes-Jordan and Andrassy, 2009). Surgical excision of the tumor is performed in conjunction with chemotherapy where commonly used drugs include: vincristine, actinomycin, cyclophosphamide/ifosfamide, and irinotecan (Hayes-Jordan and Andrassy, 2009). In high-risk cases, particularly that of alveolar subtype, there is a clear need for new treatment strategies (Morgenstern and Anderson, 2006). As nearly 25% of patients with alveolar RMS display tumors with *MYCN* deregulations, there are strong implications for targeting MYCN in this subtype.

There are several new potential treatment approaches for RMS, most of which are still at the preclinical stage (reviewed in Morgenstern and Anderson, 2006). One potential target for such therapies is IGF and its receptor. The approaches include both monoclonal antibody therapy and selective inhibitors of the receptor tyrosine kinase (Garcia-Echeverria *et al.*, 2004; Maloney *et al.*, 2003). Immunotherapy techniques comprise another potential future strategy aiming to specifically target the tumor cells and produce fewer, less severe side effects (Morgenstern and Anderson, 2006). As an example, there have been attempts at "priming" cytotoxic T lymphocytes (CTLs) into targeting the alveolar RMS-specific protein Pax3-Foxo1A (Mackall *et al.*, 2000). Similar attempts at producing CTLs specifically targeting and killing *MYCN*-amplified neuroblastoma cell lines may also present a useful alternative in RMS treatment (Sarkar and Nuchtern, 2000). The potential efficacy of a vaccination approach is also being evaluated (reviewed in Morgenstern and Anderson, 2006). Other strategies under investigation for targeting deregulated *MYCN* in tumors include the potential use of PNAs investigated for neuroblastoma treatment, small-molecule inhibitors, and antisense oligonucleotides.

IX. NOVEL THERAPIES

The most frequently used anticancer drugs, including chemotherapeutics targeting topoisomerases, DNA-damaging agents, mitotic inhibitors, antimetabolites, and nucleotide analogues, suffer the disadvantage of causing resistance development (Herr and Debatin, 2001; Luqmani, 2005). This is most likely due to a deficient apoptotic program in tumor cells together with increased efflux and decreased influx of the drug, and increased DNA repair. In addition, the adverse effects such as induction of myelotoxicity, nausea, vomiting, diarrhea, and fatigue often caused by these agents (Nieboer *et al.*, 2005) calls for novel treatments less prone to cause side effects and resistance development in the patients.

A. Rational Design and Synthetic Modeling: Successful Examples

Screenings aimed to find molecules targeting the kinase domain of tumor-associated proteins have resulted in the development of the phenylamino-pyrimidine-derivative imatinib mesylate (Gleevec™/STI-571). This compound targets the kinase domain of the fusion protein Bcr-Abl (Druker, 2002), and was later found to inhibit other kinases, such as the stem cell factor receptor c-Kit and PDGFR (Druker, 2002; Nadal and

Olavarria, 2004). Gleevec is currently used in treatment of chronic myeloid leukemia (CML) (Nicolini *et al.*, 2006) and gastrointestinal stromal tumors (GIST) (von Mehren, 2006), and is also undergoing a number of clinical trials as adjuvant treatment of refractory or metastatic solid tumors (http://www.cancer.gov/clinicaltrials).

In a more direct approach, the three-dimensional structure of crystallized protein(s) is used for modeling site-specific compounds by computer-based predictions. These compounds are then synthesized for analysis of their biological activity (Kontopidis *et al.*, 2003; McClue *et al.*, 2002). The highly specific Cdk2-Cyclin E-targeting compound R-roscovitine (Seliciclib/CYC202), currently undergoing clinical trials (Benson *et al.*, 2007), was identified using this approach (De Azevedo *et al.*, 1997; Meijer *et al.*, 1997) and has been found to significantly reduce tumor size in colorectal xenograft mouse models (McLaughlin *et al.*, 2003; Raynaud *et al.*, 2005). However, in spite of these exact measures to engineer the perfect anticancer drug, it remains difficult to find a compound for which the mechanisms of action can be exclusively specified.

Yet another strategy for identifying new treatments includes screening of low-molecular compound libraries in search for substances eliciting target-specific antiproliferative or proapoptotic effects. This method was successfully applied in identification of PRIMA-1 (p53-reactivation and induction of massive apoptosis) (Bykov *et al.*, 2002), a molecule that has been found to enhance the apoptosis of agents such as cisplatin and doxorubicin (Bykov *et al.*, 2005; Magrini *et al.*, 2008) and for which the mechanism of action is currently being elucidated (Lambert *et al.*, 2009). Similarly, molecular screens for compounds interfering with transactivation by c-MYC or MYCN or with MYC/Max heterodimerization have been successful, yielding candidate compounds awaiting further investigation (Berg *et al.*, 2002; Hueber and Evan, 1998; Lu *et al.*, 2003; Xu *et al.*, 2006; Yin *et al.*, 2003). Below, we outline some examples of novel therapies in clinical use or in preclinical studies, as well as promising approaches to bring forth MYC pathway-specific anticancer treatments.

X. TARGETED THERAPY: WHAT IS IN THE FUTURE FOR MYC?

Targeting MYC or the MYC pathway has emerged as a very attractive approach to search for cancer intervention. This is because *MYC* is frequently deregulated in human tumors and is even believed to be aberrantly expressed in a major fraction of all cancers (Hermeking, 2003; Pelengaris and Khan, 2003; Prochownik, 2004). Several new strategies are being investigated, some of which are more promising than others (Dang *et al.*, 2009; Hermeking, 2003; Johnsen *et al.*, 2009; Lu *et al.*, 2003; Pelengaris and Khan, 2003; Prochownik, 2004; Vita and Henriksson, 2006). Here, we present some of the many approaches for targeting MYC at different levels (summarized with references in Table II). However, we did not bring up implications for therapeutic interventions of MYC-mediated energy metabolism since this issue was reviewed recently by experts in the field (Dang *et al.*, 2009).

A. Substances Interfering with the MYC Pathway

The MYC pathway could be targeted either directly by tackling the MYC protein itself or indirectly by affecting upstream regulators or downstream effectors. MYC protein expression could be controlled by affecting the stability or degradation while its activity could be regulated by affecting the dimerization capacity or the DNA-binding ability (Fig. 3). Several approaches including different screening assays have been used in order to identify substances that control MYC expression or activity (Berg *et al.*, 2002; Lu *et al.*, 2003; Mo and Henriksson, 2006; Xu *et al.*, 2006; Yin *et al.*, 2003).

In search for substances affecting MYCN-mediated transactivation, Lu *et al.* utilized a luciferase screening assay in neuroblastoma cells where the MYC target gene *ODC* served as reporter (Lu *et al.*, 2003). From a library of 2800 compounds, they identified eight compounds that significantly inhibited MYC-induced luciferase activity, five of which showed MYCN-specificity. These substances are being further evaluated for their potential use as lead substances (Lu *et al.*, 2003). Our lab employed a cellular screening strategy in search for MYC pathway response agents (MYRAs) (Mo and Henriksson, 2006). Using cells with conditional c-MYC expression, we selected substances that affected viability in c-MYC-overexpressing cells. Two substances, MYRA-A and MYRA-B, were found to induce apoptosis and inhibit transformation in a MYC-dependent manner without affecting MYC/Max dimerization. Together with a third substance from the initial screen, they were also found to target MYCN overexpressing cells, suggesting their potential use in treatment of both c-MYC and MYCN overexpressing tumors (Mo *et al.*, 2006). A different approach was taken in a screen for oncogenic pathways responsive to Cdk1 inhibition (Goga *et al.*, 2007). Cells transformed with *MYC* were found to be sensitized to apoptosis in response to treatment with the Cdk1 inhibitor purvalanol in contrast to those transformed by other oncogenes. As this was independent of the p53–MDM2–ARF pathway, and appeared to be specific for MYC, Cdk1 inhibition has been suggested as a future therapeutic model for human malignancies overexpressing MYC. However, this requires identification of a Cdk1 inhibitor better suitable for

Table II Preclinical Research for MYC-Specific Therapies

MYC-specific therapy	References
MYCN antisense and antigene peptide nucleic acids (PNAs)	Sun *et al.* (2002), Tonelli *et al.* (2005), Morgenstern and Anderson (2006; review)
Disrupting MYC/Max dimerization by 10058-F4 and its analogues + 10058-F4-related molecules	Yin *et al.* (2003), Huang *et al.* (2006), Lin *et al.* (2007), Wang *et al.* (2007), Follis *et al.* (2008), Guo *et al.* (2009), Hammoudeh *et al.* (2009)
Small-molecule inhibitors other than 10058-F4 (MYRAs, IIA6B17, and others)	Berg *et al.* (2002), Lu *et al.* (2003), Mo and Henriksson (2006), Mo *et al.* (2006), Xu *et al.* (2006), Lu *et al.* (2008)
Interfering with MYC-induced energy metabolism	Dang *et al.* (2009; review)
Cdk1 or Cdk2 inhibition	Goga *et al.* (2007), Hydbring *et al.* (2009), Campaner *et al.* (2010)
Survivin targeting (such as shepherdin)	Mita *et al.* (2008), Plescia *et al.* (2005)
Histone deacetylase inhibitors (trichostatin-A and others)	McLaughlin *et al.* (2003), Albihn *et al.* (unpublished data)
G-quadruplex DNA stabilizers (TMPyP4 porphyrin, S2T1-6OTD telomestatin derivative)	Grand *et al.* (2002), Shalaby *et al.* (2010)
Targeting of miRNas (let7, miR-17-92, miR 15a, miR-34a)	He *et al.* (2005), O'Donnell *et al.* (2005), Sampson *et al.* (2007), Chang *et al.* (2008), Cole *et al.* (2008), Leucci *et al.* (2008), Shi *et al.* (2008; review), Loven *et al.* (2010)
MYCN-targeting cytotoxic T cells (CTLs), vaccination approaches	Sarkar and Nuchtern (2000), Morgenstern and Anderson (2006; review)
Gene therapy using MYC-thymidine kinase expressing adenoviral vector (adMYCTK)	Nishino *et al.* (2001)
Interfering with upstream regulation of MYC (USP28, CIP2A, PI3K/Akt)	Chesler *et al.* (2006), Popov *et al.* (2007), Johnsen *et al.* (2008), Junttila *et al.* (2007, 2008), Baryawno *et al.* (2009), Khanna *et al.* (2009)
Manipulating MYC target proteins (Bcl-2 proteins, ODC, etc.)	Mason *et al.* (2008), Nilsson *et al.* (2005), Raul (2007; review), Shantz and Levin (2007), Hogarty *et al.* (2008), Evageliou and Hogarty (2009)
Transient MYC inactivation (Tet-regulated, ER-regulated, Omomyc)	Felsher and Bishop (1999a,b), Pelengaris *et al.* (1999), Soucek *et al.* (2008)

in vivo treatment, as purvalanol does not dissolve well in aqueous solutions (Goga *et al.*, 2007). Another possibility would be to target survivin, a molecule affecting apoptosis and mitotic spindle functions and that is often overexpressed in human cancers (Mita *et al.*, 2008). There have been numerous approaches to target this inhibitor of apoptosis protein (IAP) that is partially controlled by Cdk1 and also promotes some of its functions through interaction with Hsp90 (Altieri, 2008). A small peptide, shepherdin, engineered by rational design to prevent Survivin's interaction with Hsp90, has been found to induce extensive apoptosis in tumor cells and in tumors in mice where it was well tolerated and did not induce significant signs of toxicity (Plescia *et al.*, 2005).

As previously mentioned, targeting Cdk2 is another possible future therapeutic approach for MYC-overexpressing tumors, as this kinase appears to regulate MYC protein stability by phosphorylation at the S62 site (Hydbring *et al.*, 2009). Indeed, Cdk2 was recently shown to prevent MYC-induced cellular senescence (Campaner *et al.*, 2010). In addition, it was found that pharmacological inhibition of Cdk2 induced MYC-dependent senescence independently of p53 and without enhancing MYC-driven replication stress.

Histone Deacetylase (HDAC) inhibitors became interesting as targets for cancer therapy when it was found that they could also control deacetylation of proteins other than histones (McLaughlin *et al.*, 2003) and several HDAC inhibitory compounds are already in clinical trials. In addition, we have shown that Trichostatin-A efficiently kills cells with MYC overexpression suggesting that HDAC inhibitors may be efficient in MYC-overexpressing tumors (Albihn *et al.*, unpublished data).

1. Interfering with the Upstream Signal

As tumor cell proliferation appears to require USP28-mediated MYC stabilization in many cases, this deubiquitinating enzyme (DUB) is viewed as an attractive target for future therapeutic tumor intervention (Junttila and Westermarck, 2008; Popov *et al.*, 2007). USP28 appears to be essential for MYC-induced tumorigenesis and is highly expressed in colon and breast carcinomas. Since it belongs to a class of enzymes that can be selectively targeted, small molecules could be designed to block the USP28 activity (Popov *et al.*, 2007). It has been suggested that inhibition of CIP2A would provide a possible therapeutic approach in treatment of certain cancer forms (Junttila and Westermarck, 2008; Junttila *et al.*, 2007). Indeed, CIP2A has been found overexpressed in head and neck cancer and colon cancer, and it also appeared that its depletion would cause degradation of the MYC protein. A prognostic role for CIP2A was suggested in gastric cancer where subgroups of patients, immunopositive for CIP2A, were found to have a reduced survival rate (Khanna *et al.*, 2009). The investigators identified a positive-feedback loop between CIP2A and MYC, suggesting that MYC would directly promote gene expression of CIP2A, at the same time as CIP2A stabilized the c-MYC protein. This finding highlighted the potential benefit of targeting CIP2A as a therapeutic strategy as depletion of CIP2A would prevent anchorage-independent growth as well as proliferation of the tumor cells. Similarly, inhibition of PI3K/Akt signaling has been shown to increase the degradation of MYCN in neuroblastoma (Chesler *et al.*, 2006; Johnsen *et al.*, 2008) and of c-MYC in medulloblastoma (Baryawno *et al.*, 2010).

The Mnt protein should also be considered for future therapeutic strategies as it has been proposed to be a regulator of MYC activity. The possibility to enforce the repressive effect elicited by its potential tumor suppressor activity ought to be further explored (reviewed in Wahlstrom and Henriksson, 2007).

2. DISRUPTING MYC/MAX DIMERIZATION

Several screening projects have been performed with the aim to identify substances interfering with MYC/Max dimerization. In one of these studies, 10,000 substances were screened using the yeast two-hybrid assay, and seven molecules were found to specifically disrupt c-MYC/Max dimerization (Yin *et al.*, 2003). All seven compounds were found to inhibit MYC-mediated transactivation and four of them also prevented tumor formation when cells that had been incubated with the compound for 3 days *in vitro* were inoculated into nude mice. One substance, 10058-F4 has been further studied and found to affect cellular apoptosis, differentiation, and cell-cycle progression in addition to its effect on the MYC/Max complex (Huang *et al.*, 2006; Lin *et al.*, 2007). Thus, this molecule showed great promise for further development, but turned out to be highly unstable and was rapidly degraded *in vivo* (Guo *et al.*, 2009). This problem, together with the fact that the original 10058-F4 molecule had quite low potency, called for a search for more efficient analogues. By modification of the two ring structures in the quite simple 10058-F4 backbone, Wang *et al.* created second and third generation analogues of the molecule which proved to be more stable and efficient than the original compound (Wang *et al.*, 2007). Recently, the binding site for 10058-F4 was located to the HLHZip region of the MYC protein, and it was found that three of the seven molecules indentified in the original screen could bind simultaneously to distinct sites of c-MYC without affecting the activity of the others (Follis *et al.*, 2008; Hammoudeh *et al.*, 2009).

Another technique, fluorescence resonance energy transfer (FRET) where the two proteins to be investigated are coupled to different color fluorescent proteins, is based on the color change, measured as a change in excitation wavelength, as the proteins connect. This approach was used by Berg *et al.* who identified two compounds from a library of 7000 that specifically interfered with MYC-induced oncogenic transformation of chicken embryo fibroblasts in culture (Berg *et al.*, 2002). One of those molecules (IIA6B17) was recently found to specifically disrupt the transcriptional activity of c-MYC but not that of MYCN (Lu *et al.*, 2008). This finding suggested that a cell-based MYC luciferase reporter gene assay could be used as a tool to distinguish whether the candidate molecules are specific for c-MYC MYCN, or nonselective to help select their appropriate future use. Xu *et al.* designed and synthesized a credit-card library of 285 substances and identified several compounds that disrupted c-MYC/Max dimerization (Xu *et al.*, 2006). The designation credit-card comes from the planar structure of the molecules, two of which were also found to prevent MYC-induced oncogenic transformation in cell culture.

3. ATTACKING THE MYC TARGETS

MYC-induced apoptosis is mainly mediated through the mitochondria, stimulating the release of cytochrome *c*. Proapoptotic Bcl-2 family proteins are important for permeabilization of the mitochondria and antiapoptotic members prevent this event, thus disrupting MYC-induced apoptosis. Therefore, blocking Bcl-2 activity is a possible approach to enhance MYC-driven apoptosis. ABT-737, a small molecule that similarly to BH3-only proteins binds to and inactivates some, but not all, antiapoptotic Bcl-2 proteins, was found to enhance apoptosis in lymphomas induced by MYC in combination with Bcl-2 (Mason *et al.*, 2008). This was observed *in vivo* using ABT-737 as a single agent, and the effect was further enhanced in combination with cyclophosphamide. However, there was no evident effect in lymphomas driven by MYC alone, suggesting that antagonizing Bcl-2 may be an efficient supplement to conventional therapy in treating MYC-driven lymphomas overexpressing Bcl-2 (Mason *et al.*, 2008).

Other proteins mediating the MYC effect, such as Odc, Cad, and hTERT, may also be targeted as a treatment approach to reduce the effect of MYC activation. Of these, the association between MYC and gene expression appear strongest for Odc where overexpression has been observed in human cancers such as MYC-induced lymphoma and MYCN-driven neuroblastoma (Hogarty *et al.*, 2008; Nilsson *et al.*, 2005). Potential therapeutic strategies for targeting Odc include the Odc inhibitor α-difluoromethylornithine (DFMO) in combination with conventional chemotherapy (review in Raul, 2007). Preclinical data has also proven this agent useful in treatment of lymphoma, neuroblastoma, and other malignancies (reviewed in Evageliou and Hogarty, 2009; Shantz and Levin, 2007).

4. HAMPERING WITH MYC EXPRESSION LEVEL

Antisense oligodeoxynucleotides (ASOs) targeting MYC have been successfully tested in several *in vitro* and *in vivo* models (Kutryk *et al.*, 2002 and reviewed in Morgenstern and Anderson, 2006; Pelengaris and Khan, 2003; Prochownik, 2004; Tamm, 2005). For instance, experiments have been aiming at enhancing the efficacy of cisplatin *in vitro* (reviewed in Prochownik, 2004), and reducing the MYCN protein level, resulting in reduced cell division and increased differentiation (reviewed in Morgenstern and Anderson, 2006). *In vivo* mouse data have been positive, showing that *MYCN* ASOs significantly reduced tumor incidence as well as tumor mass (Burkhart *et al.*, 2003). Similar results were shown for *c-MYC* ASOs where data from phase-I clinical trials indicate that their distribution is tolerated in healthy human subjects (Kutryk *et al.*, 2002 and reviewed in Prochownik, 2004). In addition, a third generation antisense molecule, the *c-MYC* targeting phosphorodiamidate morpholino oligomer (PMO) AVI-4126, was successfully applied in a phase-I clinical trial for prostate cancer, suggesting this as a safe and promising new therapeutic approach (Iversen *et al.*, 2003). However, the therapeutic efficacy of these molecules remains to be explored.

Cationic porphyrins such as TMPyP4 that stabilize DNA G-quadruplexes have also been shown to downregulate c-MYC expression (Grand *et al.*, 2002). Because of their additional inhibitory effect on the hTERT activity, these molecules are being studied as potential anticancer agents. A similar effect is observed in response to the telomestatin derivative S2T1-6OTD (Shalaby *et al.*, 2010).

5. MAKING USE OF THE NONCODING SEQUENCE

MicroRNAs (miRNAs), small noncoding RNA molecules that were initially identified in *Caenorhabditis elegans* (Lee *et al.*, 1993; Wightman *et al.*, 1993), measure 18–24 nucleotides and account for ~1% of known genes. They bind to and negatively regulate protein coding mRNAs and are believed to be present in all multicellular eukaryotes (Ambros, 2004; Bartel, 2004; John *et al.*, 2004; Kent and Mendell, 2006; Shi *et al.*, 2008). Several pieces of evidence suggest that many miRNAs function as tumor suppressors or oncogenes, regulating the expression of proteins important in tumorigenesis (reviewed in Kent and Mendell, 2006; Shi *et al.*, 2008). This "cancerous" feature of miRNAs makes them attractive as potential therapeutic targets, and the prospect of using them as biomarkers is also being explored. MYC has mainly been associated to two of the cancerous miRNA clusters, namely the let7 family of tumor suppressors and the oncogenic miR-17-92 cluster (Chang *et al.*, 2008; He *et al.*, 2005; Loven *et al.*, 2010; O'Donnell *et al.*, 2005; Sampson *et al.*, 2007; Shah *et al.*, 2007). The let7 family members are poorly expressed in several cancer forms and experimental evidence has shown that their ectopic expression reduced cell proliferation, inhibited tumorigenesis, and in one case even reduced metastasis (reviewed in Shi *et al.*, 2008). In addition to silencing MYC (Sampson *et al.*, 2007), let7 family members have been found to silence *Ras* and genes involved in cell-cycle and cell division control (Johnson *et al.*, 2005, 2007). This, together with the finding that let7 overexpression can reduce resistance to chemotherapy in lung cancer (Weidhaas *et al.*, 2007), suggests a future important role of this miRNA cluster in clinical use. The miR-17-92 polycistron contains seven human miRNAs (Shi *et al.*, 2008) and is strongly associated with lymphomas and several solid tumors including neuroblastoma (Dews *et al.*, 2006; Hayashita *et al.*, 2005; He *et al.*, 2005; Loven *et al.*, 2010; Ota *et al.*, 2004; Volinia *et al.*, 2006). Even though it is mostly viewed as an oncogenic cluster, the effects of individual miR-17-92 members are strictly cell type and context dependent. For instance, one study in a panel of breast cancer cell lines demonstrated a tumor suppressor function for miR-17-5p (Hossain *et al.*, 2006). The first evidence of *in vivo* oncogenic activity of the miR-17 cluster was demonstrated in Eμ-*MYC*-transgenic mice, showing cooperation between the miR-17 cluster and c-*MYC* (He *et al.*, 2005). This result was further strengthened by the finding that transcription of the miR-17 cluster is directly activated by c-MYC (O'Donnell *et al.*, 2005). In the same study the MYC target E2F1, promoting cell-cycle progression, was shown to be negatively regulated by miR-17-5p/20a of the miR-17 cluster. MYC thus has two levels of control of the proliferative signal through E2F1, directly at the transcriptional level and indirectly at the translational level by activation of the miR-17 polycistron (O'Donnell *et al.*, 2005). The miR-17-92 cluster was also found to be important in induction of B cell lymphomas, as recent experiments in mice demonstrated that miR-19 was the key oncogenic component that promoted lymphomagenesis in cooperation with c-MYC (Mu *et al.*, 2009; Olive *et al.*, 2009). This effect was partially due to repression of the tumor suppressor *pten* (Olive *et al.*, 2009), strengthening the previous notion of its association with the miR-17 cluster (Lewis *et al.*, 2003).

Another important finding is that c-MYC also represses the expression of several miRNAs (Chang *et al.*, 2008; Loven *et al.*, 2010). Among the down-regulated miRNAs were miR-15a, miR-34a, and let7; all located in genomic regions often deleted in cancer. miR-15a targets the antiapoptotic gene Bcl-2 (Cimmino *et al.*, 2005), let7 family members target Ras (Johnson *et al.*, 2005), and miR-34a has been shown to be regulated by p53 (Bommer *et al.*, 2007; Chang *et al.*, 2007). These alleged tumor suppressors were established to be directly regulated by MYC and shown to inhibit experimentally induced B cell lymphomas in mice (Chang *et al.*, 2008). There is also experimental data suggesting that the miR-34a cluster is responsible for c-MYC deregulation in cases of BL lacking the classical *c-MYC* translocation (Leucci *et al.*, 2008). In addition, it appears that miR-34a has a tumor suppressor function in neuroblastoma where it was found to regulate MYCN as well as Bcl-2 (Cole *et al.*, 2008). Taken together, these data implicate that control of miRNA expression is of great importance in MYC-mediated tumorigenesis. The fact that systemic delivery of small RNA molecules has been proven possible in animals (Soutschek *et al.*,

2004), suggests the possibility for future therapeutic strategies based on delivering MYC-repressed miRNAs to combat cancer. Furthermore, a miRNA-based therapeutic approach has the advantage over single gene therapy that it targets multiple downstream effectors and may therefore be more effective (Petrocca and Lieberman, 2009).

B. Transient Inactivation of MYC

The prospect of targeting c-MYC is complicated by its nearly ubiquitous expression in proliferating cells since such a central protein might be crucial for tissue regeneration. However, Soucek and colleagues recently showed that transgenic mice tolerated the effects of extended MYC inhibition while almost complete regression of their *K-ras*-induced lung tumors was observed (Soucek *et al.*, 2008). MYC was silenced by conditional expression of a dominant interfering MYC bHLHZip dimerization domain mutant called Omomyc. Despite strong effects on proliferating tissues in the intestinal crypts and the skin, these were rapidly reverted upon restoration of normal MYC function with no apparent damage to the animals. This study, together with other experiments where c-MYC has been transiently inactivated in more localized compartments (Felsher and Bishop, 1999a; Pelengaris *et al.*, 1999), provides evidence of the benefit of pharmacological inhibition of c-MYC. If successfully confirmed in human subjects, this may be the preferred approach in treating MYC-driven tumors in the future.

XI. CONCLUDING REMARKS

We have highlighted some of the most important MYC functions and presented an overview of current and future therapies for a few cancers with *MYC* gene activation. The need for new, more specific cancer therapies is met by an intense research activity using different approaches and strategies. In addition, aspects such as timing, cellular location, and dosing also have to be taken into account when designing novel anticancer treatments targeting MYC or the MYC pathway. Most likely a combination of different approaches, both novel and/or conventional, rather than one single agent, the magic bullet, will provide future cancer cures.

ACKNOWLEDGMENTS

We are grateful to Dr. Lars-Gunnar Larsson for critical reading of the manuscript and to our many colleagues for sharing our fascination with MYC. We apologize to colleagues whose work we were unable to cite due to the scope and space restrictions. Research from the authors' laboratories are supported by grants from the Swedish Cancer Society, the Swedish Research Council, the Swedish Childhood Cancer Society, King Gustaf V Jubilee Foundation, Karolinska Institutet, and KICancer. MAH is recipient of a Senior Investigator Award from the Swedish Cancer Society.

REFERENCES

Abal, M., Andreu, J. M., and Barasoain, I. (2003). Taxanes: Microtubule and centrosome targets, and cell cycle dependent mechanisms of action. *Curr. Cancer Drug Targets* **3**, 193–203.

Adachi, S., Obaya, A. J., Han, Z., Ramos-Desimone, N., Wyche, J. H., and Sedivy, J. M. (2001). c-Myc is necessary for DNA damage-induced apoptosis in the G(2) phase of the cell cycle. *Mol. Cell. Biol.* **21**, 4929–4937.

Adhikary, S., Marinoni, F., Hock, A., Hulleman, E., Popov, N., Beier, R., Bernard, S., Quarto, M., Capra, M., Goettig, S., Kogel, U., Scheffner, M., *et al.* (2005). The ubiquitin ligase HectH9 regulates transcriptional activation by Myc and is essential for tumor cell proliferation. *Cell* **123**, 409–421.

Albihn, A., Loven, J., Ohlsson, J., Osorio, L. M., and Henriksson, M. (2006). c-Myc-dependent etoposide-induced apoptosis involves activation of Bax and caspases, and PKCdelta signaling. *J. Cell. Biochem.* **98**, 1597–1614.

Aldoss, I. T., Weisenburger, D. D., Fu, K., Chan, W. C., Vose, J. M., Bierman, P. J., Bociek, R. G., and Armitage, J. O. (2008). Adult Burkitt lymphoma: Advances in diagnosis and treatment. *Oncology (Williston Park)* **22**, 1508–1517.

Altieri, D. C. (2008). New wirings in the survivin networks. *Oncogene* **27**, 6276–6284.

Amati, B., Alevizopoulos, K., and Vlach, J. (1998). Myc and the cell cycle. *Front. Biosci.* **3**, d250–d268.

Ambros, V. (2004). The functions of animal microRNAs. *Nature* **431**, 350–355.

Anderson, J., Gordon, A., Pritchard-Jones, K., and Shipley, J. (1999). Genes, chromosomes, and rhabdomyosarcoma. *Genes Chromosomes Cancer* **26**, 275–285.

Arnold, H. K., Zhang, X., Daniel, C. J., Tibbitts, D., Escamilla-Powers, J., Farrell, A., Tokarz, S., Morgan, C., and Sears, R. C. (2009). The Axin1 scaffold protein promotes formation of a degradation complex for c-Myc. *EMBO J.* **28**, 500–512.

Baena, E., Ortiz, M., Martinez, A. C., and de Alboran, I. M. (2007). c-Myc is essential for hematopoietic stem cell differentiation and regulates Lin(-)Sca-1(+)c-Kit(-) cell generation through p21. *Exp. Hematol.* **35**, 1333–1343.

Bahram, F., von der Lehr, N., Cetinkaya, C., and Larsson, L. G. (2000). c-Myc hot spot mutations in lymphomas result in inefficient ubiquitination and decreased proteasome-mediated turnover. *Blood* **95**, 2104–2110.

Baldwin, E. L., and Osheroff, N. (2005). Etoposide, topoisomerase II and cancer. *Curr. Med. Chem. Anticancer Agents* **5**, 363–372.

Banerjee, A., Hu, J., and Goss, D. J. (2006). Thermodynamics of protein–protein interactions of cMyc, Max, and Mad: Effect of polyions on protein dimerization. *Biochemistry* **45**, 2333–2338.

Bartel, D. P. (2004). MicroRNAs: Genomics, biogenesis, mechanism, and function. *Cell* **116**, 281–297.

Baryawno, N., Sveinbjornsson, B., Eksborg, S., Chen, C. S., Kogner, P., and Johnsen, J. I. (2010). Small-molecule inhibitors of phosphatidylinositol 3-kinase/Akt signaling inhibit Wnt/beta-catenin pathway cross-talk and suppress medulloblastoma growth. *Cancer Res.* **70**, 266–276.

Benson, C., White, J., De Bono, J., O'Donnell, A., Raynaud, F., Cruickshank, C., McGrath, H., Walton, M., Workman, P., Kaye, S., Cassidy, J., Gianella-Borradori, A., *et al.* (2007). A phase I trial of the selective oral cyclin-dependent kinase inhibitor seliciclib (CYC202; R-Roscovitine), administered twice daily for 7 days every 21 days. *Br. J. Cancer* **96,** 29–37.

Berg, T., Cohen, S. B., Desharnais, J., Sonderegger, C., Maslyar, D. J., Goldberg, J., Boger, D. L., and Vogt, P. K. (2002). Small-molecule antagonists of Myc/Max dimerization inhibit Myc-induced transformation of chicken embryo fibroblasts. *Proc. Natl. Acad. Sci. USA* **99,** 3830–3835.

Berman, D. M., Karhadkar, S. S., Hallahan, A. R., Pritchard, J. I., Eberhart, C. G., Watkins, D. N., Chen, J. K., Cooper, M. K., Taipale, J., Olson, J. M., and Beachy, P. A. (2002). Medulloblastoma growth inhibition by hedgehog pathway blockade. *Science* **297,** 1559–1561.

Billin, A. N., and Ayer, D. E. (2006). The Mlx network: Evidence for a parallel Max-like transcriptional network that regulates energy metabolism. *Curr. Top. Microbiol. Immunol.* **302,** 255–278.

Blackhall, F. H., Shepherd, F. A., and Albain, K. S. (2005). Improving survival and reducing toxicity with chemotherapy in advanced non-small cell lung cancer: A realistic goal? *Treat. Respir. Med.* **4,** 71–84.

Blackwood, E. M., and Eisenman, R. N. (1991). Max: A helix-loop-helix zipper protein that forms a sequence-specific DNA-binding complex with Myc. *Science* **251,** 1211–1217.

Blaheta, R. A., Michaelis, M., Natsheh, I., Hasenberg, C., Weich, E., Relja, B., Jonas, D., Doerr, H. W., and Cinatl, J., Jr. (2007). Valproic acid inhibits adhesion of vincristine- and cisplatin-resistant neuroblastoma tumour cells to endothelium. *Br. J. Cancer* **96,** 1699–1706.

Boehm, J. S., Hession, M. T., Bulmer, S. E., and Hahn, W. C. (2005). Transformation of human and murine fibroblasts without viral oncoproteins. *Mol. Cell. Biol.* **25,** 6464–6474.

Bommer, G. T., Gerin, I., Feng, Y., Kaczorowski, A. J., Kuick, R., Love, R. E., Zhai, Y., Giordano, T. J., Qin, Z. S., Moore, B. B., MacDougald, O. A., Cho, K. R., *et al.* (2007). p53-Mediated activation of miRNA34 candidate tumor-suppressor genes. *Curr. Biol.* **17,** 1298–1307.

Bornkamm, G. W. (2009). Epstein–Barr virus and the pathogenesis of Burkitt's lymphoma: More questions than answers. *Int. J. Cancer* **124,** 1745–1755.

Bouchalova, K., Cizkova, M., Cwiertka, K., Trojanec, R., and Hajduch, M. (2009). Triple negative breast cancer—Current status and prospective targeted treatment based on HER1 (EGFR), TOP2A and C-MYC gene assessment. *Biomed. Pap. Med. Fac. Univ. Palacky Olomouc Czech. Repub.* **153,** 13–17.

Bouchard, C., Thieke, K., Maier, A., Saffrich, R., Hanley-Hyde, J., Ansorge, W., Reed, S., Sicinski, P., Bartek, J., and Eilers, M. (1999). Direct induction of cyclin D2 by Myc contributes to cell cycle progression and sequestration of p27. *EMBO J.* **18,** 5321–5333.

Bouchard, C., Dittrich, O., Kiermaier, A., Dohmann, K., Menkel, A., Eilers, M., and Luscher, B. (2001). Regulation of cyclin D2 gene expression by the Myc/Max/Mad network: Myc-dependent TRRAP recruitment and histone acetylation at the cyclin D2 promoter. *Genes Dev.* **15,** 2042–2047.

Bousset, K., Henriksson, M., Luscher-Firzlaff, J. M., Litchfield, D. W., and Luscher, B. (1993). Identification of casein kinase II phosphorylation sites in Max: Effects on DNA-binding kinetics of Max homo- and Myc/Max heterodimers. *Oncogene* **8,** 3211–3220.

Boyd, K. E., and Farnham, P. J. (1997). Myc versus USF: Discrimination at the cad gene is determined by core promoter elements. *Mol. Cell. Biol.* **17,** 2529–2537.

Brennan, J., O'Connor, T., Makuch, R. W., Simmons, A. M., Russell, E., Linnoila, R. I., Phelps, R. M., Gazdar, A. F., Ihde, D. C., and Johnson, B. E. (1991). myc family DNA amplification in 107 tumors and tumor cell lines from patients with small cell lung cancer treated with different combination chemotherapy regimens. *Cancer Res.* **51,** 1708–1712.

Brenner, C., Deplus, R., Didelot, C., Loriot, A., Vire, E., De Smet, C., Gutierrez, A., Danovi, D., Bernard, D., Boon, T., Pelicci, P. G., Amati, B., *et al.* (2005). Myc represses transcription through recruitment of DNA methyltransferase corepressor. *EMBO J.* **24**, 336–346.

Browd, S. R., Kenney, A. M., Gottfried, O. N., Yoon, J. W., Walterhouse, D., Pedone, C. A., and Fults, D. W. (2006). N-myc can substitute for insulin-like growth factor signaling in a mouse model of sonic hedgehog-induced medulloblastoma. *Cancer Res.* **66**, 2666–2672.

Brownlie, P., Ceska, T., Lamers, M., Romier, C., Stier, G., Teo, H., and Suck, D. (1997). The crystal structure of an intact human Max–DNA complex: New insights into mechanisms of transcriptional control. *Structure* **5**, 509–520.

Brunelle, J. K., Santore, M. T., Budinger, G. R., Tang, Y., Barrett, T. A., Zong, W. X., Kandel, E., Keith, B., Simon, M. C., Thompson, C. B., Hay, N., and Chandel, N. S. (2004). c-Myc sensitization to oxygen deprivation-induced cell death is dependent on Bax/Bak, but is independent of p53 and hypoxia-inducible factor-1. *J. Biol. Chem.* **279**, 4305–4312.

Burkhart, C. A., Cheng, A. J., Madafiglio, J., Kavallaris, M., Mili, M., Marshall, G. M., Weiss, W. A., Khachigian, L. M., Norris, M. D., and Haber, M. (2003). Effects of MYCN antisense oligonucleotide administration on tumorigenesis in a murine model of neuroblastoma. *J. Natl. Cancer Inst.* **95**, 1394–1403.

Bush, N. J. (2007). Advances in hormonal therapy for breast cancer. *Semin. Oncol. Nurs.* **23**, 46–54.

Bush, A., Mateyak, M., Dugan, K., Obaya, A., Adachi, S., Sedivy, J., and Cole, M. (1998). c-myc null cells misregulate cad and gadd45 but not other proposed c-Myc targets. *Genes Dev.* **12**, 3797–3802.

Bykov, V. J., Issaeva, N., Shilov, A., Hultcrantz, M., Pugacheva, E., Chumakov, P., Bergman, J., Wiman, K. G., and Selivanova, G. (2002). Restoration of the tumor suppressor function to mutant p53 by a low-molecular-weight compound. *Nat. Med.* **8**, 282–288.

Bykov, V. J., Zache, N., Stridh, H., Westman, J., Bergman, J., Selivanova, G., and Wiman, K. G. (2005). PRIMA-1(MET) synergizes with cisplatin to induce tumor cell apoptosis. *Oncogene* **24**, 3484–3491.

Caccia, N. C., Mak, T. W., and Klein, G. (1984). c-myc involvement in chromosomal translocations in mice and men. *J. Cell. Physiol. Suppl.* **3**, 199–208.

Campaner, S., Doni, M., Hydbring, P., Verrecchia, A., Bianchi, L., Sardella, D., Schleker, T., Perna, D., Tronnersjo, S., Murga, M., Fernandez-Capetillo, O., Barbacid, M., *et al.* (2010). Cdk2 suppresses cellular senescence induced by the c-myc oncogene. *Nat. Cell Biol.* **12**, 54–59; sup pp. 51–14.

Canelles, M., Delgado, M. D., Hyland, K. M., Lerga, A., Richard, C., Dang, C. V., and Leon, J. (1997). Max and inhibitory c-Myc mutants induce erythroid differentiation and resistance to apoptosis in human myeloid leukemia cells. *Oncogene* **14**, 1315–1327.

Cartwright, P., McLean, C., Sheppard, A., Rivett, D., Jones, K., and Dalton, S. (2005). LIF/STAT3 controls ES cell self-renewal and pluripotency by a Myc-dependent mechanism. *Development* **132**, 885–896.

Chang, D. W., Claassen, G. F., Hann, S. R., and Cole, M. D. (2000). The c-Myc transactivation domain is a direct modulator of apoptotic versus proliferative signals. *Mol. Cell. Biol.* **20**, 4309–4319.

Chang, T. C., Wentzel, E. A., Kent, O. A., Ramachandran, K., Mullendore, M., Lee, K. H., Feldmann, G., Yamakuchi, M., Ferlito, M., Lowenstein, C. J., Arking, D. E., Beer, M. A., *et al.* (2007). Transactivation of miR-34a by p53 broadly influences gene expression and promotes apoptosis. *Mol. Cell* **26**, 745–752.

Chang, T. C., Yu, D., Lee, Y. S., Wentzel, E. A., Arking, D. E., West, K. M., Dang, C. V., Thomas-Tikhonenko, A., and Mendell, J. T. (2008). Widespread microRNA repression by Myc contributes to tumorigenesis. *Nat. Genet.* **40**, 43–50.

Chen, Y., and Olopade, O. I. (2008). MYC in breast tumor progression. *Expert Rev. Anticancer Ther.* **8,** 1689–1698.

Chen, C. R., Kang, Y., and Massague, J. (2001). Defective repression of c-myc in breast cancer cells: A loss at the core of the transforming growth factor beta growth arrest program. *Proc. Natl. Acad. Sci. USA* **98,** 992–999.

Chesler, L., Schlieve, C., Goldenberg, D. D., Kenney, A., Kim, G., McMillan, A., Matthay, K. K., Rowitch, D., and Weiss, W. A. (2006). Inhibition of phosphatidylinositol 3-kinase destabilizes Mycn protein and blocks malignant progression in neuroblastoma. *Cancer Res.* **66,** 8139–8146.

Chiang, Y. C., Teng, S. C., Su, Y. N., Hsieh, F. J., and Wu, K. J. (2003). c-Myc directly regulates the transcription of the NBS1 gene involved in DNA double-strand break repair. *J. Biol. Chem.* **278,** 19286–19291.

Cimmino, A., Calin, G. A., Fabbri, M., Iorio, M. V., Ferracin, M., Shimizu, M., Wojcik, S. E., Aqeilan, R. I., Zupo, S., Dono, M., Rassenti, L., Alder, H., *et al.* (2005). miR-15 and miR-16 induce apoptosis by targeting BCL2. *Proc. Natl. Acad. Sci. USA* **102,** 13944–13949.

Cohn, S. L., and Tweddle, D. A. (2004). MYCN amplification remains prognostically strong 20 years after its "clinical debut". *Eur. J. Cancer* **40,** 2639–2642.

Cohn, S. L., London, W. B., Huang, D., Katzenstein, H. M., Salwen, H. R., Reinhart, T., Madafiglio, J., Marshall, G. M., Norris, M. D., and Haber, M. (2000). MYCN expression is not prognostic of adverse outcome in advanced-stage neuroblastoma with nonamplified MYCN. *J. Clin. Oncol.* **18,** 3604–3613.

Cole, K. A., Attiyeh, E. F., Mosse, Y. P., Laquaglia, M. J., Diskin, S. J., Brodeur, G. M., and Maris, J. M. (2008). A functional screen identifies miR-34a as a candidate neuroblastoma tumor suppressor gene. *Mol. Cancer Res.* **6,** 735–742.

Conzen, S. D., Gottlob, K., Kandel, E. S., Khanduri, P., Wagner, A. J., O'Leary, M., and Hay, N. (2000). Induction of cell cycle progression and acceleration of apoptosis are two separable functions of c-Myc: Transrepression correlates with acceleration of apoptosis. *Mol. Cell. Biol.* **20,** 6008–6018.

Cory, S., and Adams, J. M. (2002). The Bcl2 family: Regulators of the cellular life-or-death switch. *Nat. Rev. Cancer* **2,** 647–656.

Cowling, V. H., Chandriani, S., Whitfield, M. L., and Cole, M. D. (2006). A conserved Myc protein domain, MBIV, regulates DNA binding, apoptosis, transformation, and G2 arrest. *Mol. Cell. Biol.* **26,** 4226–4239.

Cowling, V. H., D'Cruz, C. M., Chodosh, L. A., and Cole, M. D. (2007). c-Myc transforms human mammary epithelial cells through repression of the Wnt inhibitors DKK1 and SFRP1. *Mol. Cell. Biol.* **27,** 5135–5146.

Cully, M., You, H., Levine, A. J., and Mak, T. W. (2006). Beyond PTEN mutations: The PI3K pathway as an integrator of multiple inputs during tumorigenesis. *Nat. Rev. Cancer* **6,** 184–192.

Dadiani, M., Seger, D., Kreizman, T., Badikhi, D., Margalit, R., Eilam, R., and Degani, H. (2009). Estrogen regulation of vascular endothelial growth factor in breast cancer *in vitro* and *in vivo*: The role of estrogen receptor alpha and c-Myc. *Endocr. Relat. Cancer* **16,** 819–834.

Dai, M. S., Jin, Y., Gallegos, J. R., and Lu, H. (2006). Balance of Yin and Yang: Ubiquitylation-mediated regulation of p53 and c-Myc. *Neoplasia* **8,** 630–644.

Dalla-Favera, R., Bregni, M., Erikson, J., Patterson, D., Gallo, R. C., and Croce, C. M. (1982). Human c-myc onc gene is located on the region of chromosome 8 that is translocated in Burkitt lymphoma cells. *Proc. Natl. Acad. Sci. USA* **79,** 7824–7827.

Dang, C. V. (2009). Have you seen...?: Micro-managing and restraining pluripotent stem cells by MYC. *EMBO J.* **28,** 3065–3066.

Dang, C. V., Le, A., and Gao, P. (2009). MYC-induced cancer cell energy metabolism and therapeutic opportunities. *Clin. Cancer Res.* **15**, 6479–6483.

Davis, A. C., Wims, M., Spotts, G. D., Hann, S. R., and Bradley, A. (1993). A null c-myc mutation causes lethality before 10.5 days of gestation in homozygotes and reduced fertility in heterozygous female mice. *Genes Dev.* **7**, 671–682.

De Azevedo, W. F., Leclerc, S., Meijer, L., Havlicek, L., Strnad, M., and Kim, S. H. (1997). Inhibition of cyclin-dependent kinases by purine analogues: Crystal structure of human cdk2 complexed with roscovitine. *Eur. J. Biochem.* **243**, 518–526.

Dews, M., Homayouni, A., Yu, D., Murphy, D., Sevignani, C., Wentzel, E., Furth, E. E., Lee, W. M., Enders, G. H., Mendell, J. T., and Thomas-Tikhonenko, A. (2006). Augmentation of tumor angiogenesis by a Myc-activated microRNA cluster. *Nat. Genet.* **38**, 1060–1065.

Dorsett, Y., Robbiani, D. F., Jankovic, M., Reina-San-Martin, B., Eisenreich, T. R., and Nussenzweig, M. C. (2007). A role for AID in chromosome translocations between c-myc and the IgH variable region. *J. Exp. Med.* **204**, 2225–2232.

Druker, B. J. (2002). STI571 (Gleevec) as a paradigm for cancer therapy. *Trends Mol. Med.* **8**, S14–S18.

Dupont, J., Fernandez, A. M., Glackin, C. A., Helman, L., and LeRoith, D. (2001). Insulin-like growth factor 1 (IGF-1)-induced twist expression is involved in the anti-apoptotic effects of the IGF-1 receptor. *J. Biol. Chem.* **276**, 26699–26707.

Eberhart, C. G., Kratz, J., Wang, Y., Summers, K., Stearns, D., Cohen, K., Dang, C. V., and Burger, P. C. (2004). Histopathological and molecular prognostic markers in medulloblastoma: c-myc, N-myc, TrkC, and anaplasia. *J. Neuropathol. Exp. Neurol.* **63**, 441–449.

Ehrhardt, A., Bartels, T., Geick, A., Klocke, R., Paul, D., and Halter, R. (2001). Development of pulmonary bronchiolo-alveolar adenocarcinomas in transgenic mice overexpressing murine c-myc and epidermal growth factor in alveolar type II pneumocytes. *Br. J. Cancer* **84**, 813–818.

Eischen, C. M., Roussel, M. F., Korsmeyer, S. J., and Cleveland, J. L. (2001). Bax loss impairs Myc-induced apoptosis and circumvents the selection of p53 mutations during Myc-mediated lymphomagenesis. *Mol. Cell. Biol.* **21**, 7653–7662.

Engel, R. H., and Kaklamani, V. G. (2007). HER2-positive breast cancer: Current and future treatment strategies. *Drugs* **67**, 1329–1341.

Evageliou, N. F., and Hogarty, M. D. (2009). Disrupting polyamine homeostasis as a therapeutic strategy for neuroblastoma. *Clin. Cancer Res.* **15**, 5956–5961.

Facchini, L. M., Chen, S., Marhin, W. W., Lear, J. N., and Penn, L. Z. (1997). The Myc negative autoregulation mechanism requires Myc–Max association and involves the c-myc P2 minimal promoter. *Mol. Cell. Biol.* **17**, 100–114.

Felsher, D. W., and Bishop, J. M. (1999a). Reversible tumorigenesis by MYC in hematopoietic lineages. *Mol. Cell* **4**, 199–207.

Felsher, D. W., and Bishop, J. M. (1999b). Transient excess of MYC activity can elicit genomic instability and tumorigenesis. *Proc. Natl. Acad. Sci. USA* **96**, 3940–3944.

Feng, X. H., Liang, Y. Y., Liang, M., Zhai, W., and Lin, X. (2002). Direct interaction of c-Myc with Smad2 and Smad3 to inhibit TGF-beta-mediated induction of the CDK inhibitor p15 (Ink4B). *Mol. Cell* **9**, 133–143.

Fodde, R., and Brabletz, T. (2007). Wnt/beta-catenin signaling in cancer stemness and malignant behavior. *Curr. Opin. Cell Biol.* **19**, 150–158.

Follis, A. V., Hammoudeh, D. I., Wang, H., Prochownik, E. V., and Metallo, S. J. (2008). Structural rationale for the coupled binding and unfolding of the c-Myc oncoprotein by small molecules. *Chem. Biol.* **15**, 1149–1155.

Fredlund, E., Ringner, M., Maris, J. M., and Pahlman, S. (2008). High Myc pathway activity and low stage of neuronal differentiation associate with poor outcome in neuroblastoma. *Proc. Natl. Acad. Sci. USA* **105**, 14094–14099.

Friedman, G. K., and Castleberry, R. P. (2007). Changing trends of research and treatment in infant neuroblastoma. *Pediatr. Blood Cancer* **49**, 1060–1065.

Fulda, S., Los, M., Friesen, C., and Debatin, K. M. (1998). Chemosensitivity of solid tumor cells *in vitro* is related to activation of the CD95 system. *Int. J. Cancer* **76**, 105–114.

Gaidano, G., Ballerini, P., Gong, J. Z., Inghirami, G., Neri, A., Newcomb, E. W., Magrath, I. T., Knowles, D. M., and Dalla-Favera, R. (1991). p53 mutations in human lymphoid malignancies: Association with Burkitt lymphoma and chronic lymphocytic leukemia. *Proc. Natl. Acad. Sci. USA* **88**, 5413–5417.

Garcia-Echeverria, C., Pearson, M. A., Marti, A., Meyer, T., Mestan, J., Zimmermann, J., Gao, J., Brueggen, J., Capraro, H. G., Cozens, R., Evans, D. B., Fabbro, D., *et al.* (2004). *In vivo* antitumor activity of NVP-AEW541-A novel, potent, and selective inhibitor of the IGF-IR kinase. *Cancer Cell* **5**, 231–239.

Gartel, A. L., Ye, X., Goufman, E., Shianov, P., Hay, N., Najmabadi, F., and Tyner, A. L. (2001). Myc represses the p21(WAF1/CIP1) promoter and interacts with Sp1/Sp3. *Proc. Natl. Acad. Sci. USA* **98**, 4510–4515.

Geng, Z., Zhang, D., and Liu, Y. (2003). Expression of telomerase hTERT in human non-small cell lung cancer and its correlation with c-myc gene. *Chin. Med. J. (Engl.)* **116**, 1467–1470.

Gilladoga, A. D., Edelhoff, S., Blackwood, E. M., Eisenman, R. N., and Disteche, C. M. (1992). Mapping of MAX to human chromosome 14 and mouse chromosome 12 by *in situ* hybridization. *Oncogene* **7**, 1249–1251.

Goga, A., Yang, D., Tward, A. D., Morgan, D. O., and Bishop, J. M. (2007). Inhibition of CDK1 as a potential therapy for tumors over-expressing MYC. *Nat. Med.* **13**, 820–827.

Grand, C. L., Han, H., Munoz, R. M., Weitman, S., Von Hoff, D. D., Hurley, L. H., and Bearss, D. J. (2002). The cationic porphyrin TMPyP4 down-regulates c-MYC and human telomerase reverse transcriptase expression and inhibits tumor growth *in vivo*. *Mol. Cancer Ther.* **1**, 565–573.

Grant, S. (1998). Ara-C: Cellular and molecular pharmacology. *Adv. Cancer Res.* **72**, 197–233.

Gray, G. D., Nichol, F. R., Mickelson, M. M., Camiener, G. W., Gish, D. T., Kelly, R. C., Wechter, W. J., Moxley, T. E., and Neil, G. L. (1972). Immunosuppressive, antiviral and antitumor activities of cytarabine derivatives. *Biochem. Pharmacol.* **21**, 465–475.

Gregory, M. A., and Hann, S. R. (2000). c-Myc proteolysis by the ubiquitin–proteasome pathway: Stabilization of c-Myc in Burkitt's lymphoma cells. *Mol. Cell. Biol.* **20**, 2423–2435.

Gu, W., Bhatia, K., Magrath, I. T., Dang, C. V., and Dalla-Favera, R. (1994). Binding and suppression of the Myc transcriptional activation domain by p107. *Science* **264**, 251–254.

Guerra, L., Albihn, A., Tronnersjo, S., Yan, Q., Guidi, R., Stenerlow, B., Sterzenbach, T., Josenhans, C., Fox, J. G., Schauer, D. B., Thelestam, M., Larsson, L. G., *et al.* (2010). Myc is required for activation of the ATM-dependent checkpoints in response to DNA damage. *PLoS One* **5**, e8924.

Guo, J., Parise, R. A., Joseph, E., Egorin, M. J., Lazo, J. S., Prochownik, E. V., and Eiseman, J. L. (2009). Efficacy, pharmacokinetics, tissue distribution, and metabolism of the Myc-Max disruptor, 10058–F4 [Z, E]-5-[4-ethylbenzylidine]-2-thioxothiazolidin-4-one, in mice. *Cancer Chemother. Pharmacol.* **63**, 615–625.

Gustafson, W. C., and Weiss, W. A. (2010). Myc proteins as therapeutic targets. *Oncogene* **29**, 1249–1259.

Hahn, W. C., Counter, C. M., Lundberg, A. S., Beijersbergen, R. L., Brooks, M. W., and Weinberg, R. A. (1999). Creation of human tumour cells with defined genetic elements. *Nature* **400**, 464–468.

Hambardzumyan, D., Becher, O. J., and Holland, E. C. (2008a). Cancer stem cells and survival pathways. *Cell Cycle* **7**, 1371–1378.

Hambardzumyan, D., Becher, O. J., Rosenblum, M. K., Pandolfi, P. P., Manova-Todorova, K., and Holland, E. C. (2008b). PI3K pathway regulates survival of cancer stem cells residing in the perivascular niche following radiation in medulloblastoma *in vivo*. *Genes Dev.* **22,** 436–448.

Hammoudeh, D. I., Follis, A. V., Prochownik, E. V., and Metallo, S. J. (2009). Multiple independent binding sites for small-molecule inhibitors on the oncoprotein c-Myc. *J. Am. Chem. Soc.* **131,** 7390–7401.

Hande, K. R. (1998). Clinical applications of anticancer drugs targeted to topoisomerase II. *Biochim. Biophys. Acta* **1400,** 173–184.

Hann, S. R. (2006). Role of post-translational modifications in regulating c-Myc proteolysis, transcriptional activity and biological function. *Semin. Cancer Biol.* **16,** 288–302.

Hann, S. R., Dixit, M., Sears, R. C., and Sealy, L. (1994). The alternatively initiated c-Myc proteins differentially regulate transcription through a noncanonical DNA-binding site. *Genes Dev.* **8,** 2441–2452.

Harrington, E. A., Bennett, M. R., Fanidi, A., and Evan, G. I. (1994). c-Myc-induced apoptosis in fibroblasts is inhibited by specific cytokines. *EMBO J.* **13,** 3286–3295.

Hartmann, W., Digon-Sontgerath, B., Koch, A., Waha, A., Endl, E., Dani, I., Denkhaus, D., Goodyer, C. G., Sorensen, N., Wiestler, O. D., and Pietsch, T. (2006). Phosphatidylinositol 3′-kinase/AKT signaling is activated in medulloblastoma cell proliferation and is associated with reduced expression of PTEN. *Clin. Cancer Res.* **12,** 3019–3027.

Hatton, K. S., Mahon, K., Chin, L., Chiu, F. C., Lee, H. W., Peng, D., Morgenbesser, S. D., Horner, J., and DePinho, R. A. (1996). Expression and activity of L-Myc in normal mouse development. *Mol. Cell. Biol.* **16,** 1794–1804.

Hatton, B. A., Knoepfler, P. S., Kenney, A. M., Rowitch, D. H., de Alboran, I. M., Olson, J. M., and Eisenman, R. N. (2006). N-myc is an essential downstream effector of Shh signaling during both normal and neoplastic cerebellar growth. *Cancer Res.* **66,** 8655–8661.

Hayashita, Y., Osada, H., Tatematsu, Y., Yamada, H., Yanagisawa, K., Tomida, S., Yatabe, Y., Kawahara, K., Sekido, Y., and Takahashi, T. (2005). A polycistronic microRNA cluster, miR-17-92, is overexpressed in human lung cancers and enhances cell proliferation. *Cancer Res.* **65,** 9628–9632.

Hayes-Jordan, A., and Andrassy, R. (2009). Rhabdomyosarcoma in children. *Curr. Opin. Pediatr.* **21,** 373–378.

He, T. C., Sparks, A. B., Rago, C., Hermeking, H., Zawel, L., da Costa, L. T., Morin, P. J., Vogelstein, B., and Kinzler, K. W. (1998). Identification of c-MYC as a target of the APC pathway. *Science* **281,** 1509–1512.

He, L., Thomson, J. M., Hemann, M. T., Hernando-Monge, E., Mu, D., Goodson, S., Powers, S., Cordon-Cardo, C., Lowe, S. W., Hannon, G. J., and Hammond, S. M. (2005). A microRNA polycistron as a potential human oncogene. *Nature* **435,** 828–833.

Hecht, J. L., and Aster, J. C. (2000). Molecular biology of Burkitt's lymphoma. *J. Clin. Oncol.* **18,** 3707–3721.

Henriksson, M., and Luscher, B. (1996). Proteins of the Myc network: Essential regulators of cell growth and differentiation. *Adv. Cancer Res.* **68,** 109–182.

Henriksson, M., Bakardjiev, A., Klein, G., and Luscher, B. (1993). Phosphorylation sites mapping in the N-terminal domain of c-myc modulate its transforming potential. *Oncogene* **8,** 3199–3209.

Herbst, A., Salghetti, S. E., Kim, S. Y., and Tansey, W. P. (2004). Multiple cell-type-specific elements regulate Myc protein stability. *Oncogene* **23,** 3863–3871.

Herbst, A., Hemann, M. T., Tworkowski, K. A., Salghetti, S. E., Lowe, S. W., and Tansey, W. P. (2005). A conserved element in Myc that negatively regulates its proapoptotic activity. *EMBO Rep.* **6,** 177–183.

Hermeking, H. (2003). The MYC oncogene a cancer drug target. *Curr. Cancer Drug Targets* **3**, 163–175.

Hermeking, H., and Eick, D. (1994). Mediation of c-Myc-induced apoptosis by p53. *Science* **265**, 2091–2093.

Hermeking, H., Rago, C., Schuhmacher, M., Li, Q., Barrett, J. F., Obaya, A. J., O'Connell, B. C., Mateyak, M. K., Tam, W., Kohlhuber, F., Dang, C. V., Sedivy, J. M., *et al.* (2000). Identification of CDK4 as a target of c-MYC. *Proc. Natl. Acad. Sci. USA* **97**, 2229–2234.

Herms, J., Neidt, I., Luscher, B., Sommer, A., Schurmann, P., Schroder, T., Bergmann, M., Wilken, B., Probst-Cousin, S., Hernaiz-Driever, P., Behnke, J., Hanefeld, F., *et al.* (2000). C-MYC expression in medulloblastoma and its prognostic value. *Int. J. Cancer* **89**, 395–402.

Hernan, R., Fasheh, R., Calabrese, C., Frank, A. J., Maclean, K. H., Allard, D., Barraclough, R., and Gilbertson, R. J. (2003). ERBB2 up-regulates S100A4 and several other prometastatic genes in medulloblastoma. *Cancer Res.* **63**, 140–148.

Herold, S., Wanzel, M., Beuger, V., Frohme, C., Beul, D., Hillukkala, T., Syvaoja, J., Saluz, H. P., Haenel, F., and Eilers, M. (2002). Negative regulation of the mammalian UV response by Myc through association with Miz-1. *Mol. Cell* **10**, 509–521.

Herr, I., and Debatin, K. M. (2001). Cellular stress response and apoptosis in cancer therapy. *Blood* **98**, 2603–2614.

Higgins, M. J., and Ettinger, D. S. (2009). Chemotherapy for lung cancer: The state of the art in 2009. *Expert Rev. Anticancer Ther.* **9**, 1365–1378.

Hirose, M., and Kuroda, Y. (1998). p53 may mediate the mdr-1 expression via the WT1 gene in human vincristine-resistant leukemia/lymphoma cell lines. *Cancer Lett.* **129**, 165–171.

Ho, K. S., and Scott, M. P. (2002). Sonic hedgehog in the nervous system: Functions, modifications and mechanisms. *Curr. Opin. Neurobiol.* **12**, 57–63.

Hoang, A. T., Cohen, K. J., Barrett, J. F., Bergstrom, D. A., and Dang, C. V. (1994). Participation of cyclin A in Myc-induced apoptosis. *Proc. Natl. Acad. Sci. USA* **91**, 6875–6879.

Hoang, A. T., Lutterbach, B., Lewis, B. C., Yano, T., Chou, T. Y., Barrett, J. F., Raffeld, M., Hann, S. R., and Dang, C. V. (1995). A link between increased transforming activity of lymphoma-derived MYC mutant alleles, their defective regulation by p107, and altered phosphorylation of the c-Myc transactivation domain. *Mol. Cell. Biol.* **15**, 4031–4042.

Hogarty, M. D. (2003). The requirement for evasion of programmed cell death in neuroblastomas with MYCN amplification. *Cancer Lett.* **197**, 173–179.

Hogarty, M. D., Norris, M. D., Davis, K., Liu, X., Evageliou, N. F., Hayes, C. S., Pawel, B., Guo, R., Zhao, H., Sekyere, E., Keating, J., Thomas, W., *et al.* (2008). ODC1 is a critical determinant of MYCN oncogenesis and a therapeutic target in neuroblastoma. *Cancer Res.* **68**, 9735–9745.

Hossain, A., Kuo, M. T., and Saunders, G. F. (2006). Mir-17-5p regulates breast cancer cell proliferation by inhibiting translation of AIB1 mRNA. *Mol. Cell. Biol.* **26**, 8191–8201.

Huang, M. J., Cheng, Y. C., Liu, C. R., Lin, S., and Liu, H. E. (2006). A small-molecule c-Myc inhibitor, 10058-F4, induces cell-cycle arrest, apoptosis, and myeloid differentiation of human acute myeloid leukemia. *Exp. Hematol.* **34**, 1480–1489.

Hueber, A. O., and Evan, G. I. (1998). Traps to catch unwary oncogenes. *Trends Genet.* **14**, 364–367.

Hui, M. K., Wu, W. K., Shin, V. Y., So, W. H., and Cho, C. H. (2006). Polysaccharides from the root of Angelica sinensis protect bone marrow and gastrointestinal tissues against the cytotoxicity of cyclophosphamide in mice. *Int. J. Med. Sci.* **3**, 1–6.

Hydbring, P., Bahram, F., Su, Y., Tronnersjo, S., Hogstrand, K., von der Lehr, N., Sharifi, H. R., Lilischkis, R., Hein, N., Wu, S., Vervoorts, J., Henriksson, M., *et al.* (2009). Phosphorylation by Cdk2 is required for Myc to repress Ras-induced senescence in cotransformation. *Proc. Natl. Acad. Sci. USA***107**, 58–63.

Iversen, P. L., Arora, V., Acker, A. J., Mason, D. H., and Devi, G. R. (2003). Efficacy of antisense morpholino oligomer targeted to c-myc in prostate cancer xenograft murine model and a Phase I safety study in humans. *Clin. Cancer Res.* **9**, 2510–2519.

Izumi, H., Molander, C., Penn, L. Z., Ishisaki, A., Kohno, K., and Funa, K. (2001). Mechanism for the transcriptional repression by c-Myc on PDGF beta-receptor. *J. Cell Sci.* **114**, 1533–1544.

Jacobs, J. J., Scheijen, B., Voncken, J. W., Kieboom, K., Berns, A., and van Lohuizen, M. (1999). Bmi-1 collaborates with c-Myc in tumorigenesis by inhibiting c-Myc-induced apoptosis via INK4a/ARF. *Genes Dev.* **13**, 2678–2690.

Jakacki, R. I., Hamilton, M., Gilbertson, R. J., Blaney, S. M., Tersak, J., Krailo, M. D., Ingle, A. M., Voss, S. D., Dancey, J. E., and Adamson, P. C. (2008). Pediatric phase I and pharmacokinetic study of erlotinib followed by the combination of erlotinib and temozolomide: A Children's Oncology Group Phase I Consortium Study. *J. Clin. Oncol.* **26**, 4921–4927.

Jeruss, J. S., Sturgis, C. D., Rademaker, A. W., and Woodruff, T. K. (2003). Down-regulation of activin, activin receptors, and Smads in high-grade breast cancer. *Cancer Res.* **63**, 3783–3790.

John, B., Enright, A. J., Aravin, A., Tuschl, T., Sander, C., and Marks, D. S. (2004). Human MicroRNA targets. *PLoS Biol.* **2**, e363.

Johnsen, J. I., Segerstrom, L., Orrego, A., Elfman, L., Henriksson, M., Kagedal, B., Eksborg, S., Sveinbjornsson, B., and Kogner, P. (2008). Inhibitors of mammalian target of rapamycin downregulate MYCN protein expression and inhibit neuroblastoma growth *in vitro* and *in vivo*. *Oncogene* **27**, 2910–2922.

Johnsen, J. I., Kogner, P., Albihn, A., and Henriksson, M. A. (2009). Embryonal neural tumours and cell death. *Apoptosis* **14**, 424–438.

Johnson, S. M., Grosshans, H., Shingara, J., Byrom, M., Jarvis, R., Cheng, A., Labourier, E., Reinert, K. L., Brown, D., and Slack, F. J. (2005). RAS is regulated by the let-7 microRNA family. *Cell* **120**, 635–647.

Johnson, C. D., Esquela-Kerscher, A., Stefani, G., Byrom, M., Kelnar, K., Ovcharenko, D., Wilson, M., Wang, X., Shelton, J., Shingara, J., Chin, L., Brown, D., *et al.* (2007). The let-7 microRNA represses cell proliferation pathways in human cells. *Cancer Res.* **67**, 7713–7722.

Johnson, N. A., Savage, K. J., Ludkovski, O., Ben-Neriah, S., Woods, R., Steidl, C., Dyer, M. J., Siebert, R., Kuruvilla, J., Klasa, R., Connors, J. M., Gascoyne, R. D., *et al.* (2009). Lymphomas with concurrent BCL2 and MYC translocations: The critical factors associated with survival. *Blood* **114**, 2273–2279.

Jordan, M. A., Thrower, D., and Wilson, L. (1991). Mechanism of inhibition of cell proliferation by Vinca alkaloids. *Cancer Res.* **51**, 2212–2222.

Judson, R. L., Babiarz, J. E., Venere, M., and Blelloch, R. (2009). Embryonic stem cell-specific microRNAs promote induced pluripotency. *Nat. Biotechnol.* **27**, 459–461.

Juin, P., Hueber, A. O., Littlewood, T., and Evan, G. (1999). c-Myc-induced sensitization to apoptosis is mediated through cytochrome *c* release. *Genes Dev.* **13**, 1367–1381.

Juin, P., Hunt, A., Littlewood, T., Griffiths, B., Swigart, L. B., Korsmeyer, S., and Evan, G. (2002). c-Myc functionally cooperates with Bax to induce apoptosis. *Mol. Cell. Biol.* **22**, 6158–6169.

Junttila, M. R., and Westermarck, J. (2008). Mechanisms of MYC stabilization in human malignancies. *Cell Cycle* **7**, 592–596.

Junttila, M. R., Puustinen, P., Niemela, M., Ahola, R., Arnold, H., Bottzauw, T., Ala-aho, R., Nielsen, C., Ivaska, J., Taya, Y., Lu, S. L., Lin, S., *et al.* (2007). CIP2A inhibits PP2A in human malignancies. *Cell* **130**, 51–62.

Karpinich, N. O., Tafani, M., Rothman, R. J., Russo, M. A., and Farber, J. L. (2002). The course of etoposide-induced apoptosis from damage to DNA and p53 activation to mitochondrial release of cytochrome *c*. *J. Biol. Chem.* **277,** 16547–16552.

Kent, O. A., and Mendell, J. T. (2006). A small piece in the cancer puzzle: MicroRNAs as tumor suppressors and oncogenes. *Oncogene* **25,** 6188–6196.

Khanna, A., Bockelman, C., Hemmes, A., Junttila, M. R., Wiksten, J. P., Lundin, M., Junnila, S., Murphy, D. J., Evan, G. I., Haglund, C., Westermarck, J., and Ristimaki, A. (2009). MYC-dependent regulation and prognostic role of CIP2A in gastric cancer. *J. Natl. Cancer Inst.* **101,** 793–805.

Kim, S. Y., Herbst, A., Tworkowski, K. A., Salghetti, S. E., and Tansey, W. P. (2003). Skp2 regulates Myc protein stability and activity. *Mol. Cell* **11,** 1177–1188.

Kim, J., Chu, J., Shen, X., Wang, J., and Orkin, S. H. (2008). An extended transcriptional network for pluripotency of embryonic stem cells. *Cell* **132,** 1049–1061.

Kime, L., and Wright, S. C. (2003). Mad4 is regulated by a transcriptional repressor complex that contains Miz-1 and c-Myc. *Biochem. J.* **370,** 291–298.

Kleine-Kohlbrecher, D., Adhikary, S., and Eilers, M. (2006). Mechanisms of transcriptional repression by Myc. *Curr. Top. Microbiol. Immunol.* **302,** 51–62.

Klinakis, A., Szabolcs, M., Politi, K., Kiaris, H., Artavanis-Tsakonas, S., and Efstratiadis, A. (2006). Myc is a Notch1 transcriptional target and a requisite for Notch1-induced mammary tumorigenesis in mice. *Proc. Natl. Acad. Sci. USA* **103,** 9262–9267.

Knoepfler, P. S., and Kenney, A. M. (2006). Neural precursor cycling at sonic speed: N-Myc pedals, GSK-3 brakes. *Cell Cycle* **5,** 47–52.

Knoepfler, P. S., Cheng, P. F., and Eisenman, R. N. (2002). N-myc is essential during neurogenesis for the rapid expansion of progenitor cell populations and the inhibition of neuronal differentiation. *Genes Dev.* **16,** 2699–2712.

Kontopidis, G., Andrews, M. J., McInnes, C., Cowan, A., Powers, H., Innes, L., Plater, A., Griffiths, G., Paterson, D., Zheleva, D. I., Lane, D. P., Green, S., *et al.* (2003). Insights into cyclin groove recognition: Complex crystal structures and inhibitor design through ligand exchange. *Structure* **11,** 1537–1546.

Kutryk, M. J., Foley, D. P., van den Brand, M., Hamburger, J. N., van der Giessen, W. J., deFeyter, P. J., Bruining, N., Sabate, M., and Serruys, P. W. (2002). Local intracoronary administration of antisense oligonucleotide against c-myc for the prevention of in-stent restenosis: Results of the randomized investigation by the Thoraxcenter of antisense DNA using local delivery and IVUS after coronary stenting (ITALICS) trial. *J. Am. Coll. Cardiol.* **39,** 281–287.

Lacasce, A., Howard, O., Lib, S., Fisher, D., Weng, A., Neuberg, D., and Shipp, M. (2004). Modified magrath regimens for adults with Burkitt and Burkitt-like lymphomas: Preserved efficacy with decreased toxicity. *Leuk. Lymphoma* **45,** 761–767.

Lambert, J. M., Gorzov, P., Veprintsev, D. B., Soderqvist, M., Segerback, D., Bergman, J., Fersht, A. R., Hainaut, P., Wiman, K. G., and Bykov, V. J. (2009). PRIMA-1 reactivates mutant p53 by covalent binding to the core domain. *Cancer Cell* **15,** 376–388.

Land, H., Parada, L. F., and Weinberg, R. A. (1983). Tumorigenic conversion of primary embryo fibroblasts requires at least two cooperating oncogenes. *Nature* **304,** 596–602.

Laurenti, E., Varnum-Finney, B., Wilson, A., Ferrero, I., Blanco-Bose, W. E., Ehninger, A., Knoepfler, P. S., Cheng, P. F., MacDonald, H. R., Eisenman, R. N., Bernstein, I. D., and Trumpp, A. (2008). Hematopoietic stem cell function and survival depend on c-Myc and N-Myc activity. *Cell Stem Cell* **3,** 611–624.

Lee, R. C., Feinbaum, R. L., and Ambros, V. (1993). The *C. elegans* heterochronic gene lin-4 encodes small RNAs with antisense complementarity to lin-14. *Cell* **75,** 843–854.

Lee, T. C., Li, L., Philipson, L., and Ziff, E. B. (1997). Myc represses transcription of the growth arrest gene gas1. *Proc. Natl. Acad. Sci. USA* **94,** 12886–12891.

Leucci, E., Cocco, M., Onnis, A., De Falco, G., van Cleef, P., Bellan, C., van Rijk, A., Nyagol, J., Byakika, B., Lazzi, S., Tosi, P., van Krieken, H., *et al.* (2008). MYC translocation-negative classical Burkitt lymphoma cases: An alternative pathogenetic mechanism involving miRNA deregulation. *J. Pathol.* **216,** 440–450.

Levine, M. N., and Whelan, T. (2006). Adjuvant chemotherapy for breast cancer—30 years later. *N. Engl. J. Med.* **355,** 1920–1922.

Lewis, B. P., Shih, I. H., Jones-Rhoades, M. W., Bartel, D. P., and Burge, C. B. (2003). Prediction of mammalian microRNA targets. *Cell* **115,** 787–798.

Li, Z., and Hann, S. R. (2009). The Myc–nucleophosmin–ARF network: A complex web unveiled. *Cell Cycle* **8,** 2703–2707.

Li, L. H., Nerlov, C., Prendergast, G., MacGregor, D., and Ziff, E. B. (1994). c-Myc represses transcription *in vivo* by a novel mechanism dependent on the initiator element and Myc box II. *EMBO J.* **13,** 4070–4079.

Li, H., Lee, T. H., and Avraham, H. (2002). A novel tricomplex of BRCA1, Nmi, and c-Myc inhibits c-Myc-induced human telomerase reverse transcriptase gene (hTERT) promoter activity in breast cancer. *J. Biol. Chem.* **277,** 20965–20973.

Lin, C. P., Liu, J. D., Chow, J. M., Liu, C. R., and Liu, H. E. (2007). Small-molecule c-Myc inhibitor, 10058-F4, inhibits proliferation, downregulates human telomerase reverse transcriptase and enhances chemosensitivity in human hepatocellular carcinoma cells. *Anticancer Drugs* **18,** 161–170.

Liu, X., Mazanek, P., Dam, V., Wang, Q., Zhao, H., Guo, R., Jagannathan, J., Cnaan, A., Maris, J. M., and Hogarty, M. D. (2008). Deregulated Wnt/beta-catenin program in high-risk neuroblastomas without MYCN amplification. *Oncogene* **27,** 1478–1488.

Longo-Sorbello, G. S., and Bertino, J. R. (2001). Current understanding of methotrexate pharmacology and efficacy in acute leukemias. Use of newer antifolates in clinical trials. *Haematologica* **86,** 121–127.

Loven, J., Zinin, N., Wahlstrom, T., Muller, I., Brodin, P., Fredlund, E., Ribacke, U., Pivarcsi, A., Pahlman, S., and Henriksson, M. (2010). MYCN-regulated microRNAs repress estrogen receptor-{alpha} (ESR1) expression and neuronal differentiation in human neuroblastoma. *Proc. Natl. Acad. Sci. USA* **107,** 1553–1558.

Lu, X., Pearson, A., and Lunec, J. (2003). The MYCN oncoprotein as a drug development target. *Cancer Lett.* **197,** 125–130.

Lu, X., Vogt, P. K., Boger, D. L., and Lunec, J. (2008). Disruption of the MYC transcriptional function by a small-molecule antagonist of MYC/MAX dimerization. *Oncol. Rep.* **19,** 825–830.

Luqmani, Y. A. (2005). Mechanisms of drug resistance in cancer chemotherapy. *Med. Princ. Pract.* **14**(Suppl. 1), 35–48.

Luscher, B., and Larsson, L. G. (1999). The basic region/helix-loop-helix/leucine zipper domain of Myc proto-oncoproteins: Function and regulation. *Oncogene* **18,** 2955–2966.

Ma, L., Krishnamachary, N., Perbal, B., and Center, M. S. (1992). HL-60 cells isolated for resistance to vincristine are defective in 12-O-tetradecanoylphorbol-13-acetate induced differentiation and the formation of a functional AP-1 complex. *Oncol. Res.* **4,** 291–298.

Mackall, C., Berzofsky, J., and Helman, L. J. (2000). Targeting tumor specific translocations in sarcomas in pediatric patients for immunotherapy. *Clin. Orthop. Relat. Res.* 25–31.

Magrath, I. (1990). The pathogenesis of Burkitt's lymphoma. *Adv. Cancer Res.* **55,** 133–270.

Magrini, R., Russo, D., Ottaggio, L., Fronza, G., Inga, A., and Menichini, P. (2008). PRIMA-1 synergizes with adriamycin to induce cell death in non-small cell lung cancer cells. *J. Cell. Biochem.* **104,** 2363–2373.

Mai, S., Hanley-Hyde, J., Rainey, G. J., Kuschak, T. I., Paul, J. T., Littlewood, T. D., Mischak, H., Stevens, L. M., Henderson, D. W., and Mushinski, J. F. (1999). Chromosomal and extrachromosomal instability of the cyclin D2 gene is induced by Myc overexpression. *Neoplasia* **1,** 241–252.

Maloney, E. K., McLaughlin, J. L., Dagdigian, N. E., Garrett, L. M., Connors, K. M., Zhou, X. M., Blattler, W. A., Chittenden, T., and Singh, R. (2003). An anti-insulin-like growth factor I receptor antibody that is a potent inhibitor of cancer cell proliferation. *Cancer Res.* **63,** 5073–5083.

Maris, J. M., Hogarty, M. D., Bagatell, R., and Cohn, S. L. (2007). Neuroblastoma. *Lancet* **369,** 2106–2120.

Mason, K. D., Vandenberg, C. J., Scott, C. L., Wei, A. H., Cory, S., Huang, D. C., and Roberts, A. W. (2008). *In vivo* efficacy of the Bcl-2 antagonist ABT-737 against aggressive Myc-driven lymphomas. *Proc. Natl. Acad. Sci. USA* **105,** 17961–17966.

Matthay, K. K., Reynolds, C. P., Seeger, R. C., Shimada, H., Adkins, E. S., Haas-Kogan, D., Gerbing, R. B., London, W. B., and Villablanca, J. G. (2009). Long-term results for children with high-risk neuroblastoma treated on a randomized trial of myeloablative therapy followed by 13-cis-retinoic acid: A children's oncology group study. *J. Clin. Oncol.* **27,** 1007–1013.

McClue, S. J., Blake, D., Clarke, R., Cowan, A., Cummings, L., Fischer, P. M., MacKenzie, M., Melville, J., Stewart, K., Wang, S., Zhelev, N., Zheleva, D., *et al.* (2002). *In vitro* and *in vivo* antitumor properties of the cyclin dependent kinase inhibitor CYC202 (R-roscovitine). *Int. J. Cancer* **102,** 463–468.

McLaughlin, F., Finn, P., and La Thangue, N. B. (2003). The cell cycle, chromatin and cancer: Mechanism-based therapeutics come of age. *Drug Discov. Today* **8,** 793–802.

McMahon, S. B., Van Buskirk, H. A., Dugan, K. A., Copeland, T. D., and Cole, M. D. (1998). The novel ATM-related protein TRRAP is an essential cofactor for the c-Myc and E2F oncoproteins. *Cell* **94,** 363–374.

McMahon, S. B., Wood, M. A., and Cole, M. D. (2000). The essential cofactor TRRAP recruits the histone acetyltransferase hGCN5 to c-Myc. *Mol. Cell. Biol.* **20,** 556–562.

Mead, G. M., Sydes, M. R., Walewski, J., Grigg, A., Hatton, C. S., Pescosta, N., Guarnaccia, C., Lewis, M. S., McKendrick, J., Stenning, S. P., and Wright, D. (2002). An international evaluation of CODOX-M and CODOX-M alternating with IVAC in adult Burkitt's lymphoma: Results of United Kingdom Lymphoma Group LY06 study. *Ann. Oncol.* **13,** 1264–1274.

Meijer, L., Borgne, A., Mulner, O., Chong, J. P., Blow, J. J., Inagaki, N., Inagaki, M., Delcros, J. G., and Moulinoux, J. P. (1997). Biochemical and cellular effects of roscovitine, a potent and selective inhibitor of the cyclin-dependent kinases cdc2, cdk2 and cdk5. *Eur. J. Biochem.* **243,** 527–536.

Meyer, N., and Penn, L. Z. (2008). Reflecting on 25 years with MYC. *Nat. Rev. Cancer* **8,** 976–990.

Meyer, G. E., Chesler, L., Liu, D., Gable, K., Maddux, B. A., Goldenberg, D. D., Youngren, J. F., Goldfine, I. D., Weiss, W. A., Matthay, K. K., and Rosenthal, S. M. (2007). Nordihydroguaiaretic acid inhibits insulin-like growth factor signaling, growth, and survival in human neuroblastoma cells. *J. Cell. Biochem.* **102,** 1529–1541.

Miltenberger, R. J., Sukow, K. A., and Farnham, P. J. (1995). An E-box-mediated increase in cad transcription at the G1/S-phase boundary is suppressed by inhibitory c-Myc mutants. *Mol. Cell. Biol.* **15,** 2527–2535.

Mita, A. C., Mita, M. M., Nawrocki, S. T., and Giles, F. J. (2008). Survivin: Key regulator of mitosis and apoptosis and novel target for cancer therapeutics. *Clin. Cancer Res.* **14,** 5000–5005.

Mitchell, K. O., Ricci, M. S., Miyashita, T., Dicker, D. T., Jin, Z., Reed, J. C., and El-Deiry, W. S. (2000). Bax is a transcriptional target and mediator of c-myc-induced apoptosis. *Cancer Res.* **60,** 6318–6325.

Mo, H., and Henriksson, M. (2006). Identification of small molecules that induce apoptosis in a Myc-dependent manner and inhibit Myc-driven transformation. *Proc. Natl. Acad. Sci. USA* **103,** 6344–6349.

Mo, H., Vita, M., Crespin, M., and Henriksson, M. (2006). Myc overexpression enhances apoptosis induced by small molecules. *Cell Cycle* **5,** 2191–2194.

Momota, H., Shih, A. H., Edgar, M. A., and Holland, E. C. (2008). c-Myc and beta-catenin cooperate with loss of p53 to generate multiple members of the primitive neuroectodermal tumor family in mice. *Oncogene* **27,** 4392–4401.

Morgenstern, D. A., and Anderson, J. (2006). MYCN deregulation as a potential target for novel therapies in rhabdomyosarcoma. *Expert Rev. Anticancer Ther.* **6,** 217–224.

Moshier, J. A., Dosescu, J., Skunca, M., and Luk, G. D. (1993). Transformation of NIH/3 T3 cells by ornithine decarboxylase overexpression. *Cancer Res.* **53,** 2618–2622.

Mosse, Y. P., Wood, A., and Maris, J. M. (2009). Inhibition of ALK signaling for cancer therapy. *Clin. Cancer Res.* **15,** 5609–5614.

Mu, P., Han, Y. C., Betel, D., Yao, E., Squatrito, M., Ogrodowski, P., de Stanchina, E., D'Andrea, A., Sander, C., and Ventura, A. (2009). Genetic dissection of the miR-17 92 cluster of microRNAs in Myc-induced B-cell lymphomas. *Genes Dev.* **23,** 2806–2811.

Mueller, S., and Chang, S. (2009). Pediatric brain tumors: Current treatment strategies and future therapeutic approaches. *Neurotherapeutics* **6,** 570–586.

Mukherjee, S., and Conrad, S. E. (2005). c-Myc suppresses p21WAF1/CIP1 expression during estrogen signaling and antiestrogen resistance in human breast cancer cells. *J. Biol. Chem.* **280,** 17617–17625.

Mulholland, D. J., Dedhar, S., Wu, H., and Nelson, C. C. (2006). PTEN and GSK3beta: Key regulators of progression to androgen-independent prostate cancer. *Oncogene* **25,** 329–337.

Musgrove, E. A., Sergio, C. M., Loi, S., Inman, C. K., Anderson, L. R., Alles, M. C., Pinese, M., Caldon, C. E., Schutte, J., Gardiner-Garden, M., Ormandy, C. J., McArthur, G., *et al.* (2008). Identification of functional networks of estrogen- and c-Myc-responsive genes and their relationship to response to tamoxifen therapy in breast cancer. *PLoS One* **3,** e2987.

Nadal, E., and Olavarria, E. (2004). Imatinib mesylate (Gleevec/Glivec) a molecular-targeted therapy for chronic myeloid leukaemia and other malignancies. *Int. J. Clin. Pract.* **58,** 511–516.

Nair, S. K., and Burley, S. K. (2003). X-ray structures of Myc-Max and Mad-Max recognizing DNA. Molecular bases of regulation by proto-oncogenic transcription factors. *Cell* **112,** 193–205.

Nakagawa, M., Koyanagi, M., Tanabe, K., Takahashi, K., Ichisaka, T., Aoi, T., Okita, K., Mochiduki, Y., Takizawa, N., and Yamanaka, S. (2008). Generation of induced pluripotent stem cells without Myc from mouse and human fibroblasts. *Nat. Biotechnol.* **26,** 101–106.

Nau, M. M., Brooks, B. J., Battey, J., Sausville, E., Gazdar, A. F., Kirsch, I. R., McBride, O. W., Bertness, V., Hollis, G. F., and Minna, J. D. (1985). L-myc, a new myc-related gene amplified and expressed in human small cell lung cancer. *Nature* **318,** 69–73.

Nau, M. M., Brooks, B. J., Jr., Carney, D. N., Gazdar, A. F., Battey, J. F., Sausville, E. A., and Minna, J. D. (1986). Human small-cell lung cancers show amplification and expression of the N-myc gene. *Proc. Natl. Acad. Sci. USA* **83,** 1092–1096.

Nesbit, C. E., Tersak, J. M., and Prochownik, E. V. (1999). MYC oncogenes and human neoplastic disease. *Oncogene* **18,** 3004–3016.

Nicolini, F. E., Corm, S., Le, Q. H., Sorel, N., Hayette, S., Bories, D., Leguay, T., Roy, L., Giraudier, S., Tulliez, M., Facon, T., Mahon, F. X., *et al.* (2006). Mutation status and clinical outcome of 89 imatinib mesylate-resistant chronic myelogenous leukemia patients: A retrospective analysis from the French intergroup of CML (Fi(varphi)-LMC GROUP). *Leukemia* **20,** 1061–1066.

Nieboer, P., de Vries, E. G., Mulder, N. H., and van der Graaf, W. T. (2005). Relevance of high-dose chemotherapy in solid tumours. *Cancer Treat. Rev.* **31,** 210–225.

Nilsson, J. A., and Cleveland, J. L. (2003). Myc pathways provoking cell suicide and cancer. *Oncogene* **22**, 9007–9021.

Nilsson, J. A., Keller, U. B., Baudino, T. A., Yang, C., Norton, S., Old, J. A., Nilsson, L. M., Neale, G., Kramer, D. L., Porter, C. W., and Cleveland, J. L. (2005). Targeting ornithine decarboxylase in Myc-induced lymphomagenesis prevents tumor formation. *Cancer Cell* **7**, 433–444.

Nishino, K., Osaki, T., Kumagai, T., Kijima, T., Tachibana, I., Goto, H., Arai, T., Kimura, H., Funakoshi, T., Takeda, Y., Tanio, Y., and Hayashi, S. (2001). Adenovirus-mediated gene therapy specific for small cell lung cancer cells using a Myc-Max binding motif. *Int. J. Cancer* **91**, 851–856.

Nishioka, C., Ikezoe, T., Yang, J., and Yokoyama, A. (2009). Inhibition of MEK signaling enhances the ability of cytarabine to induce growth arrest and apoptosis of acute myelogenous leukemia cells. *Apoptosis* **14**, 1108–1120.

O'Donnell, K. A., Wentzel, E. A., Zeller, K. I., Dang, C. V., and Mendell, J. T. (2005). c-Myc-regulated microRNAs modulate E2F1 expression. *Nature* **435**, 839–843.

Oh, S., Song, Y. H., Kim, U. J., Yim, J., and Kim, T. K. (1999). *In vivo* and *in vitro* analyses of Myc for differential promoter activities of the human telomerase (hTERT) gene in normal and tumor cells. *Biochem. Biophys. Res. Commun.* **263**, 361–365.

Oh, S., Song, Y. H., Yim, J., and Kim, T. K. (2000). Identification of Mad as a repressor of the human telomerase (hTERT) gene. *Oncogene* **19**, 1485–1490.

Okita, K., Ichisaka, T., and Yamanaka, S. (2007). Generation of germline-competent induced pluripotent stem cells. *Nature* **448**, 313–317.

Okubo, T., Knoepfler, P. S., Eisenman, R. N., and Hogan, B. L. (2005). Nmyc plays an essential role during lung development as a dosage-sensitive regulator of progenitor cell proliferation and differentiation. *Development* **132**, 1363–1374.

Olive, V., Bennett, M. J., Walker, J. C., Ma, C., Jiang, I., Cordon-Cardo, C., Li, Q. J., Lowe, S. W., Hannon, G. J., and He, L. (2009). miR-19 is a key oncogenic component of mir-17-92. *Genes Dev.* **23**, 2839–2849.

Opel, D., Poremba, C., Simon, T., Debatin, K. M., and Fulda, S. (2007). Activation of Akt predicts poor outcome in neuroblastoma. *Cancer Res.* **67**, 735–745.

Oster, S. K., Ho, C. S., Soucie, E. L., and Penn, L. Z. (2002). The myc oncogene: MarvelouslY Complex. *Adv. Cancer Res.* **84**, 81–154.

Ota, A., Tagawa, H., Karnan, S., Tsuzuki, S., Karpas, A., Kira, S., Yoshida, Y., and Seto, M. (2004). Identification and characterization of a novel gene, C13orf25, as a target for 13q31-q32 amplification in malignant lymphoma. *Cancer Res.* **64**, 3087–3095.

Ozaki, S., Ikeda, S., Ishizaki, Y., Kurihara, T., Tokumoto, N., Iseki, M., Arihiro, K., Kataoka, T., Okajima, M., and Asahara, T. (2005). Alterations and correlations of the components in the Wnt signaling pathway and its target genes in breast cancer. *Oncol. Rep.* **14**, 1437–1443.

Packham, G., and Cleveland, J. L. (1994). Ornithine decarboxylase is a mediator of c-Myc-induced apoptosis. *Mol. Cell. Biol.* **14**, 5741–5747.

Pawar, S. A., Szentirmay, M. N., Hermeking, H., and Sawadogo, M. (2004). Evidence for a cancer-specific switch at the CDK4 promoter with loss of control by both USF and c-Myc. *Oncogene* **23**, 6125–6135.

Pearson, A. D., Pinkerton, C. R., Lewis, I. J., Imeson, J., Ellershaw, C., and Machin, D. (2008). High-dose rapid and standard induction chemotherapy for patients aged over 1 year with stage 4 neuroblastoma: A randomised trial. *Lancet Oncol.* **9**, 247–256.

Pelengaris, S., and Khan, M. (2003). The c-MYC oncoprotein as a treatment target in cancer and other disorders of cell growth. *Expert Opin. Ther. Targets* **7**, 623–642.

Pelengaris, S., Littlewood, T., Khan, M., Elia, G., and Evan, G. (1999). Reversible activation of c-Myc in skin: Induction of a complex neoplastic phenotype by a single oncogenic lesion. *Mol. Cell* **3**, 565–577.

Pelengaris, S., Khan, M., and Evan, G. (2002). c-MYC: More than just a matter of life and death. *Nat. Rev. Cancer* **2,** 764–776.

Perez-Roger, I., Kim, S. H., Griffiths, B., Sewing, A., and Land, H. (1999). Cyclins D1 and D2 mediate myc-induced proliferation via sequestration of p27(Kip1) and p21(Cip1). *EMBO J.* **18,** 5310–5320.

Petrocca, F., and Lieberman, J. (2009). Micromanipulating cancer: MicroRNA-based therapeutics? *RNA Biol.* **6,** 335–340.

Peukert, K., Staller, P., Schneider, A., Carmichael, G., Hanel, F., and Eilers, M. (1997). An alternative pathway for gene regulation by Myc. *EMBO J.* **16,** 5672–5686.

Plescia, J., Salz, W., Xia, F., Pennati, M., Zaffaroni, N., Daidone, M. G., Meli, M., Dohi, T., Fortugno, P., Nefedova, Y., Gabrilovich, D. I., Colombo, G., *et al.* (2005). Rational design of shepherdin, a novel anticancer agent. *Cancer Cell* **7,** 457–468.

Pomeroy, S. L., Tamayo, P., Gaasenbeek, M., Sturla, L. M., Angelo, M., McLaughlin, M. E., Kim, J. Y., Goumnerova, L. C., Black, P. M., Lau, C., Allen, J. C., Zagzag, D., *et al.* (2002). Prediction of central nervous system embryonal tumour outcome based on gene expression. *Nature* **415,** 436–442.

Ponzielli, R., Katz, S., Barsyte-Lovejoy, D., and Penn, L. Z. (2005). Cancer therapeutics: Targeting the dark side of Myc. *Eur. J. Cancer* **41,** 2485–2501.

Popov, N., Wanzel, M., Madiredjo, M., Zhang, D., Beijersbergen, R., Bernards, R., Moll, R., Elledge, S. J., and Eilers, M. (2007). The ubiquitin-specific protease USP28 is required for MYC stability. *Nat. Cell Biol.* **9,** 765–774.

Prat, A., and Baselga, J. (2008). The role of hormonal therapy in the management of hormonal-receptor-positive breast cancer with co-expression of HER2. *Nat. Clin. Pract. Oncol.* **5,** 531–542.

Prochownik, E. V. (2004). c-Myc as a therapeutic target in cancer. *Expert Rev. Anticancer Ther.* **4,** 289–302.

Rabbani, A., Finn, R. M., and Ausio, J. (2005). The anthracycline antibiotics: Antitumor drugs that alter chromatin structure. *Bioessays* **27,** 50–56.

Rajagopalan, H., Jallepalli, P. V., Rago, C., Velculescu, V. E., Kinzler, K. W., Vogelstein, B., and Lengauer, C. (2004). Inactivation of hCDC4 can cause chromosomal instability. *Nature* **428,** 77–81.

Rao, G., Pedone, C. A., Del Valle, L., Reiss, K., Holland, E. C., and Fults, D. W. (2004). Sonic hedgehog and insulin-like growth factor signaling synergize to induce medulloblastoma formation from nestin-expressing neural progenitors in mice. *Oncogene* **23,** 6156–6162.

Rapp, U. R., Korn, C., Ceteci, F., Karreman, C., Luetkenhaus, K., Serafin, V., Zanucco, E., Castro, I., and Potapenko, T. (2009). MYC is a metastasis gene for non-small-cell lung cancer. *PLoS One* **4,** e6029.

Raul, F. (2007). Revival of 2-(difluoromethyl)ornithine (DFMO), an inhibitor of polyamine biosynthesis, as a cancer chemopreventive agent. *Biochem. Soc. Trans.* **35,** 353–355.

Raynaud, F. I., Whittaker, S. R., Fischer, P. M., McClue, S., Walton, M. I., Barrie, S. E., Garrett, M. D., Rogers, P., Clarke, S. J., Kelland, L. R., Valenti, M., Brunton, L., *et al.* (2005). *In vitro* and *in vivo* pharmacokinetic–pharmacodynamic relationships for the trisubstituted aminopurine cyclin-dependent kinase inhibitors olomoucine, bohemine and CYC202. *Clin. Cancer Res.* **11,** 4875–4887.

Reynolds, C. P., and Lemons, R. S. (2001). Retinoid therapy of childhood cancer. *Hematol. Oncol. Clin. North Am.* **15,** 867–910.

Rich, T., Allen, R. L., and Wyllie, A. H. (2000). Defying death after DNA damage. *Nature* **407,** 777–783.

Richardson, G. E., and Johnson, B. E. (1993). The biology of lung cancer. *Semin. Oncol.* **20,** 105–127.

Rodrik, V., Gomes, E., Hui, L., Rockwell, P., and Foster, D. A. (2006). Myc stabilization in response to estrogen and phospholipase D in MCF-7 breast cancer cells. *FEBS Lett.* **580,** 5647–5652.

Rom, W. N., and Tchou-Wong, K. M. (2003). Molecular and genetic aspects of lung cancer. *Methods Mol. Med.* **75,** 3–26.

Romer, J., and Curran, T. (2005). Targeting medulloblastoma: Small-molecule inhibitors of the Sonic Hedgehog pathway as potential cancer therapeutics. *Cancer Res.* **65,** 4975–4978.

Rossi, A., Caracciolo, V., Russo, G., Reiss, K., and Giordano, A. (2008). Medulloblastoma: From molecular pathology to therapy. *Clin. Cancer Res.* **14,** 971–976.

Roughan, J. E., and Thorley-Lawson, D. A. (2009). The intersection of Epstein–Barr virus with the germinal center. *J. Virol.* **83,** 3968–3976.

Roy, A. L., Meisterernst, M., Pognonec, P., and Roeder, R. G. (1991). Cooperative interaction of an initiator-binding transcription initiation factor and the helix-loop-helix activator USF. *Nature* **354,** 245–248.

Rudin, C. M., Hann, C. L., Laterra, J., Yauch, R. L., Callahan, C. A., Fu, L., Holcomb, T., Stinson, J., Gould, S. E., Coleman, B., LoRusso, P. M., Von Hoff, D. D., *et al.* (2009). Treatment of medulloblastoma with hedgehog pathway inhibitor GDC-0449. *N. Engl. J. Med.* **361,** 1173–1178.

Rustighi, A., Tiberi, L., Soldano, A., Napoli, M., Nuciforo, P., Rosato, A., Kaplan, F., Capobianco, A., Pece, S., Di Fiore, P. P., and Del Sal, G. (2009). The prolyl-isomerase Pin1 is a Notch1 target that enhances Notch1 activation in cancer. *Nat. Cell Biol.* **11,** 133–142.

Salahshor, S., and Woodgett, J. R. (2005). The links between axin and carcinogenesis. *J. Clin. Pathol.* **58,** 225–236.

Salghetti, S. E., Kim, S. Y., and Tansey, W. P. (1999). Destruction of Myc by ubiquitin-mediated proteolysis: Cancer-associated and transforming mutations stabilize Myc. *EMBO J.* **18,** 717–726.

Salter, K. H., Acharya, C. R., Walters, K. S., Redman, R., Anguiano, A., Garman, K. S., Anders, C. K., Mukherjee, S., Dressman, H. K., Barry, W. T., Marcom, K. P., Olson, J., *et al.* (2008). An integrated approach to the prediction of chemotherapeutic response in patients with breast cancer. *PLoS One* **3,** e1908.

Sampson, V. B., Rong, N. H., Han, J., Yang, Q., Aris, V., Soteropoulos, P., Petrelli, N. J., Dunn, S. P., and Krueger, L. J. (2007). MicroRNA let-7a down-regulates MYC and reverts MYC-induced growth in Burkitt lymphoma cells. *Cancer Res.* **67,** 9762–9770.

Santen, R. J., Song, R. X., Zhang, Z., Kumar, R., Jeng, M. H., Masamura, S., Yue, W., and Berstein, L. (2003). Adaptive hypersensitivity to estrogen: Mechanism for superiority of aromatase inhibitors over selective estrogen receptor modulators for breast cancer treatment and prevention. *Endocr. Relat. Cancer* **10,** 111–130.

Sarkar, A. K., and Nuchtern, J. G. (2000). Lysis of MYCN-amplified neuroblastoma cells by MYCN peptide-specific cytotoxic T lymphocytes. *Cancer Res.* **60,** 1908–1913.

Satou, A., Taira, T., Iguchi-Ariga, S. M., and Ariga, H. (2001). A novel transrepression pathway of c-Myc. Recruitment of a transcriptional corepressor complex to c-Myc by MM-1, a c-Myc-binding protein. *J. Biol. Chem.* **276,** 46562–46567.

Schulein, C., and Eilers, M. (2009). An unsteady scaffold for Myc. *EMBO J.* **28,** 453–454.

Schuller, U., Heine, V. M., Mao, J., Kho, A. T., Dillon, A. K., Han, Y. G., Huillard, E., Sun, T., Ligon, A. H., Qian, Y., Ma, Q., Alvarez-Buylla, A., *et al.* (2008). Acquisition of granule neuron precursor identity is a critical determinant of progenitor cell competence to form Shh-induced medulloblastoma. *Cancer Cell* **14,** 123–134.

Schuster, C., Fernbach, N., Rix, U., Superti-Furga, G., Holy, M., Freissmuth, M., Sitte, H. H., and Sexl, V. (2007). Selective serotonin reuptake inhibitors—A new modality for the treatment of lymphoma/leukaemia? *Biochem. Pharmacol.* **74,** 1424–1435.

Scionti, I., Michelacci, F., Pasello, M., Hattinger, C. M., Alberghini, M., Manara, M. C., Bacci, G., Ferrari, S., Scotlandi, K., Picci, P., and Serra, M. (2008). Clinical impact of the methotrexate resistance-associated genes C-MYC and dihydrofolate reductase (DHFR) in high-grade osteosarcoma. *Ann. Oncol.* **19**, 1500–1508.

Scotting, P. J., Walker, D. A., and Perilongo, G. (2005). Childhood solid tumours: A developmental disorder. *Nat. Rev. Cancer* **5**, 481–488.

Sears, R., Nuckolls, F., Haura, E., Taya, Y., Tamai, K., and Nevins, J. R. (2000). Multiple Ras-dependent phosphorylation pathways regulate Myc protein stability. *Genes Dev.* **14**, 2501–2514.

Seoane, J., Le, H. V., and Massague, J. (2002). Myc suppression of the p21(Cip1) Cdk inhibitor influences the outcome of the p53 response to DNA damage. *Nature* **419**, 729–734.

Serra, J. M., Gutierrez, A., Alemany, R., Navarro, M., Ros, T., Saus, C., Gines, J., Sampol, A., Amat, J. C., Serra-Moises, L., Martin, J., Galmes, A., *et al.* (2008). Inhibition of c-Myc down-regulation by sustained extracellular signal-regulated kinase activation prevents the antimetabolite methotrexate- and gemcitabine-induced differentiation in non-small-cell lung cancer cells. *Mol. Pharmacol.* **73**, 1679–1687.

Shah, Y. M., Morimura, K., Yang, Q., Tanabe, T., Takagi, M., and Gonzalez, F. J. (2007). Peroxisome proliferator-activated receptor alpha regulates a microRNA-mediated signaling cascade responsible for hepatocellular proliferation. *Mol. Cell. Biol.* **27**, 4238–4247.

Shalaby, T., von Bueren, A. O., Hurlimann, M. L., Fiaschetti, G., Castelletti, D., Masayuki, T., Nagasawa, K., Arcaro, A., Jelesarov, I., Shin-ya, K., and Grotzer, M. (2010). Disabling c-Myc in childhood medulloblastoma and atypical teratoid/rhabdoid tumor cells by the potent G-quadruplex interactive agent S2T1-6OTD. *Mol. Cancer Ther.* **9**, 167–179.

Shanafelt, T. D., Lin, T., Geyer, S. M., Zent, C. S., Leung, N., Kabat, B., Bowen, D., Grever, M. R., Byrd, J. C., and Kay, N. E. (2007). Pentostatin, cyclophosphamide, and rituximab regimen in older patients with chronic lymphocytic leukemia. *Cancer* **109**, 2291–2298.

Shantz, L. M., and Levin, V. A. (2007). Regulation of ornithine decarboxylase during oncogenic transformation: Mechanisms and therapeutic potential. *Amino Acids* **33**, 213–223.

Sharma, S. V., Bell, D. W., Settleman, J., and Haber, D. A. (2007). Epidermal growth factor receptor mutations in lung cancer. *Nat. Rev. Cancer* **7**, 169–181.

Sheiness, D., and Bishop, J. M. (1979). DNA and RNA from uninfected vertebrate cells contain nucleotide sequences related to the putative transforming gene of avian myelocytomatosis virus. *J. Virol.* **31**, 514–521.

Shen-Li, H., O'Hagan, R. C., Hou, H., Jr., Horner, J. W., II, Lee, H. W., and DePinho, R. A. (2000). Essential role for Max in early embryonic growth and development. *Genes Dev.* **14**, 17–22.

Sherr, C. J., and Weber, J. D. (2000). The ARF/p53 pathway. *Curr. Opin. Genet. Dev.* **10**, 94–99.

Shi, X. B., Tepper, C. G., and deVere White, R. W. (2008). Cancerous miRNAs and their regulation. *Cell Cycle* **7**, 1529–1538.

Shusterman, S., Grupp, S. A., Barr, R., Carpentieri, D., Zhao, H., and Maris, J. M. (2001). The angiogenesis inhibitor tnp-470 effectively inhibits human neuroblastoma xenograft growth, especially in the setting of subclinical disease. *Clin. Cancer Res.* **7**, 977–984.

Singh, A. M., and Dalton, S. (2009). The cell cycle and Myc intersect with mechanisms that regulate pluripotency and reprogramming. *Cell Stem Cell* **5**, 141–149.

Sirohi, B., and Smith, K. (2008). Bevacizumab in the treatment of breast cancer. *Expert Rev. Anticancer Ther.* **8**, 1559–1568.

Sorlie, T., Perou, C. M., Tibshirani, R., Aas, T., Geisler, S., Johnsen, H., Hastie, T., Eisen, M. B., van de Rijn, M., Jeffrey, S. S., Thorsen, T., Quist, H., *et al.* (2001). Gene expression patterns of breast carcinomas distinguish tumor subclasses with clinical implications. *Proc. Natl. Acad. Sci. USA* **98**, 10869–10874.

Soucek, L., Whitfield, J., Martins, C. P., Finch, A. J., Murphy, D. J., Sodir, N. M., Karnezis, A. N., Swigart, L. B., Nasi, S., and Evan, G. I. (2008). Modelling Myc inhibition as a cancer therapy. *Nature* **455**, 679–683.

Soucie, E. L., Annis, M. G., Sedivy, J., Filmus, J., Leber, B., Andrews, D. W., and Penn, L. Z. (2001). Myc potentiates apoptosis by stimulating Bax activity at the mitochondria. *Mol. Cell. Biol.* **21**, 4725–4736.

Soutschek, J., Akinc, A., Bramlage, B., Charisse, K., Constien, R., Donoghue, M., Elbashir, S., Geick, A., Hadwiger, P., Harborth, J., John, M., Kesavan, V., *et al.* (2004). Therapeutic silencing of an endogenous gene by systemic administration of modified siRNAs. *Nature* **432**, 173–178.

Staller, P., Peukert, K., Kiermaier, A., Seoane, J., Lukas, J., Karsunky, H., Moroy, T., Bartek, J., Massague, J., Hanel, F., and Eilers, M. (2001). Repression of p15INK4b expression by Myc through association with Miz-1. *Nat. Cell Biol.* **3**, 392–399.

Stanton, B. R., Perkins, A. S., Tessarollo, L., Sassoon, D. A., and Parada, L. F. (1992). Loss of N-myc function results in embryonic lethality and failure of the epithelial component of the embryo to develop. *Genes Dev.* **6**, 2235–2247.

Stojanova, A., and Penn, L. Z. (2009). The role of INI1/hSNF5 in gene regulation and cancer. *Biochem. Cell Biol.* **87**, 163–177.

Strieder, V., and Lutz, W. (2002). Regulation of N-myc expression in development and disease. *Cancer Lett.* **180**, 107–119.

Stylianou, S., Clarke, R. B., and Brennan, K. (2006). Aberrant activation of notch signaling in human breast cancer. *Cancer Res.* **66**, 1517–1525.

Sun, L., Fuselier, J. A., Murphy, W. A., and Coy, D. H. (2002). Antisense peptide nucleic acids conjugated to somatostatin analogs and targeted at the n-myc oncogene display enhanced cytotoxity to human neuroblastoma IMR32 cells expressing somatostatin receptors. *Peptides* **23**, 1557–1565.

Sweetenham, J. W., Pearce, R., Taghipour, G., Blaise, D., Gisselbrecht, C., and Goldstone, A. H. (1996). Adult Burkitt's and Burkitt-like non-Hodgkin's lymphoma—Outcome for patients treated with high-dose therapy and autologous stem-cell transplantation in first remission or at relapse: Results from the European Group for Blood and Marrow Transplantation. *J. Clin. Oncol.* **14**, 2465–2472.

Takahashi, K., and Yamanaka, S. (2006). Induction of pluripotent stem cells from mouse embryonic and adult fibroblast cultures by defined factors. *Cell* **126**, 663–676.

Tamm, I. (2005). Antisense therapy in clinical oncology: Preclinical and clinical experiences. *Methods Mol. Med.* **106**, 113–134.

Tang, X. X., Zhao, H., Kung, B., Kim, D. Y., Hicks, S. L., Cohn, S. L., Cheung, N. K., Seeger, R. C., Evans, A. E., and Ikegaki, N. (2006). The MYCN enigma: Significance of MYCN expression in neuroblastoma. *Cancer Res.* **66**, 2826–2833.

Taub, R., Kirsch, I., Morton, C., Lenoir, G., Swan, D., Tronick, S., Aaronson, S., and Leder, P. (1982). Translocation of the c-myc gene into the immunoglobulin heavy chain locus in human Burkitt lymphoma and murine plasmacytoma cells. *Proc. Natl. Acad. Sci. USA* **79**, 7837–7841.

Teitz, T., Wei, T., Valentine, M. B., Vanin, E. F., Grenet, J., Valentine, V. A., Behm, F. G., Look, A. T., Lahti, J. M., and Kidd, V. J. (2000). Caspase 8 is deleted or silenced preferentially in childhood neuroblastomas with amplification of MYCN. *Nat. Med.* **6**, 529–535.

Thomas, D. A., Faderl, S., O'Brien, S., Bueso-Ramos, C., Cortes, J., Garcia-Manero, G., Giles, F. J., Verstovsek, S., Wierda, W. G., Pierce, S. A., Shan, J., Brandt, M., *et al.* (2006). Chemoimmunotherapy with hyper-CVAD plus rituximab for the treatment of adult Burkitt and Burkitt-type lymphoma or acute lymphoblastic leukemia. *Cancer* **106**, 1569–1580.

Tonelli, R., Purgato, S., Camerin, C., Fronza, R., Bologna, F., Alboresi, S., Franzoni, M., Corradini, R., Sforza, S., Faccini, A., Shohet, J. M., Marchelli, R., *et al.* (2005). Anti-gene peptide nucleic acid specifically inhibits MYCN expression in human neuroblastoma cells leading to cell growth inhibition and apoptosis. *Mol. Cancer Ther.* **4**, 779–786.

Tworkowski, K. A., Salghetti, S. E., and Tansey, W. P. (2002). Stable and unstable pools of Myc protein exist in human cells. *Oncogene* **21**, 8515–8520.

Van Roy, N., De Preter, K., Hoebeeck, J., Van Maerken, T., Pattyn, F., Mestdagh, P., Vermeulen, J., Vandesompele, J., and Speleman, F. (2009). The emerging molecular pathogenesis of neuroblastoma: Implications for improved risk assessment and targeted therapy. *Genome Med.* **1**, 74.

Vennström, B., Sheiness, D., Zabielski, J., and Bishop, J. M. (1982). Isolation and characterization of c-myc, a cellular homolog of the oncogene (v-myc) of avian myelocytomatosis virus strain 29. *J. Virol.* **42**, 773–779.

Verneris, M. R., and Wagner, J. E. (2007). Recent developments in cell-based immune therapy for neuroblastoma. *J. Neuroimmune Pharmacol.* **2**, 134–139.

Vita, M., and Henriksson, M. (2006). The Myc oncoprotein as a therapeutic target for human cancer. *Semin. Cancer Biol.* **16**, 318–330.

Volinia, S., Calin, G. A., Liu, C. G., Ambs, S., Cimmino, A., Petrocca, F., Visone, R., Iorio, M., Roldo, C., Ferracin, M., Prueitt, R. L., Yanaihara, N., *et al.* (2006). A microRNA expression signature of human solid tumors defines cancer gene targets. *Proc. Natl. Acad. Sci. USA* **103**, 2257–2261.

von Bueren, A. O., Shalaby, T., Oehler-Janne, C., Arnold, L., Stearns, D., Eberhart, C. G., Arcaro, A., Pruschy, M., and Grotzer, M. A. (2009). RNA interference-mediated c-MYC inhibition prevents cell growth and decreases sensitivity to radio- and chemotherapy in childhood medulloblastoma cells. *BMC Cancer* **9**, 10.

von der Lehr, N., Johansson, S., Wu, S., Bahram, F., Castell, A., Cetinkaya, C., Hydbring, P., Weidung, I., Nakayama, K., Nakayama, K. I., Söderberg, O., Kerppola, T. K., *et al.* (2003). The F-box protein Skp2 participates in c-Myc proteasomal degradation and acts as a cofactor for c-Myc-regulated transcription. *Mol. Cell* **11**, 1189–1200.

von Mehren, M. (2006). Imatinib-refractory gastrointestinal stromal tumors: The clinical problem and therapeutic strategies. *Curr. Oncol. Rep.* **8**, 192–197.

Wagner, A. J., Kokontis, J. M., and Hay, N. (1994). Myc-mediated apoptosis requires wild-type p53 in a manner independent of cell cycle arrest and the ability of p53 to induce p21waf1/cip1. *Genes Dev.* **8**, 2817–2830.

Wahlstrom, T., and Henriksson, M. (2007). Mnt takes control as key regulator of the myc/max/mxd network. *Adv. Cancer Res.* **97**, 61–80.

Wang, J., Xie, L. Y., Allan, S., Beach, D., and Hannon, G. J. (1998). Myc activates telomerase. *Genes Dev.* **12**, 1769–1774.

Wang, H., Hammoudeh, D. I., Follis, A. V., Reese, B. E., Lazo, J. S., Metallo, S. J., and Prochownik, E. V. (2007). Improved low molecular weight Myc-Max inhibitors. *Mol. Cancer Ther.* **6**, 2399–2408.

Weidhaas, J. B., Babar, I., Nallur, S. M., Trang, P., Roush, S., Boehm, M., Gillespie, E., and Slack, F. J. (2007). MicroRNAs as potential agents to alter resistance to cytotoxic anticancer therapy. *Cancer Res.* **67**, 11111–11116.

Weinstein, J. L., Katzenstein, H. M., and Cohn, S. L. (2003). Advances in the diagnosis and treatment of neuroblastoma. *Oncologist* **8**, 278–292.

Weiss, W. A., Aldape, K., Mohapatra, G., Feuerstein, B. G., and Bishop, J. M. (1997). Targeted expression of MYCN causes neuroblastoma in transgenic mice. *EMBO J.* **16**, 2985–2995.

Welcker, M., Orian, A., Jin, J., Grim, J. E., Harper, J. W., Eisenman, R. N., and Clurman, B. E. (2004). The Fbw7 tumor suppressor regulates glycogen synthase kinase 3 phosphorylation-dependent c-Myc protein degradation. *Proc. Natl. Acad. Sci. USA* **101**, 9085–9090.

Wernig, M., Meissner, A., Foreman, R., Brambrink, T., Ku, M., Hochedlinger, K., Bernstein, B. E., and Jaenisch, R. (2007). *In vitro* reprogramming of fibroblasts into a pluripotent ES-cell-like state. *Nature* **448**, 318–324.

Westermann, F., Muth, D., Benner, A., Bauer, T., Henrich, K. O., Oberthuer, A., Brors, B., Beissbarth, T., Vandesompele, J., Pattyn, F., Hero, B., Konig, R., *et al.* (2008). Distinct transcriptional MYCN/c-MYC activities are associated with spontaneous regression or malignant progression in neuroblastomas. *Genome Biol.* **9,** R150.

White, J., and Dalton, S. (2005). Cell cycle control of embryonic stem cells. *Stem Cell Rev.* **1,** 131–138.

Wightman, B., Ha, I., and Ruvkun, G. (1993). Posttranscriptional regulation of the heterochronic gene lin-14 by lin-4 mediates temporal pattern formation in *C. elegans. Cell* **75,** 855–862.

Williamson, D., Lu, Y. J., Gordon, T., Sciot, R., Kelsey, A., Fisher, C., Poremba, C., Anderson, J., Pritchard-Jones, K., and Shipley, J. (2005). Relationship between MYCN copy number and expression in rhabdomyosarcomas and correlation with adverse prognosis in the alveolar subtype. *J. Clin. Oncol.* **23,** 880–888.

Wilson, A., Murphy, M. J., Oskarsson, T., Kaloulis, K., Bettess, M. D., Oser, G. M., Pasche, A. C., Knabenhans, C., Macdonald, H. R., and Trumpp, A. (2004). c-Myc controls the balance between hematopoietic stem cell self-renewal and differentiation. *Genes Dev.* **18,** 2747–2763.

Wolf, M., Tebbe, S., and Fink, T. (2004). First-line chemotherapy in metastatic small-cell lung cancer (SCLC). *Lung Cancer* **45**(Suppl. 2), S223–S234.

Wong, A. J., Ruppert, J. M., Eggleston, J., Hamilton, S. R., Baylin, S. B., and Vogelstein, B. (1986). Gene amplification of c-myc and N-myc in small cell carcinoma of the lung. *Science* **233,** 461–464.

Wu, K. J., Grandori, C., Amacker, M., Simon-Vermot, N., Polack, A., Lingner, J., and Dalla-Favera, R. (1999). Direct activation of TERT transcription by c-MYC. *Nat. Genet.* **21,** 220–224.

Wu, S., Cetinkaya, C., Munoz-Alonso, M. J., von der Lehr, N., Bahram, F., Beuger, V., Eilers, M., Leon, J., and Larsson, L. G. (2003). Myc represses differentiation-induced p21CIP1 expression via Miz-1-dependent interaction with the p21 core promoter. *Oncogene* **22,** 351–360.

Xu, D., Popov, N., Hou, M., Wang, Q., Bjorkholm, M., Gruber, A., Menkel, A. R., and Henriksson, M. (2001). Switch from Myc/Max to Mad1/Max binding and decrease in histone acetylation at the telomerase reverse transcriptase promoter during differentiation of HL60 cells. *Proc. Natl. Acad. Sci. USA* **98,** 3826–3831.

Xu, Y., Shi, J., Yamamoto, N., Moss, J. A., Vogt, P. K., and Janda, K. D. (2006). A credit-card library approach for disrupting protein-protein interactions. *Bioorg. Med. Chem.* **14,** 2660–2673.

Yada, M., Hatakeyama, S., Kamura, T., Nishiyama, M., Tsunematsu, R., Imaki, H., Ishida, N., Okumura, F., Nakayama, K., and Nakayama, K. I. (2004). Phosphorylation-dependent degradation of c-Myc is mediated by the F-box protein Fbw7. *EMBO J.* **23,** 2116–2125.

Yang, W., Shen, J., Wu, M., Arsura, M., FitzGerald, M., Suldan, Z., Kim, D. W., Hofmann, C. S., Pianetti, S., Romieu-Mourez, R., Freedman, L. P., and Sonenshein, G. E. (2001). Repression of transcription of the p27(Kip1) cyclin-dependent kinase inhibitor gene by c-Myc. *Oncogene* **20,** 1688–1702.

Yang, Z. J., Ellis, T., Markant, S. L., Read, T. A., Kessler, J. D., Bourboulas, M., Schuller, U., Machold, R., Fishell, G., Rowitch, D. H., Wainwright, B. J., and Wechsler-Reya, R. J. (2008). Medulloblastoma can be initiated by deletion of Patched in lineage-restricted progenitors or stem cells. *Cancer Cell* **14,** 135–145.

Yauch, R. L., Dijkgraaf, G. J., Alicke, B., Januario, T., Ahn, C. P., Holcomb, T., Pujara, K., Stinson, J., Callahan, C. A., Tang, T., Bazan, J. F., Kan, Z., *et al.* (2009). Smoothened mutation confers resistance to a Hedgehog pathway inhibitor in medulloblastoma. *Science* **326,** 572–574.

Yeh, E., Cunningham, M., Arnold, H., Chasse, D., Monteith, T., Ivaldi, G., Hahn, W. C., Stukenberg, P. T., Shenolikar, S., Uchida, T., Counter, C. M., Nevins, J. R., *et al.* (2004). A signalling pathway controlling c-Myc degradation that impacts oncogenic transformation of human cells. *Nat. Cell Biol.* **6,** 308–318.

Yin, X., Giap, C., Lazo, J. S., and Prochownik, E. V. (2003). Low molecular weight inhibitors of Myc–Max interaction and function. *Oncogene* **22,** 6151–6159.

Young, S. D., Whissell, M., Noble, J. C., Cano, P. O., Lopez, P. G., and Germond, C. J. (2006). Phase II clinical trial results involving treatment with low-dose daily oral cyclophosphamide, weekly vinblastine, and rofecoxib in patients with advanced solid tumors. *Clin. Cancer Res.* **12,** 3092–3098.

Yu, Q., Ciemerych, M. A., and Sicinski, P. (2005). Ras and Myc can drive oncogenic cell proliferation through individual D-cyclins. *Oncogene* **24,** 7114–7119.

Zhang, H., Fan, S., and Prochownik, E. V. (1997). Distinct roles for MAX protein isoforms in proliferation and apoptosis. *J. Biol. Chem.* **272,** 17416–17424.

Zhuang, D., Mannava, S., Grachtchouk, V., Tang, W. H., Patil, S., Wawrzyniak, J. A., Berman, A. E., Giordano, T. J., Prochownik, E. V., Soengas, M. S., and Nikiforov, M. A. (2008). C-MYC overexpression is required for continuous suppression of oncogene-induced senescence in melanoma cells. *Oncogene* **27,** 6623–6634.

Zimmerman, K. A., Yancopoulos, G. D., Collum, R. G., Smith, R. K., Kohl, N. E., Denis, K. A., Nau, M. M., Witte, O. N., Toran-Allerand, D., Gee, C. E., *et al.* (1986). Differential expression of myc family genes during murine development. *Nature* **319,** 780–783.

Index

M

N

O

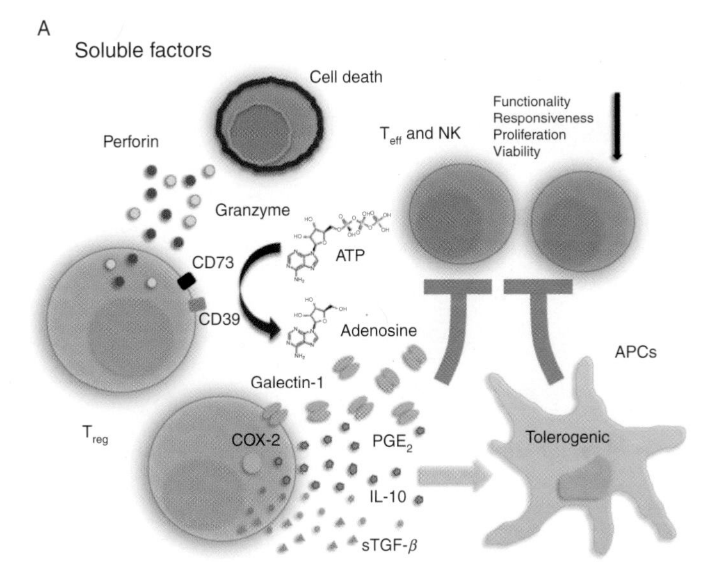
A
Soluble factors
Cell death
Perforin
Granzyme
T_{eff} and NK
Functionality
Responsiveness
Proliferation
Viability
CD73
CD39
ATP
Adenosine
APCs
Galectin-1
T_{reg}
COX-2
PGE_2
IL-10
sTGF-β
Tolerogenic

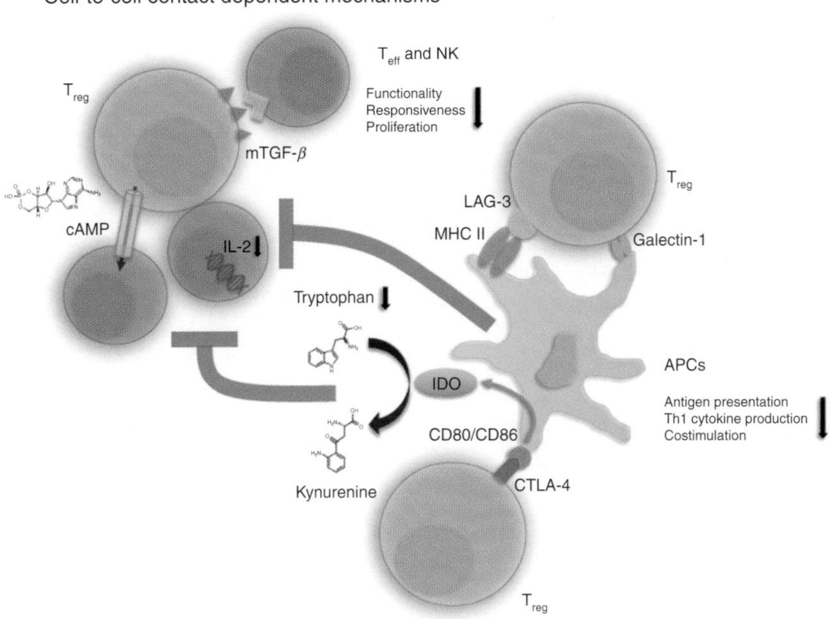
B
Cell-to-cell contact dependent mechanisms
T_{eff} and NK
T_{reg}
Functionality
Responsiveness
Proliferation
mTGF-β
T_{reg}
LAG-3
MHC II
Galectin-1
cAMP
IL-2
Tryptophan
IDO
APCs
Antigen presentation
Th1 cytokine production
Costimulation
CD80/CD86
CTLA-4
Kynurenine
T_{reg}

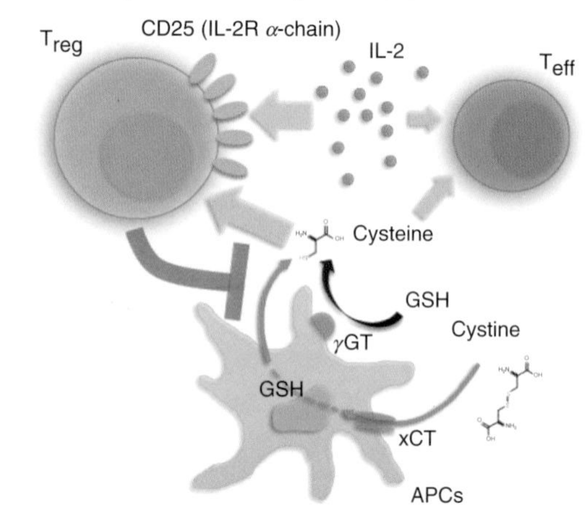

Fig. 1, Dimitrios Mougiakakos *et al.* (See Page 68 of this volume.)

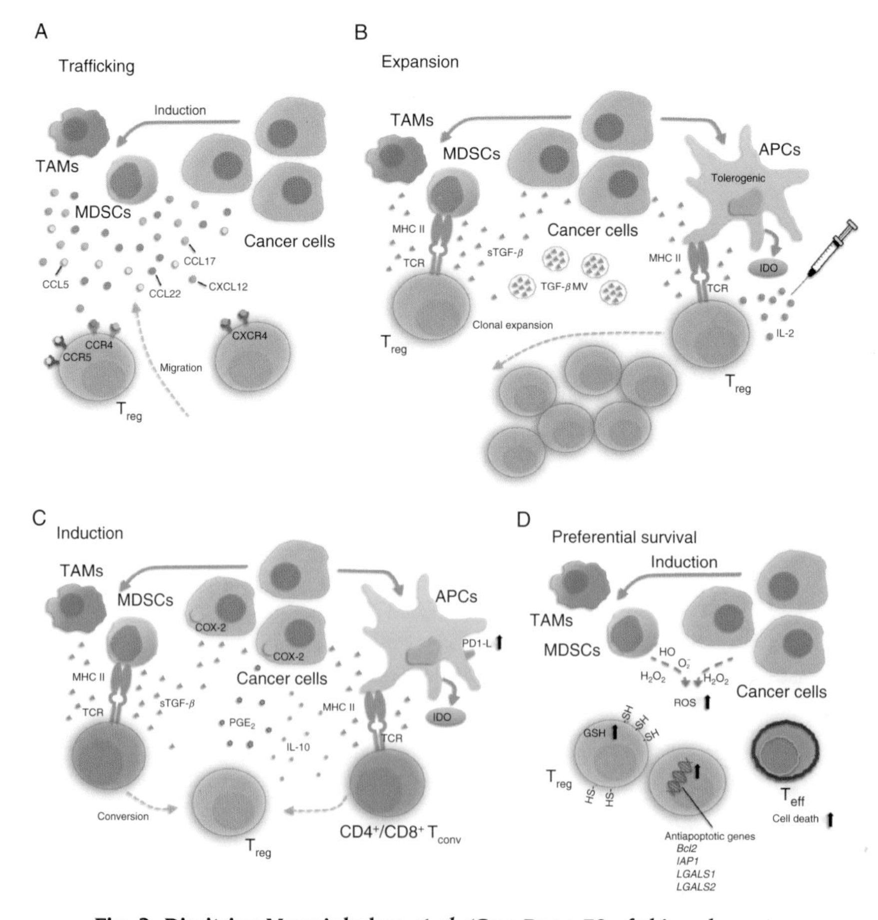

Fig. 2, Dimitrios Mougiakakos *et al.* (See Page 79 of this volume.)

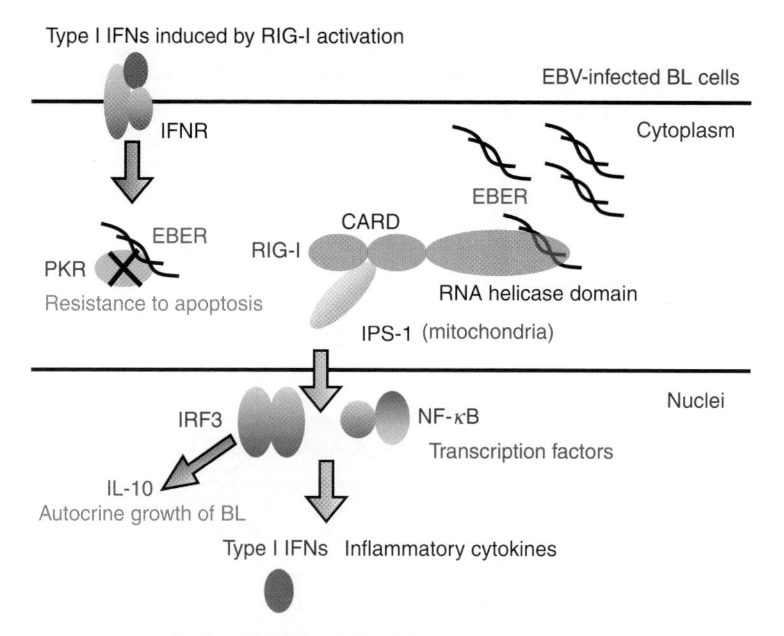

Fig. 3, Dai Iwakiri and Kenzo Takada (See Page 129 of this volume.)

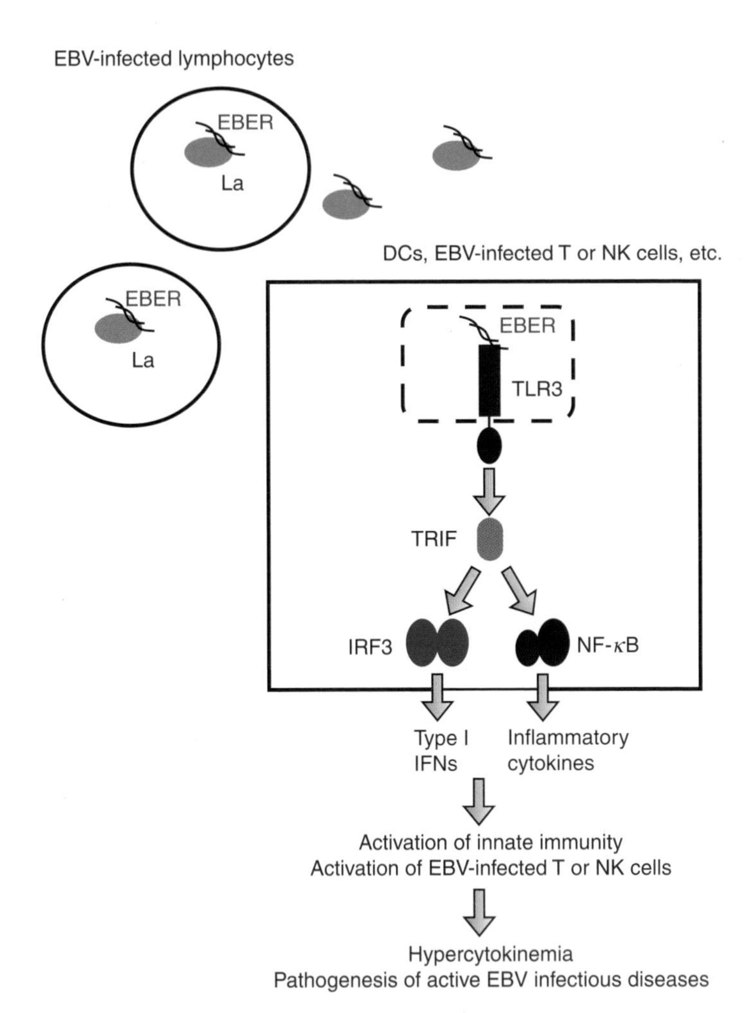

Fig. 4, Dai Iwakiri and Kenzo Takada (See Page 130 of this volume.)